# GLOBALIZATION AND SOCIAL CHANGE

## People and places in a divided world

*Diane Perrons*

Routledge
Taylor & Francis Group

LONDON AND NEW YORK

First published 2004
by Routledge
2 Park Square, Milton Park, Abingdon, Oxon OX14 4RN

Simultaneously published in the USA and Canada
by Routledge
270 Madison Ave, New York, NY 10016

Reprinted 2005

*Routledge is an imprint of the Taylor & Francis Group*

© 2004 Diane Perrons

Typeset in Garamond by
Keystroke, Jacaranda Lodge, Wolverhampton
Printed and bound in Great Britain by
TJ International Ltd, Padstow, Cornwall

*British Library Cataloguing in Publication Data*
A catalogue record for this book is available from the British Library

*Library of Congress Cataloging in Publication Data*
Perrons, Diane.
Globalization and social change: people and places in a divided world /
Diane Perrons.–1st ed.
p. cm.
Simultaneously published in the USA and Canada.
Includes bibliographical references and index.
1. Social change. 2. Globalization. 3. Globalization–Economic aspects.
I. Title: Globalization and social change. II. Title.
HM836.P47 2004
303.4–dc22

ISBN 0–415–26695–5 (hbk)
ISBN 0–415–26696–3 (pbk)

To Lotte and Robin

# CONTENTS

# CONTENTS

# PREFACE

*Globalization and Social Change* analyses the effects of globalization and the new economy on people living and working in different places. Emphasis is placed on socio-economic aspects of change, particularly on the development of information and communication technologies and changes in working patterns and living arrangements. A theoretical approach is combined with detailed comparative case studies in order to provide a social scientific interpretation of contemporary socio-economic change.

Reference is made to existing theoretical explanations and popular conceptualizations but a key aspect of the book is to provide a coherent account of these changes based on the author's own synthesis of a number of theoretical approaches arising from the French Regulation theory, Beckian risk analysis and Gösta Esping-Andersen's welfare regimes in the context of a feminist historical materialist understanding of social change. The book also provides detailed comparative empirical analyses to illustrate how societies have responded to similar external pressures in different ways in order to identify more socially inclusive patterns of development. The approach followed is that of the 'reformist tinkerer' rather than the 'utopian visionary' (Harvey 2000) in conformity with the author's belief that marginal improvements are worthwhile as they have a real and immediate impact on people's lives. By so doing the book seeks to counter the determinism and lack of optimism found in some accounts of globalization without diminishing the significance of increasing inequalities found in the contemporary divided world.

# ACKNOWLEDGEMENTS

The author and publishers would like to thank the following for granting permission to reproduce material in this work: Oxford University Press for Box 3.3; Barbara Ehrenrich and Rogers, Coleridge and White Ltd for the contents of Box 3.5; UNICEF Innocenti Research Centre, Florence, Italy, Innocenti Report Card No.1 June 2000 for 'A league table of child poverty in rich nations' reproduced in Figure 5.1; The National Glass Centre for Figure 5.5; Babel Media Limited for Figure 6.1; Alexa Koller for Figure 6.5; War on Want/Ben Blackall for Figure 8.2; Silvia Posocco for Figure 9.2; ETI for Box 9.4; Eleanor Phant for Figure 9.4. Every effort has been made to contact copyright holders for their permission to reprint material in this book. The publishers would be grateful to hear from any copyright holder who is not here acknowledged and will undertake to rectify any errors or omissions in future editions of this book.

I would like also to give thanks to friends and colleagues who taught on the Basic Social Science, Introduction to Political Economy and Women and Business courses at what was at one time City of London Polytechnic and whose ideas have shaped my thoughts, in particular, to Les Budd, Judy Klein, Anne Phillips, Frank Rodriguez, Sam Whimster, Caroline Woodhead, the late Eva Colorni and Michael Cowen. To students on my courses at the LSE, Silvia Posocco who helped me design a Web Ct version of a course which helped structure ideas for this book and to Róisín Ryan Flood and Inge Strüder who did my teaching while I was on leave. Special thanks go to the reviewers of the proposal and the first draft for their encouragement; to Melanie Attridge, Andrew Mould, Angie Doran at Routledge, Sarah Cahill and Letitia Grant for their copy-editing and proofreading work and Jo Underwood at the LSE for administrative assistance, to Mina Moshkeri for the graphics, and Sylvia Chant for comments on Chapter 4, both at the LSE. I would also like to acknowledge financial support from the LSE, Leverhulme and the ESRC for various pieces of my own research commented on in the book. Very special thanks to Stephanie and Armando Barrientos for many discussions and dinners and to Lotte, Robin and Mick Dunford for living with me.

# 1

# ANALYSING GLOBALIZATION AND SOCIAL CHANGE

Globalization and the new economy encapsulate the transformation of economic and social relations across the globe. People and places are increasingly inter-linked through the organization of work, the flows of goods and services and the exchange of ideas. Even so the contemporary world is characterized by differ-ence rather than uniformity and widening rather than narrowing inequality but the spatial pattern is complex; while some people and places are involved in highly interactive global networks others are largely excluded, creating new and reinforcing old patterns of uneven development. Despite the enormous advances in human ingenuity and technology that have created unparalleled wealth and an economically more integrated world, social and spatial divisions are widening. This book illustrates and explains some of the divisions between countries, between regions and within cities, emphasizing how they provide quite different opportunities within which people live their lives, even though they are increasingly linked within the global economy.

Globalization was a term first used towards the end of the last century. It became the subject of academic conferences, TV programmes, bestselling books, websites and papers in learned journals[1] all around the globe such that it is a concept 'with which one argues but about which one does not argue' (Bourdieu and Wacquant 1999: 41). Descriptively globalization refers to the growing interconnectedness and interdependencies between countries on a global scale as in the World Bank's definition: 'Globalization can be summarized as the global circulation of goods, services and capital but also of information, ideas and people' (World Bank 2000: 3). This definition implies that the world has become increasingly interconnected leading some writers, but few geographers, to sug-gest that geography has become irrelevant.[2] Space has been compressed by fast modes of communication making it possible for money, ideas, goods and people to flow around the world ever more quickly with significant implications for the organization of economic activities and the security and stability of employment. Money can be transferred instantaneously from one part of the globe to another inducing financial crises with real effects on people's lives. Commodities are designed, made and marketed between a range of countries with correspondingly different job opportunities in the different locations. People have also become increasingly mobile; international travel is commonplace for people in wealthier

1

countries and both legal and illegal migration are significant.[3] Similarly ideas, news, films and music flow instantaneously from one place to another creating world audiences for events such as film premieres, pop concerts and football matches.[4] Furthermore, following the transformations in Eastern Europe, the vast majority of countries currently subscribe to some version of political democracy and market economics.

In some ways these global flows have generated a certain homogenization of ideas, cultures, political and economic systems, and the vision of the global village – 'where tribes people in remote rain forests tap away on lap top computers, Sicilian grandmothers conduct e-business, and global teens share a world wide style culture' (Klein 1999: xvii) has a certain resonance even though such iconic instances of globalization are realized only by a minority. Important differences remain between places so geography does matter as these differences are often built upon to increase corporate profitability reinforcing and creating new patterns of uneven development. International trade has always built upon world climate differences to increase the range of goods available in particular locations but the development of global supply chains has led to complex patterns of production and distribution to ensure the continual year round stocking of particular commodities such as seedless grapes in UK supermarkets.[5] Different wage zones have been used to lower the costs of manufactured goods and now different time and wage zones are drawn upon to lower costs in the service sector through the re-routeing of phone calls for airline reservations, credit card inquiries and fast transcription so overall the interconnections between places are becoming more complex. Despite this fluidity, however, all activities, even virtual ones, have to take place somewhere and many activities gain from geographical proximity, leading to clusters of activity in some locations as others are bypassed.

Technical developments in transportation and the rapid diffusion of information and communication technologies, especially the Internet, have facilitated these flows but the causes of increased integration are found in economic, social and political changes, in particular the growing dominance of capitalism as a system of social and economic regulation. The World Bank for example argues that in addition to technology, a second factor promoting globalization is the 'shift in policy orientation as governments everywhere have reduced barriers that had curbed the development of domestic markets and their links to the international economy' (World Bank 2000: 1). Effectively the Bank is referring to the prevalence of neo-liberal ideology without indicating how it has itself been instrumental in this process by making increased openness a condition for financial assistance. Such political and economic harmonization is also a precondition for realizing Castells's (1996: 92) definition of globalization as 'an economy with the capacity to work as a unit in real time on a planetary scale'. For firms, individuals and organizations to be interconnected in a 'network society' that stretches across national borders and within which people engage in communications and transactions in real time, they must have accepted broadly the same economic and social institutions and ways of working.

Just as some writers have suggested that increasing interconnectedness has ended geography others have suggested that increasing openness has effectively ended politics or at least undermined the power of nation states to exercise their economic and political autonomy (Ohmae 1995), while others (Castells 1997; Hirst and Thompson 1996, 2002) are more sceptical. As development is uneven so too is the power of nation states with some states, such as the United States, having unprecedented power, while the powers of others are constrained by supra national institutions. Whatever the level and wherever power lies, however, it is crucial to recognize the significance of economic, social and political processes in shaping the contemporary world and to highlight the fact that globalization is a product of political decision making or choice and correspondingly open to change and modification through human decision making. By so doing the use of globalization as a noun, that is, as a seemingly unstoppable process, almost independent of human will, can be challenged, and a space is created for thinking about the ways in which globalization can and indeed is being developed differently in different countries sometimes to create more inclusive outcomes.

Globalization is sometimes used to explain contemporary events but it is a summary term for processes that require explanation not an analytical concept. In this book globalization is taken as a given[6] and used descriptively to reflect the increasingly interconnected nature of the contemporary world and correspondingly the need to contextualize analyses of people and places within the economic, social, political and cultural processes currently shaping the global economy. This chapter provides a framework for analysing these processes and for connecting theories of economic and social change at different levels. In some ways the framework could be construed as a new meta-narrative, but it is neither singular nor deterministic but rather provides an intellectual space for thinking about the connectedness of processes shaping contemporary change and within which to situate events affecting people in different places. If as Anthony Giddens argues 'distant events, whether economic or not, affect everyone more directly and immediately than ever before' (Giddens 1999:31), then it is important to situate specific analyses within a framework which recognizes this degree of interconnectedness. This rather daunting task is not intended to impose any singular interpretation of events but reflects a genuine desire to understand and articulate the nature of contemporary connections between people and places. Why for example does a nurse leave her family in South Africa to work in a local Brighton hospital, why are jobs being transferred from a major employer in the same city to new call centres in India, why do 53 per cent of children in inner London, one of the richest cities in the world, live in poverty, why is there a Starbucks inside the Forbidden City in Beijing, why do western governments subsidize their agricultural producers while urging open markets on other countries, why do some people pay large sums of money and risk their lives to travel illegally to other countries, only to find themselves in jobs which barely cover their survival needs, why in every country do women on average fare less well than men, why do children suffer from undernourishment in some countries and obesity in others, and more generally why in the context of growing opulence, does inequality appear to be increasing at every spatial scale? In order to answer these questions different theoretical perspectives are drawn upon, because theories identify general processes and therefore help to understand the context within which people ranging from corporate managers to economic migrants make choices, but theories are not substitutes for explanation; each event will always have a unique justification, so the framework needs to be applied in specific contexts. The book draws upon this framework to illustrate and illuminate aspects of globalization and social change at a world, regional and city scale, and considers some of the responses made by people trying to influence the trajectory of development in different ways.

## THEORIZING PEOPLE AND PLACES IN THE NEW ECONOMY

Beginnings are always difficult but given the complexity and interdependency of the contemporary world and its rapidly evolving nature the problem here is immense because it is almost as if in order to understand anything it is necessary to understand everything. The framework outlined in Figure 1.1 is designed to provide a way of breaking through this complexity in order to understand the contemporary world and in particular the growing social and spatial divisions. Figure 1.1 contains different elements that are central to understanding and specifies interactions between them. In any specific context the elements are mutually constitutive or interdependent so each and every outcome will in some ways always be unique. Thus the framework itself is not an explanation but provides a set of tools that can be applied in specific contexts to produce an understanding or at least insights towards an understanding of specific events. It goes beyond the common sense view that everything affects everything because it specifies a directional logic, albeit with feedback loops, which helps provide a starting point for analysis.

The starting point lies with the economy and so reflects a historical materialist perspective, but not a deterministic one (see Box 1.1 for a brief summary of the different theoretical perspectives referred to in the book). There are two reasons for this starting point. First, the material reproduction of everyday life, or how people obtain their needs and wants on a day to day basis,[7] irrespective of whether they are physical necessities or cultural preferences, is fundamental to human life and existence on this planet. Second, the ways in which people do this are increasingly shaped by capitalism, which is a dynamic system. Understanding this dynamic helps to explain change, both the apparently ceaseless search for new markets and correspondingly the spread of this system across the globe, as well as the constant tendency for change in terms of the range of goods and services produced, the way they are produced, where they are produced and so on and these changes profoundly shape the context in which people live their daily lives. This privileging of the economy may seem rather

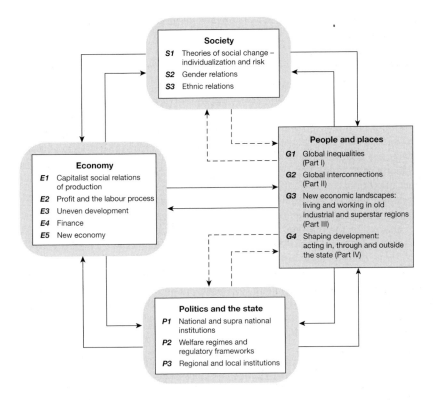

*Figure 1.1* People and places: a framework for analysis.

old-fashioned in comparison to contemporary perspectives that have privileged individuals and identities but my concern is to explain continuing social and spatial inequalities. To begin analyses with individuals and their identities, with diversity rather than commonalities, is to overlook the origins of inequality, that is, it focuses too much attention on finding out who people are rather than on how they came to be located in the socially hierarchical, gendered and racialized space where they reside (Mirza 1997). Correspondingly solutions are sought more in terms of changing personal behaviour, recognizing diversity and widening participation, rather than challenging the wider structures that promote and intensify social and spatial divisions. Knowledge or knowledges of these processes is crucial in order to challenge globalization in its contemporary unequal form.[8]

This materialist approach however does not in any way overlook the significance of human agency. Social change results from the interaction of two types of causal mechanism: ideal and material causality. Ideal causality refers to the way that people conceptualize their future or elements of their future and either individually or collectively seek to bring their ideas into being. People are not however entirely free agents because their choices are made in existing

*Box 1.1* Theoretical perspectives

*Historical materialism*
Historical materialism is associated with Karl Marx (1973b) and is a per-spective that foregrounds the reproduction of everyday life. Economic and social analyses begin by explaining how material needs and wants have been and continue to be produced and distributed in any given society. It focuses on how people obtain their daily survival, in terms of what and how things are produced and consumed, that is with the material and social relations of reproduction and in class societies how a surplus is produced and appropriated. Understanding how life is reproduced on a day to day basis is regarded as the key to a wider understanding of how any given society works. Within this perspective people's ideas and social institutions are shaped by the material circumstances and social relations within which they live. Historical materialism has often been mistakenly conflated with economic determinism, but this neglects the double sided nature of change central to historical materialism; specifically the notion that social change is the outcome of ideal and material causality. Human beings are the active agents and seek to bring their ideas into being so their ideas shape their histories, but not in circumstances of their own choosing.

*Feminism*
A perspective which foregrounds gender relations in the issues that are being analysed. It also pays attention to how relations of gender, social class, race, ethnicity, ability, and sexuality are mutually constituted.

*Feminist historical materialism*
A perspective concerned with analysing how societies reproduce themselves but highlights the gendered nature of social relations.

These perspectives or standpoints are at quite a high level of abstraction or generality. They specify an approach to research by identifying what sorts of processes should be examined. The perspectives below lie between the abstract world of ideas and the concrete world of things and develop a set of intermediate concepts that relate explicitly to capitalist society and can be drawn upon to highlight fields for empirical enquiry.

*French Regulation School*
This perspective, associated with Michel Aglietta (1979) and taken forward by a number of writers including Alain Lipietz (1992) and Bob Jessop (1990), sought to explain how there were periods of relative stability and growth in capitalism, which, following Marx, they believed to be an anarchic and crisis ridden system. To do so they developed a series of intermediate concepts including the regime of accumulation and the mode of regulation which link economic and social reproduction. In particular they used the term Fordism to refer to the post-Second World War economic boom, which they

argue was founded on a new labour process with unprecedented increases in labour productivity. This labour process was based on a combination of scientific management and the flow line principle of Henry Ford. The increases in productivity allowed wages and profits to rise simultaneously but changes in the mode of regulation or social institutions were necessary in order to realize the potential gains and allow a mass production/mass consumption society to come into being. Many countries introduced some form of Keynesian demand management and expanded the welfare state. It was a national system of economic and social regulation. The contradictions of capitalist society re-emerged, as the productivity increases could not be sustained and competition increased as an increasing range of countries industrialized. From the 1970s onwards the era of post-Fordism has been characterized by slower and less certain economic growth and many of the social institutions of Fordism, in particular trade unions and the welfare state, have ceased to be so influential. Different countries initially took different pathways out of the crisis of Fordism, and there are some parallels with the welfare regime perspective discussed below, so the analysis of post-Fordism is more varied. The central contribution of this perspective is the idea of the balance or links between production and reproduction, which during Fordism were established at the national level. In the contemporary global economy such a balance has yet to be secured on a global scale.

*Risk Society*
Ulrich Beck (1992, 2000a) developed the idea of the risk society. His periodization of economic and social change has parallels with the French Regulation approach. Rather than Fordism and post-Fordism he refers to modernity and the risk society or reflexive modernization (Beck 1992). More recently he has termed this second phase the second modernity (Beck 2000a). In contrast to the French Regulation approach Beck foregrounds the social impact of these changes. He uses the concept of individualization to refer to the way that many traditional social supports, such as regular employment, trade unions and the family, have been eroded creating more insecure times or risks. At the same time people are freer to shape their own biographies but because structural inequalities have not disappeared people have different levels of resources with which to shape their futures, so inequalities expand. Individual and social risks, including environmental problems, are heightened by globalization.

*Welfare regimes*
Gösta Esping-Andersen (1990) identified the existence of three different welfare regimes that characterized the relationship between the state and society specifically in relation to social policy and the nature of the welfare state. He differentiates these regimes according to the degree of de-commodification or the extent to which the provision of goods such as health, education and housing are not supplied through the market but more directly

continued . . .

supplied by the state as social rights. His initial perspective was criticized by feminists for not recognizing that de-commodification can depend on unpaid female labour rather than state provision, and that the transfer of domestic work to the market can relieve domestic burdens and create paid employment, which despite being low paid, can be more empowering. Esping-Andersen has responded by developing the idea of defamilialization or the extent to which certain tasks are removed from the family and supplied either by the state or through the market. This refinement however does not affect his basic categorization of states into three welfare regimes: social democratic, conservative corporatist and liberal market. I draw upon his analysis in order to highlight the existence of different variants of capitalist economic and social development within the prevailing neo-liberal order.

*New economy*
This term has many meanings. In this book I use it to refer to the con- temporary era which is shaped by globalization, defined as the growing interconnections between people and places across the globe facilitated by new ICTs, and by growing social and spatial divisions. These social divisions are shaped by the increasing polarization of work between people working in high paid knowledge sectors and others working in low paid caring and reproductive work. These different forms of work have quite different economic properties which in part accounts for their different market valuations but not why societies accept these market valuations. This perspective draws on the work of Danny Quah (1996, 2003) in terms of his analysis of social divisions but takes this analysis further by drawing on the work of Nancy Folbre and Julie Nelson (2000) to explain why these divisions take a gendered form. It also develops these ideas geographically and considers why these social and gendered divisions also take a spatial form.

*Neo-liberalism*
This refers to the prevailing economic orthodoxy, which believes in the efficacy of the market. It involves a number of interconnected policies including: financial and trade liberalization, flexible and deregulated labour markets, privatization of the economy including public utilities, macro- economic stability, fiscal discipline or low levels of public debt and public sector expenditure directed towards 'productive' economic expenditure rather than social welfare. These policies have been recommended to and at times imposed upon countries experiencing economic transition and on countries falling into debt and seeking assistance from the supra national institutions.

*Global cities and superstar regions*
Global cities and superstar regions are not theoretical perspectives in the same way as the others above as they are one of the spatial forms associated with the new economy. Global cities are where the pinnacles of the market

economy are found in the form of the stock markets, financial institutions and corporations which regulate or manage the 'free global economy'. At the same time global cities or global city regions also attract people from all over the world employed in the major institutions but also in the low paid service work that caters to the needs of the high paid knowledge workers as well as work in the low paid manufacturing employment that remains. I use the term superstar region, drawing an analogy with the economic concept of superstar, which refers to sectors where it is possible for some people, the superstars, to capture such a high proportion of the market that the income gap between themselves and others working in the sector is extremely wide. Likewise with contemporary ICTs global cities are able to capture a large share of high level activity and supply the world from their headquarters. Thus in these cities or city regions the widest income gaps are likely to be found.

*Synthetic approach*
The perspective followed in this book is consistent with a historical materialist approach as it foregrounds the reproduction of everyday life from a feminist standpoint, that is it seeks to highlight gender inequalities associated with existing forms of development. It also draws upon the ideas of individualization associated with Beck, while recognizing that the extent of individualization is not uniform and that the state and other social institutions, considered within the Regulation School and welfare regimes perspectives, can still affect the way that different parts of the world can influence their trajectories, albeit within the prevailing neo-liberal global agenda. In this way while being implicitly critical of the current capitalist and patriarchal social system it nevertheless seeks to identify possibilities of change from within to move towards a more inclusive society.

natural, technological, social and cultural conditions and material causality refers to the way these existing circumstances shape decisions.[9] In the contemporary global economy where for example Coca-Cola, Nike trainers and Microsoft software can be bought almost everywhere though not by everyone, these circumstances are increasingly shaped by capitalist social relations of production (*E1* in Figure 1.1), so to understand the lives of different people and places it is important to understand how capitalism works. People may not be engaged in capitalist social relations of production all of the time, they may work in agribusiness and then return to smallholdings or be self-employed and then work as employees, and there are small groups of people who so far remain almost entirely untouched by capitalism, but the lives of the majority of the world's population are shaped by their direct or indirect connections with the capitalist economy, though how they act in these circumstances will vary. Outcomes are the result of both ideal and material causality or human action and their context thus they can never be predicted with certainty – only tendencies can be identified.

Within capitalist societies the profit motive (*E2* in Figure 1.1) underlies the production and distribution of social output. Indeed this motive is the internal dynamic of capitalist society and underlies the decision making strategies of small firms as well as the major capitalist organizations that continue to shape the trajectory of social development within and between localities. There are always exceptions – some people are entirely self-subsistent and some small firms and independent operators are more concerned with 'moulding their own lives rather than conquering world markets' (Beck 2000a: 54–5) – but as long as they sell their products on world markets their actions will be constrained by competitiveness and profit. This motive also underlies the related processes of globalization, the organization of work and relations between employers and employees or the labour process and so is crucial to an analysis of social change.

Profit arises in the production of commodities. See Box 1.2, which illustrates the circuit of capital. Money is advanced to buy equipment, components and labour, which are then combined in the production process to produce commodities that contain more value than the parts of which they are composed. That is, labour adds value by applying skills; the value of any equipment or components used is simply transferred to the new commodity. When the commodity is sold this added value is realized; part is used to pay for the labour, materials, equipment, land or factory costs and the rest is profit.

There are two main lines of struggle or contestation in capitalist societies. One is between firms or between capitals to capture the market; that is, each firm tries to ensure that it is its products and not its rival's that capture the market. Firms or capitals are constantly seeking for new ways to lower the costs of production of existing products and to create markets for new ones. The second struggle is between firms and their employees or between capital and labour. There is always a tension between how much goes to pay the employees, including all those in design, production, marketing, managing, advertising and any other functions that contribute to the marketability or exchange value of the commodity, and how much can be profit.

Contemporary production in the global economy is highly complex. Box 1.3 illustrates this complexity by tracing the production pattern through the distribution of the revenue from a car sold by General Motors. Some activities will be carried out by branch plants of General Motors, some by joint ventures between General Motors and local firms and others may be entirely independent firms. Each firm will try to remain competitive and profitable by constantly reviewing its production processes and negotiating the best prices, constrained by the knowledge of competition from other firms. Moreover, within each firm the struggle between pay and profit will occur, constrained by the knowledge that the firm could move on to lower cost labour elsewhere.[10] The way economic activity is organized and located across the globe lies at the heart of uneven development (*E3*), which is discussed in greater depth in Chapter 3.

The struggle between firms and between capital and labour underscores the dynamic nature of capitalist society. It explains why the organization of work

Box 1.2 The origins of profit in capitalist societies

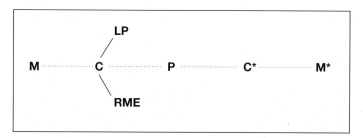

Money (M) is used to buy labour power (LP), raw materials and equipment (RME) which are combined in the production process (P) to produce new commodities (C*) which contain more value than the material of which they are composed because labour adds value, thus C* > C and M* >M.

This is why production or the labour process is also referred to as the valorization process. More specifically workers or labour add value to and release the value already contained in the raw materials and equipment by making new commodities. Some of this value contributes to the maintenance or social reproduction of the workers and is paid to them as wages – the rest is profit.

Because the labour process is simultaneously the valorization process or process through which profit is generated, it is carefully organized and constantly scrutinized. Changes are continually made in order to maintain and increase the rate of profit. Thus capitalism is a dynamic system technologically. The profit is only realized if and when the commodities are sold and converted to money, that is as C* becomes M*.

is constantly scrutinized and day to day adjustments as well as longer-term changes linked with new technologies continually made. There are different ways of increasing profit: extracting more value added from the existing workforce by increasing the intensity of work or introducing new technologies which increase labour productivity; transferring work to lower cost locations; or expanding the overall scale of operations. These processes lower the cost of production and increase the competitiveness of individual firms, thus making it more likely that they will capture a greater share of the market and increase profits. There is a contradiction however between the interests of individual capitalists and capitalists in general because although profits will rise if wage costs fall, other things being equal, capitalists also need markets for their products, i.e. people with income, and so capitalists benefit from other capitalists who pay higher wages and from the expansion of incomes more generally.

---

*Box 1. 3* The global web of value

When an American buys a Pontiac le mans from General Motors, for example, he or she engages unwittingly in an international transaction. Of the $10,000 paid to GM, about $3000 goes to South Korea for routine assembly opera- tions, $1750 to Japan for advanced components (engines, transaxles and electronics), $750 to West Germany for styling and design engineering, $400 to Taiwan, Singapore and Japan for small components, $250 to Britain for advertising and marketing services and about $50 to Ireland and Barbados for data processing. The rest – less than $4000 – goes to strategists in Detroit, lawyers and bankers in New York, lobbyists in Washington, insurance and health care workers all over the country, and General Motors shareholders – most of whom live in the United States, but an increasing number are foreign nationals.

*Source*: See Reich (1991).

---

In the early years of the twentieth century the introduction of Taylorism or scientific management allowed a more extensive fragmentation of work including a division between mental and manual labour. Combined with the conveyor belt of Henry Ford these new labour processes led to enormous increases in labour productivity, the full benefits of which could however only be realized by increasing the consumption standards of working people. Thus it was not until the middle of the century when the United States and Europe introduced Keynesian macroeconomic management, which increased economic stability, and expanded the welfare state, which improved education and housing standards, that the potential benefits from increased productivity were realized. The French Regulation School describe the ensuing era of unprecedented productivity increases, rising profits and rising wages as a new regime of accumulation, which they termed Fordism (see Box 1.1).[11] Without this theoretical perspective Robert Reich (2001a) points out that one of the ironies associated with the contemporary era, the new economy (*E5*, discussed below), is that the sharp polarization of incomes means that many working people cannot afford to consume the goods and services they are producing and suggests that some equivalent of Keynesian economic and social regulation which recognizes the global nature of contemporary economies be reintroduced to restore social tranquillity. These issues are discussed further in Part IV. The point here is that the economic analysis of capitalist society has to be combined with analyses of social and political (*S* and *P* in Figure 1.1) developments discussed later on. First though some other aspects of economic change are considered.

As the organization of production has become more functionally fragmented, people are needed to recombine the different activities, manage the accounts and organize their financing. In Box 1.3, all the activities were indirectly connected

12

to car making but there is an increasing range of financial activities (*E4*) where connections with physical goods are increasingly remote. Firms raise money by issuing shares on financial markets but there is a whole range of secondary markets in share options, derivatives (futures, options and swaps), government bonds and foreign currencies where individuals and institutional investors, some of whom control vast fortunes linked with pension funds, make profits through trading.[12] No new value is produced, but distributional changes occur as funds are rapidly switched between these markets in response to small changes in prices. People working at high levels in finance are among the highest wage earners and although they deal in a virtual world they have very real effects on lives elsewhere. These traders typically hold and act upon very prescriptive and orthodox ideas about how organizations and national economies should be run and if organizations or countries deviate from these expectations then they transfer funds, which can undermine companies and whole countries if speculation takes place against their currency. The speed of switching has increased with information/communication technology ICT and this has led to financial instability as financial shocks or crises can rapidly reverberate across the globe. Furthermore the proliferation of financial markets and their separation from the real economy have undermined long-term investment, which is often necessary for modernization. The Asian model of development, for example, was based on longer-term connections between finance and industry, but the IMF and World Bank insisted on changes to this model as a condition for receiving aid during the Asian financial crisis in the late 1990s (see Chapter 8 and Box 8.3).

The widening prevalence of neo-liberalism is one of the reasons why the traditional powers of nation states have been undermined, that is while 'democracy' may be spreading across the world, the range of issues over which 'the people' have effective choice is simultaneously narrowed by the preferences and prejudices of people managing the supra national institutions and global financial markets. Thus the Asian model of development mentioned above has to some degree been undermined as a consequence of the conditions attached to IMF aid, and the capacity of the left wing government elected in Brazil in 2002 in response to falling living standards to reverse economic decline is constrained by fears of financial instability should it deviate too far from the norms and expectations of the IMF and the World Bank. The IMF economic orthodoxy has affected many countries that sought its assistance, making the 1980s a 'lost decade' as rates of growth became negative (UNDP 1997). Argentina, once the richest country in Latin America, has experienced dramatic economic decline following compliance with the neo-liberal economic orthodoxy, such that by 2002, 50 per cent of the population were living below the poverty line (Lewis 2002). Structural adjustment policies are discussed in some detail in Chapter 3 (see Box 3.1) in the context of a review of neo-liberal development theory. Not all countries have suffered to the same degree, which suggests that some discretion for economic management remains at the national level if only in terms of the extent to which these policies are adhered to in practice. Moreover some states are more able or more will-ing to implement policies with more inclusive outcomes and the precise conditions in every country will vary such that similar policies can have different impacts.

Referring back to the car industry example illustrated in Box 1.3, just over 50 per cent of the revenue went to work concerned with the physical pro-duction of the car, the rest going to strategists, lawyers, bankers and health care workers, and this typifies contemporary employment trends especially in the already industrialized countries where new employment has been increasingly concentrated in the service sector. The service sector is more polarized than manufacturing, with high paid jobs among the knowledge workers in finance, marketing, communications and information technology largely concentrated in more developed cities and regions, and low paid, more generic work in services and reproductive activities such as caring, cleaning and catering having a much wider spatial distribution. This division is central to 'the new economy' (E5) which is discussed in some detail below because it provides interesting insights into the nature and continuation of widening social and spatial divisions in this divided world.

## THE NEW ECONOMY

The 'new economy' is a concept that has rapidly entered academic and media discourse in the last few years and like globalization is a widely used term but with several different meanings. The new economy is characterized by global-ization, increasing use of computing and information technologies, growth of knowledge goods and employment polarization, feminization and new patterns of working. Optimists such as Alan Greenspan (1998) refer to the almost unprecedented coexistence of economic growth and a booming stock market with low inflation, tight labour markets and low wage pressures. Pessimists refer to the development of polarized and more precarious forms of work (Sennett 1998; Beck 2000a) associated with globalization and deregulation, which, in turn, have generated problems for the sustainability of families and com-munities.[13] Danny Quah (1996 and 2003) and Robert Reich (2001a) link these different dimensions arguing that the positive and negative aspects of the new economy are opposite sides of the same coin and form part of an emerging digital divide. That is, some of the essential characteristics of the knowledge based economy that contribute to economic growth simultaneously exacerbate social division,[14] putting increasing pressure on the individual work–life balance and the maintenance of overall social sustainability.[15] By combining their ideas with those of feminist economists, in particular Nancy Folbre and Julie Nelson (2000), it is possible to explain not only the paradox of increasing social divisions at many different spatial scales but also why this takes a gendered and often an ethnic form.

For Quah, the defining feature of the new economy is the increasing range of goods and services, from business computer software to music, that take the form and properties of knowledge goods, namely: weightlessness, infinite expansibility (that is very low production costs) and non-rivalry (that is where one individual's consumption does not prevent another's, so two people can use the same software program but cannot eat the same chocolate hobnob (Quah 1999). Weightlessness or dematerialization, for example, mean that 'international trade becomes not a matter of shipping wine and textiles from one country to the next, but of bouncing bits off satellites' (Quah 1996: 7), creating a disrespect for physical distance, potentially an infinite global reach and tendencies towards more even development socially and spatially. However, in practice, as Quah points out, the opposite tendencies can be observed leading to increasing inequality. So how can this paradox be explained?

Infinite expansibility means that knowledge goods are subject to increasing economies of scale and therefore a tendency to monopoly. Although the costs of producing the first copy of any knowledge good can be enormous, replication costs are minuscule. Thus, firms price to recoup their initial research and development outlay, but the low marginal cost means that they can always lower their prices to eliminate potential competition. Furthermore, some firms come to dominate the market even though they may not produce the best technical products, as ICT software and equipment are often linked to other products, locking consumers into particular networks which generate externalities. The potential consumer is also likely to require certain kinds of equipment to be able to receive and make use of the product. Thus while 'dematerialized content is freely reproducible by the originating agent, it can be costly for the receiving one to use' (Quah 1996: 7). Furthermore, the existence of dematerialized products contributes to the 'superstar effect', which helps to explain increasing income polarization. Quah (1996) demonstrates this by explaining why the income differential between opera singers is greater than that between shoemakers. He argues that it makes little difference to the singers' effort whether they are singing to 2 or 20,000 people but consumers generally prefer to listen to singers of greater rather than lesser renown, even though for the majority of listeners there is probably minimal perceptible difference. The almost cost-less replication of products means that market size, market share and correspondingly income taken by top singers is unlimited by distance. So 'the winner takes all', hence the wider income dispersion for the growing range of goods and services where dematerialized replication is possible.

Quah (1996) suggests that one reason why people accept widening inequalities is because of increasing social mobility. That is the poor tolerate the rich because they can see a greater opportunity for becoming rich themselves. However it is also clear that there are specific features associated with the nature of work in the new economy that make the chances of becoming rich through work systematically uneven. Further, the uneven spatial distribution of different activities with different earnings potential including clustering, together with 'risk sorting', also means that the chances of becoming rich are likewise spatially differentiated.[16]

Reich's (2001a and 2001b) analysis of increasing economic inequality and insecurity in the new economy also relates to both sides of the digital divide. He argues that the new economy is characterized by intense competition and increased risk and uncertainty for employers caused partly by the ease of switching between suppliers. These risks are passed on to employees by the use of subcontracting, contingent pay and contracts and longer working hours. Knowledge is a key asset, and firms are often prepared to pay high sums for innovative people, albeit on short-term contracts. Buying people in or hiring freelancers for specific projects is often preferable to in house training, given uncertainty about future skill requirements. Employers seek to maximize the amount of 'billable' work they get out of these people, consequently work is increasingly intense with frequent deadlines. As well as being pressed to do so by employers, high level employees or virtual employees work long hours because of the high short-term opportunity cost of not working, given the high current pay and the uncertainty of future contracts, and because they often enjoy their work, seeing no clear distinction between work and life.

In the US the dual adult household worked an equivalent of seven additional weeks a year in 2000 compared to 1990, each person often working 50–60 hours and referred to as DINS ('dual income no sex') by the media (Reich 2001b). Similarly in the UK over 25 per cent of the workforce worked more hours than permitted under the EU Working Time Directive (TUC 2002). Not everyone considers long working hours a problem because they often reflect worker preferences. Richard Reeves (2001) for example argues that time at work increasingly involves doing interesting things in attractive physical and social environments and so may be preferred to watching a TV soap, carrying out domestic work or looking after children. In part following the ideas of Arlie Hochschild (1997) Reeves argues that:

> while the workplace is growing in attractiveness for many people home, or 'life' is looking a bit gloomy. For dual-earner couples with children, life outside work is one of fixed timetables (childcare), conflict (whose turn is it to pick up the kids?), low-skill work (cooking, cleaning, nappy disposal) and thankless masters and mistresses (the kids). As work enters the post-industrial era, home life has become industrial.
>
> (Reeves 2001: 128)

These long hours may be self-chosen and there may be some truth in this illustration for some people, or on occasions for many but Reeves (2001) pays little attention to the terms and conditions of employment for those who might provide childcare (see Chapter 4 on the global care chain) and domestic services or whether they similarly would welcome increased working hours.

Increased time pressures together with the increasing feminization of employment are leading to growing demands for marketed services whose inherent characteristics lead to low pay. Some major companies are providing concierge services and 'lifestyle fixers' for their elite employees, such as meals, shopping and dry cleaning services, as well as organizing childcare, dog walking and house maintenance.[17] These companies are advanced in that they recognize there is rarely a 'wife' at home to play this role and they doubtless also aim to relieve the strain on these workers. They are also self-interested and the services designed to increase their employees' productivity by ensuring their work time is focused on the job as well as facilitating the long working hours. The people working in the sectors supplying these services, disproportionately women and people from ethnic minorities, are sometimes overlooked in discussions focused on sectors or clusters in the new economy yet they clearly play a vital role in sustaining development. They are however generally low paid and have little chance of becoming rich themselves, because of the nature of the work they do and because of its low social valuation, despite being employed by large-scale corporations that have become increasingly involved in supplying cleaning services to hospitals, offices and private homes. For these people the divisions in the new economy are likely to be permanent.

Care/reproductive work is highly labour intensive and in contrast to the opera singer discussed by Quah (1996) is intrinsically not infinitely expansible or non-rival and in William Baumol's (1967) terms, inherently technologically unprogressive. For example, although a professional childcare worker can care for more than one child simultaneously, there is a fixed and relatively small limit, constraining productivity, market share and earnings. This is analogous to but not quite the same as Baumol's conundrum about how to increase the productivity of a string quintet. It is impossible to increase the productivity in a live performance without changing the work being played, but music does share the properties of knowledge goods, and CDs or Internet relays would expand output by creating a remote audience. These properties simply do not apply to care work. Furthermore, good quality childcare is probably associated with positive externalities in the form of better motivated, trustworthy workers, less crime and so on. But those providing the services cannot realize these gains. Furthermore, women disproportionately carry out caring work and their skills are frequently taken to be inherent characteristics of womanhood and rarely rewarded equivalently in monetary terms to stereotypically male skills in for example car maintenance. Pay for caring for the elderly, despite the strength, patience and skills that are often required, is rarely much above the minimum wage, especially in neo-liberal economies[18] which suggests that pay is deter-mined more by who does the job and also perhaps for whom it is done[19] rather than by material competencies, an issue which affects gender segregation and pay determination more generally.[20] Thus issues of power and gender relations, labour organization and the extent to which the state is involved in the provision of collective services and in the negotiation of labour relations all influence pay determination and well being so it is important to combine these abstract analyses of economic processes with theories which explain prevailing political and social relations (P and S in Figure 1.1).

This theoretical discussion of economic processes is important for identifying processes or tendencies of change or the partial dynamics of contemporary life but they reflect market logic, which while almost universal, in reality materializes in specific national and local settings. These settings are in turn influenced by prevailing political and social structures and traditions, and ongoing political struggles which can and indeed do influence the expected roles of women and men, the existence and scale of a minimum wage, the degree of poverty and inequality, the form of service delivery, including education and health care and whether they are provided by the state or through the market. These issues in turn offer feedback to and influence the content of material reproduction and by so doing shape specific geographies, hence the need to move forwards and backwards between different theories and between theories and empirical analysis in specific contexts.

## POLITICS AND THE STATE

Economic theories can explain the tendency towards widening social and spatial divisions but the extent of inequality varies. Countries with similar levels of economic wealth measured by gross domestic product (GDP) per capita can have quite different levels of development as measured by the United Nations Human Development Index (HDI) and different degrees of internal inequality and poverty, differences which are illustrated in Chapter 2. These differences will in part reflect past patterns of development, but also indicate that even within contemporary global capitalism with its pressure to harmonize economic and social policies, it is still possible for different states to follow different development trajectories (*P1*), determined by internal political processes and choices. Gösta Esping-Andersen (1990 and 1999) has defined three different welfare regimes (*P2*), social democratic, conservative corporatist and liberal market, according to the extent to which the state provides resources to people as social rights (decommodification and defamilialization)[21] rather than relying on market provision or unpaid female labour in the home (see Box 1.1 on p. 6). The analysis has mainly been applied to already industrialized countries but as models or ideal types they represent quite different outlooks on the role of the state and as such provide valuable insights into the ways that societies in general can and indeed are organized differently with differing implications for the extent of social divisions. The social democratic and liberal market regimes are discussed in more detail below as they represent quite contrasting traditions.

Within the social democratic tradition the state provides social resources, health and education to enable people to be effective citizens, that is social policy is not only valued for itself but considered a precondition for social cohesion and economic efficiency. In this model, the state provides services of the highest standards to all citizens so the market is effectively 'crowded out', i.e. there is little to gain from going private. The state also socializes the cost of child and elder care through collective provision to maximize capacities for individual independence. By eradicating poverty, unemployment and complete wage dependency, political capacities are increased, social divisions are diminished and barriers to political unity reduced.[22] Interestingly this perspective considers social rights necessary for citizenship or political capacity and effective participation rather than simply seeking the latter by opening up discussions with currently marginalized groups, something currently very much in vogue and often a precondition for international funding, issues discussed further in Chapter 9. To sustain this high quality and extensive cover, however, high taxation and full employment are necessary.

The liberal market model by contrast places much greater emphasis on individual responsibility and market provision, which are equated with greater freedom and efficiency. State provision of services or social rights is thought to undermine work commitment and encourage people to become scroungers. The welfare state is therefore something of a residual providing only a minimum level of service. There is commitment to formal equality but no recognition of the structural barriers that may prevent apparently equal opportunities from being realized substantively. Tessa Jowell, at the time the UK Labour government's minister for public health, illustrates the limitations of this liberal conceptualization of equal opportunities in the following way:

> When I had my baby, there were five babies lined up in their cots like runners in a race – but the most important thing had already happened to them, the circumstances they were born in. One was going straight into care, because his brother had been sexually abused at home; one was going back to a bed and breakfast; the father of one had just lost his job with the closure of a steel works; and two were going cosily back to well provided homes.
>
> (Jowell cited by Whitehorn, *Observer* 1997)

Clearly if these babies are treated by the state in the same way then they are scarcely being provided with equal opportunities in terms of their life chances. In relation to child and elder care, some provision is made for the destitute but the remainder buy care or rely on unpaid care within the family. Thus within liberal market economies the majority of the population are differentiated through their market earning capacity while those dependent on social welfare will be more or less equal with each other, but they will be equally poor, generating wide social divisions.

Esping-Andersen's (1990, 1999) work relates mainly to social policy but his welfare regime types correspond to the regulatory frameworks identified by Danielle Leborgne and Alain Lipietz (1991), which are defined according to the degree to which relationships between the state, firms and employees are negotiated or determined through the market. Specifically they identified five pathways taken by countries in response to the crisis of Fordism[23] which involve increasing degrees of negotiation between the social partners, i.e. the state, capital and labour: neo-liberal/neo-Taylorist, Californian/individualist (which refers to the way in which contracts between firm and employees are individualized), Toyotism (which refers to the significance of the firm in providing for the general well being of employees), corporatist (West German) and social democratic (Kalmarism). Focusing again on the two extremes, in the neo-liberal model relations between firms and between firms and employees are predominantly market ones. Between firms, relations are likely to be adversarial and short term with few functional linkages between local and external firms, moreover relations with the outside world will be open and control over domestic markets may be lost. Relations between capital and labour also take place largely through the market and where regulations exist, they are increasingly dismantled or bypassed by new contingent forms of working, such as short-term contracts or use of agency workers, where regulations are difficult to apply. Social protection is regarded as archaic and against the assumed common interest of firms and their employees for greater competitiveness. The state has a distant relation with the economy and is confined to maintaining a stable macroeconomic framework. Any specific problems are left to the affected areas to resolve or met by ad hoc responses. Countries following this kind of model may set up export processing zones, where regulations are even fewer and which are occupied by firms attracted by low labour costs or tax concessions with few links to each other or to domestic firms.

In between the neo-liberal and social democratic models relations between individual firms and the state are more likely to be negotiated.[24] The social democratic model is characterized by strong vertical near integration between firms, i.e. stable agreements with networks for exchanging information, allowing an inter-sectoral diffusion of knowledge giving local firms greater control over domestic markets. This kind of model has parallels with the networked firm described by Castells (1996) or in value chain analysis (discussed in Chapter 3 – see Box 3.6). Within the regulation perspective, however, rather than just referring to relations between firms, the model applies to prevailing cultural and social norms about the relations between all the social partners, and thus to the economic, social and political environment. The relations between labour and capital are characterized by more stable wage contracts and mutually advantageous compromises are negotiated in relation to modernization which occurs through training and retraining and encouraging workers to be adaptable through becoming multiskilled rather than by being permanently mobile. In this context the state brokers or mediates agreements between firms and between firms and workers. As a consequence more functionally integrated and stable clusters of activity are more likely to form and stimulate endogenous growth.

Clearly these models are ideal types and different countries are unlikely to match any exactly, but the purpose of models is to provide a representation of reality that reflects *some* of its properties, which in turn provide guidelines for investigation in specific contexts. In contrast to the prevailing neo-liberal orthodoxy however it would seem that where and indeed when states have played a more involved role they have generally performed better economically.

Taking countries at similar levels of economic development, Scandinavian countries with a social democratic tradition have tended to perform better economically and more inclusively than the neo-liberal United States or UK; similarly the newly industrializing countries of Asia have performed better than those in Latin America, despite the Asian financial crisis. Moreover for OECD countries, if the period of 1960–1973, when much greater regulation existed both internally and externally in relation to capital controls and labour markets and when the state owned key sectors, is compared to the period between 1979 and 1993, which is characterized by much greater liberalization and privatization, growth in GDP and productivity were twice as fast in the former (Singh 1999a). For European countries unemployment was three times lower in the former period and although the United States has a record of lower unemployment in the second period this has occurred in the context of zero increases in real wages as well as a very high rate of male imprisonment, especially for young black males.

The advantage of analyses that include the state and labour movements as well as firms is that policies which reflect the interests of the people and places can be more easily identified. Linking the analyses of Leborgne and Lipietz (1991) with those of Esping-Andersen (1990) also allows questions of reproduction and care to be brought into the analyses. In the context of globalization, however, where markets are effectively global, the capacity of nation states to follow a more social democratic route has been questioned. Some of the powers of nation states may have been lost upwards to supra national institutions and downwards to regions and cities and non-elected institutions may be playing a greater role in the organization of economic and social affairs, i.e. governments may to some degree have been displaced by governance, issues discussed further in Chapter 8. Furthermore different states have different degrees of power, depending on their level of development, their size and their geo-political position, as well as their ability to secure support from their population either by consent or coercion, which is also reflected in their power within international institutions. Nevertheless while analyses conducted at an aggregate level can indicate how states have lost powers, finer comparative analyses between countries indicate the continuing significance of national policies in terms of comparative well being, issues that again can be illustrated in specific geographical settings but within an understanding of broader processes generating changes which may constrain these powers. The state also plays a part in influencing and being influenced by social and personal relationships, which have also changed in the contemporary era and are discussed in more detail below.

## SOCIETY

As economists write about the new economy, sociologists use terms such as the risk society, reflexive modernization or the second modernity (Beck 1992, 2000a) to encapsulate the changing times (see Box 1.1). Ulrich Beck identifies two main forms of risk: those associated with the unpredictable effects of science, especially on the environment; and social, biographical and cultural risks in everyday life arising from the erosion of traditional social structures in work and the family, in particular the increase in flexible and contingent forms of working, which he terms the 'Brazilianization of the West', and the increase in divorce and new styles of family. These processes are collectively termed individualization (*S1*), which refers to the way that people have been liberated from traditional roles and structures and so have greater freedom to author their lives but at the same time have much less social support within the workplace, family or community to do so. People have to construct their own work biographies as traditional career structures have declined and the temporal and contractual fragmentation of work loosens social ties between people and places; 'family, neighbourhood, even friendship as well as ties to a regional culture and landscape contradict individual mobility and the mobile individual required by the labour market' (Beck 1992: 88; see also Sennett 1998). At the same time, and partly because of these labour market changes, the traditional family is being displaced by a 'post-family' or a 'new negotiated provisional family composed of multiple relationships' (Beck 2002: 202–3) and a growing number of female headed and single person households.[25] Thus individualization is associated with risks as well as opportunities because new life patterns 'lie outside the classical employee's biography, outside union agreements and statutory pay scales, outside collective bargaining and home mortgage contracts' (Beck 2000a: 5, citing C. Clermont and J. Goebel). Furthermore structural inequalities remain, so different people have very different levels of resources with which to either shape their biography or to confront the risks. Thus:

> instead of the promised classless society the fine old distinctions are suddenly changing into intense social polarization. Instead of an elevator effect for all layers in society, a revolving door effect admits a few winners and casts out many losers.
>
> (Beck, 2000a: 53, citing T. Westphal)

These circumstances have quite significant political implications because inequality is perceived as the consequence of individual success or failure rather than deriving from wider social and economic processes linked with the region, locality or social class. That is, 'problems of the system are lessened politically and transformed into personal failure' (Beck 1992: 89) so individuals feel responsible for developing their own solutions. In the case of the family, for example, Beck (2000a) argues that struggles over time within dual earning households appear to be individual ones but in fact reflect the imbalance between the extent of collective care services and the new patterns of life and work.

Clearly the extent to which traditional structures have broken down or actually provided economic security is variable between places and between people. Even in societies where they existed the proportion of women covered by collective bargaining agreements was far lower than for men, and likewise neither mortgage contracts nor the nuclear family were universal even within western style societies and even less so elsewhere. Moreover the extent to which job insecurity is actually increasing has also been challenged.[26] Further, while Beck (1992) argues that the freeing of traditional gender relations places women at greater risk of poverty because they lose the protection of the spouse's income, it is important to recognize that household incomes were not necessarily shared, so some female headed households are more secure after separation because at least then they have control over the income they do receive. Statistically though, women are likely to be poorer, as on average women earn less than men and are underrepresented at the higher levels in the employment hierarchy in all societies where data exists (see Chapters 2 and 4). These statistics also shed doubt on a simple association between feminization of employment and increasing gender equality, though positive benefits in terms of self-esteem, empowerment and socialization are reported in qualitative research carried out in a wide range of countries, even when the terms and conditions of work are arduous and paid work is added to domestic work.[27]

Many of these issues are contingent and need to be explored empirically but ideas about family breakdown, the crisis of the family, the end of the male breadwinner model[28] or the end of patriarchalism,[29] the development of new family forms and expansion in the number of female headed households[30] are discussed in a number of countries across the globe. There is particular concern about social sustainability and the welfare of children brought up in an insecure environment in terms of paid work and the family and where time pressures induced by the juggling of paid work and caring responsibilities intensify.[31]

One of the driving forces behind the changes is the increasingly competitive nature of the global economy, which is associated with work fragmentation and increasing mobility, and the other is the feminization of paid employment, partly linked to economic changes but also leading to growing female economic independence which challenges the nuclear family, marriage and sexuality which have all become freer and less traditional. Women have always worked in one way or another but they are increasingly working outside the home, which as discussed in greater detail in Chapter 4, gives rise to a certain sense of empowerment. This engagement in paid work, once a demand of feminist movements, has now become a social expectation in many countries. It is built into IMF and World Bank structural adjustment and economic reform programmes and forms a key part of the European Union's strategy for economic growth and competitiveness. Specifically, the European Union seeks to expand the overall employment rate from just over 60 per cent to 70 per cent by 2010 and the female employment rate from 54 per cent to 60 per cent (European Commission 2001). Reference is made to the need for improved childcare but the motivation for expanding female employment is as much linked to the economic strategy as to equal opportunities. Esping-Andersen (1999) also regards expanded female employment as conducive to both micro and macro welfare but points out that while women are becoming more engaged in the labour markets there has been no parallel increase in men's care and reproductive work, leaving a gap he suggests be filled by the state or market as men are unlikely to change, and, in his view, it would be inefficient for them to do so. Quite often, especially in northern and western European countries, this gap is filled by women taking paid work on a part time basis and doing an unequal share of housework and childcare, but this partial engagement in the labour market is not always recognized by those who argue that gender equality is increasing or by policies which increasingly assume that all adults are involved in income generating activities. Both of these assumptions are ahead of reality and thereby fail to appreciate the processes that sustain inequalities between women and men especially in the long term (see Lewis 2001). Other ways of filling this gap are simply by women working a double shift, in paid work and then in the home, or by grandparents or older siblings, but sometimes the gap is left empty, leaving children to fend for themselves. When the state or the market fills the care gap employees are disproportionately female and low paid, generally more so when employed in the private market.[32]

There are some common trends across the globe in terms of changing social relations which are linked with the increased involvement of women in paid work, but the outcome of these changes will vary as people start from very different social and gender relations (S2). Even in the Nordic countries, which appear to have proceeded furthest along this route, and where women's earnings and employment rates are within 15 per cent or so of men's,[33] horizontal and vertical segregation continue in paid work and domestic labour and child-care continue to be unequally divided. A detailed study carried out in 1990 of the total work load (TWL), which includes paid and unpaid work, of over 1000 women and men in Sweden, matched for age, family situation, education and occupational level, found that women's TWL exceeded men's. The study was repeated ten years later (2000) and found only very small changes had been made towards equality, even though here, as elsewhere, the role of fatherhood is socially valued.[34]

Family structure may be changing and women are more likely to be in paid employment, but in no society where statistics are gathered do women fare better than men. It is possible to explain analytically why caring work tends to be low paid, but this does not explain why it is that women are over-represented in this sector, or why people across the globe choose to accept this market logic. So gender continues to be an important organizing principle or structure even though there is significant variation between women in terms of social class, ethnicity, age, qualifications and so on. Explaining these continuing gender divisions is complex and a number of different categories and concepts have been used, including gender regime, gender order, gender arrangements and patriarchy[35] but these terms only identify and describe the form of gender relations in any particular country, rather than explaining why gender continues to play a role in shaping the structure of societies. Within neo-classical economics, new household economics explains gender differences in terms of comparative advantage, but this argument is rather circular. Men's comparative advantage in waged work is attributed to their higher pay, a consequence of their higher productivity, in turn explained by their skill, which is created through investment in education and training and on the job experience; that is, men specialize in waged work because of their higher human capital. Not only are the links between wages and productivity contentious but also the reason given for men's greater investment in human capital is that they expect an uninterrupted working life. Women by contrast are argued to invest less in themselves because they expect to take a break for raising children. As this division of labour between the family and work is *explained* by comparative advantage, demonstrated by men's higher wages, the argument is circular. There are many other theories of gender inequality in the labour market[36] but in Chapter 4 the cooperative conflict model is discussed in some detail because it links power relations in the household with external market factors within which household decisions are made.

Societies are also increasingly differentiated by race and ethnicity (S3) and this differentiation often takes a hierarchical form, with different ethnic and racial groups in practice being over- or underrepresented in different roles, spaces and levels of well being. These differences are most stark in South Africa, where economic inequalities are forming along similar lines to the former system of apartheid, which ended only in the 1990s. In the United States, where formally citizens have had equal rights for a long time, economic inequalities also fragment along racial lines with the Afro American population being disproportionately poor as well as being overrepresented in the prison population. Afro Americans constitute 13 per cent of the population but half of all prison inmates, and young black men are particularly vulnerable with one in three being either locked up, on probation or on parole (Parkin 2002).[37] Similarly to gender, there are many differences within ethnic and racial groups and just as not all women are oppressed neither are all minority individuals, but they face a higher probability of being so and therefore in racially and ethnically mixed societies this also forms a key determination of life chances.

Having identified simple determinations or cornerstones of social processes it is crucial to consider how they intermesh with each other and with the prevailing social, historical and geographical context. Figure 1.1 illustrates these relationships and in some ways is a contemporary adaptation of the methodology proposed by Marx in the *Grundrisse*, where he argued that:

> It seems to be correct to begin with the real and the concrete, with the real precondition, and thus to begin, in economics, with e.g. the population, which is the foundation and the subject of the entire social act of production. However, on closer examination this proves false. The population is an abstraction if I leave out, for example, the classes of which it is composed. These classes in turn are an empty phrase if I am not familiar with the elements on which they rest, e.g. wage labour, capital etc. These latter in turn presuppose exchange, division of labour, prices, etc. For example, capital is

nothing without wage labour, without value, money, price etc. Thus, if I were to begin with the population, this would be a chaotic conception of the whole and I would then, by means of further determination move analytically towards even more simple concepts, from the imagined concrete towards ever thinner abstractions until I had arrived at the simplest determinations. From *there the journey would have to be retraced until I had finally arrived at the population again, but this time not as the chaotic conception of the whole, but as a rich totality of many determinations and relations. The concrete is concrete because it is the synthesis of many determinations, hence the unity of the diverse.*

(Marx 1973a: 100–1, my emphasis)

Marx's simple concepts all relate to the capitalist economy and although he considered the connections between capital and labour as a social relation, other social relations, especially gender and ethnicity, are not specified in his approach, though if they were, they would not be inconsistent with the underlying methodology. What is important about this quotation however is the insistence that analytical, theoretical or in his words 'simple' concepts are necessary to understand the world. Having been identified, however, they have to be employed in particular contexts, that is 'the journey would have to be retraced' in order to understand reality but as a 'rich totality of many deter-minations and relations' rather than a 'chaotic conception of the whole'. Thus the spatial world is a synthesis of many determinations (not determinisms) or the outcome of a multiplicity of social dynamics operating at different levels and these need to be articulated and understood but they are not substitutes for detailed analysis, which has to take place in specific settings. For practical and logistical purposes there has to be some division of intellectual labour but it is important that all the determinations and processes are on the agenda so that an intellectual space is created for thinking about the connectedness of processes shaping change.

Some social theories try to encapsulate the interrelations between the economy, society, politics and geography. For example regulation theory links the economy and society and politics through the relations between the regime of accumulation and the mode of regulation but has tended to overlook the way such systems are also shaped by gender relations and how they will be experienced differently by different people according to their gender, ethnicity, age and stage in the life course.[38] Beck's (2000a) analysis of individualization and the risk society also links different elements. However, though tendencies towards insecurity and individualization may be universal the nature and pace of change are likely to differ between nation states and different types of firms depending on a range of other factors including the legislative framework and social and cultural norms leading to different regional and local outcomes. The welfare regime approach emphasizes different state practices at the national level but pays less attention to changes in global capitalism, which has generated

26

homogenizing tendencies, making the preservation of nationally differentiated economic and social policies difficult. Moreover, disaggregated analyses of sectors, organizations and localities indicate that common state policies are not experienced uniformly and so this approach needs to be incorporated within a framework which takes account of the broader global context as well as local institutions both of which influence specific outcomes. Likewise ideas about , the new economy outline a tendency towards widening social and spatial divisions but not how these take different forms in different social and political contexts.

The framework outlined in Figure 1.1 tries to indicate how the general processes that shape capitalist development are interconnected and how they are activated by and materialize within particular social and spatial forms. How things work out in practice will depend on how the many determinations, including those initiated by people in specific places, are synthesized or unified. Detailed comparative analysis on these lines is crucial if real 'spaces of hope' are to be identified within capitalism. This perspective may be dismissed as the ideas of a 'reformist tinkerer' rather than a 'utopian visionary' (see Harvey 2000) but it is perhaps more likely to provide an understanding upon which realistic alternative trajectories could be designed rather than starting from utopian visions where the complexities and inequalities of contemporary economic and social relations are assumed away.

## CONCLUSION

At the heart of globalization lies the organization of production and distribution on a world scale. These developments have been made possible by the widespread acceptance or domination of capitalism as a system of social and economic regulation, new technologies in transport and communications, and the development of global financial institutions and global companies. Corresponding with the globalization of production there is global geography consisting of localized concentrations of global power, for example global cities, localized production districts scattered throughout the world as well as places that, superficially at least, are overlooked by these developments. One of the purposes of this book is to provide a series of vignettes of people and places in different parts of the globe that are connected through this global context within the framework outlined in this chapter in order to try to explain some of the complexities of the contemporary world.

The rest of the book is divided into four parts. Part I outlines concepts and measures of inequality on an international scale and examines some theories of uneven development that aim to explain these inequalities. Part II deals with more concrete illustrations, focusing on how people are connected through these divisions at an international and regional level. It also focuses on the social implications of the feminization of employment and considers the extent to which such changes can be said to have equalized gender relations. Part III focuses on people and places in the new economy more empirically, looking at its uneven architecture and the widening spatial and social divisions between the 'superstar' regions and the rest as well as within the superstar regions where some of the greatest inequalities are found. Part IV considers ways of challenging the form of contemporary globalization through traditional state forms, NGOs and protest movements. While anti-globalization protesters have highlighted the intensity of inequalities and the corruption of many corporations and NGOs have been very active in criticizing the existing model of development and proposing alternatives, action through nation states singly and in combination may be the most effective way of peacefully challenging contemporary social and spatial divisions, provided enough people wish to do so. In documenting and explaining some of the contemporary processes leading to inequalities my intention is to convince more people of the need to do so.

## Further reading

P. Dicken (2003) *Global Shift: Reshaping the Global Economy in the 21ˢᵗ Century*, London: Sage.
J. Peck and H. Yeung (2003) *Remaking the Global Economy*, London: Sage.
D. Quah (1996) The invisible hand and the weightless economy, Centre for Economic Performance Occasional paper No. 12, London: LSE.

## Websites

Eldis Gateway to Development Information: http://www.eldis.org/.
Global Exchange: http://www.globalexchange.org/.
United Nations Human Development Programme: http://www.undp.org/.

## Notes

1  There were 9832 references listed on BIDS between 1998 and 2002; 2202 between 1994 and 1998 and only 148 between 1990 and 1994.
2  Writers referring to the death of geography include Ohmae (1990) and Robertson (1992).
3  Although proportionately the scale of migration does not exceed the nineteenth century flows from Europe to the United States (Hirst and Thompson 2002).
4  See Appadurai (2001).
5  See Barrientos and Perrons (1999).
6  There is an extensive literature which debates the meanings and measuring of globalization; see for example Hirst and Thompson (1996, 2002); Held *et al.* (1999).
7  The starting point is therefore material reproduction of everyday life. These material needs and wants vary in different societies at different points in time and 'whether

they arise, from the stomach or the *imagination* [my emphasis] makes no difference'
(Marx 1976: 125).

8  Knowledge is necessary to formulate effective challenges not simply to control and
can be derived from a range of sources.

9  Thus paraphrasing Karl Marx human beings make history but not in circumstances
of their own choosing. Any reader of Marx will note the closeness of this portrayal
of a historical materialist perspective to Marx's own writings (see especially Marx
1973b). Elsewhere I (Perrons 1999a) and with Mick Dunford (Dunford and Perrons
1983) draw more explicitly on his ideas. Here I wanted to give them a contemporary
rendition, reflecting what I consider to be the spirit of his ideas so as not to deter
any readers who may have unshakeable adverse preconceptions. I have not referred
to Giddens's ideas about structuration because I think they are in part responsible
for unwarranted deterministic readings of Marx.

10  In some theories workers are assumed to be paid according to how much they
contribute to the value of the product, but this is highly contentious because it is
clear that people in different parts of the world using the same technologies and
correspondingly contributing similar amounts of value added get quite different
levels of pay. It is much more likely that pay is determined by prevailing norms
and expectations about the worth of different kinds of work, the perceived value
of different kinds of labourers – women, for example, are quite often paid less than
men for doing very similar work – and the real costs of living in different locations.
In Saudi Arabia, where there are a large number of migrant workers, nationality is
a key determinant of pay – 'in most other parts of the world it would be incon-
ceivable to pay a supervisor less than a subordinate simply because of nationality,
but it is common in Saudi Arabia' (Mahdi and Barrientos 2003). Power relations
also influence pay and entry into certain kinds of work is controlled not simply
through qualifications but by restrictive hiring practices, which seem to persist
even in those countries where equal opportunities policies operate.

11  See Chapter 5 and Lipietz (1992), as well as Dunford and Perrons (1983) where the
issues are discussed in detail.

12  See Pollard (2000).

13  See for example Beck (2000a); Carnoy (2000); Hochschild (1997); Reich (2001a).

14  See Quah (1996).

15  See Reich (2001a).

16  See Reich (2001a) and Chapter 7.

17  See Chaudhuri (2000), Denny (2001) and Chapter 7.

18  For example in Brighton and Hove a Grade B auxiliary nurse would receive
marginally above the minimum wage (approximately 56 per cent of the wage paid
to the predominantly male street cleaners and refuse collectors) for doing 'routine
work like leg ulcers, blood sugar, urine analysis, blood pressure and bladder
washout' (Prism Research 2000: 44).

19  For example the fitness trainer for Cherie Booth, top lawyer and wife of the UK
prime minister, was paid £5000 a month while a worker in a care home for the
elderly who in practice is given a wide range of responsibilities will be paid barely
above the minimum wage, i.e. the former would earn roughly seven times as
much as the latter assuming that the trainer worked 35 hours a week for this one
client.

20  See Phillips and Taylor (1980).

21  A defamilializing regime is one that seeks to unburden the household and diminish

individual welfare dependence on kinship. A familialistic system (not pro-family) is one 'where public policy assumes – indeed insists – that households must carry the principal responsibility for their member's welfare' (Esping-Andersen 1999: 51).

22  See Esping-Andersen (1990).

23  The regime of intensive accumulation based on Fordism came to an end in the late 1960s and early 1970s. Productivity increases had reached their limits within this technology, social resistance to increased intensification of work was growing and competition from newly industrializing countries was increasing, thereby breaking the link between mass production and mass consumption within national markets.

24  One of the contradictions of purely market relations is low worker commitment and high turnover, which create insecurity and costs for the firms especially for employees with high market values. Consequently core workers may be employed on more permanent contracts but in an individualized way as in the Californian model. Toyotism relates to the Japanese employment for life model which by no means ever applied to all workers but did generate commitment and provide a degree of security for some; moreover state institutions collaborated with firms to modernize the economy. The corporatist state took this further as agreements were made between whole sectors, the state and the unions such that they would also cover smaller firms.

25  See Chant (1997).

26  See Doogan (2001).

27  For example see Grossman (1979); Kabeer (2000); Denman (2002); Perrons (1999b).

28  See Lewis (2001).

29  See Castells (1997) and Chapter 4.

30  See Chant (2002).

31  See Carnoy (2000).

32  See Ehrenrich and Hochschild (2003).

33  In 1999 the gender pay gap for Denmark was 11 per cent, and for Sweden 12 per cent. The gender employment rate gap for both countries was less than 5 per cent (EC 2001).

34  See Lundberg and Berntsson (2002).

35  There is a huge literature providing models of gender relations some of which provide gendered interpretations, critiques and alternatives to the welfare regimes perspective; see for example Connell (2002); Duncan (1996); Hirdman (1990); Lewis (1997, 2001); Walby (1997).

36  These theories include dual labour market theory, the Marxist industrial reserve army, patriarchy, and gendered moral rationalities; see Geske and Plantenga (1997) for a review.

37  While 1 in 210 white men are in prison, the figure for black men is 1 in 21 (US Bureau of Justice 1996). These figures vary geographically; thus in 1996 1 in 13 black men in Washington DC were in prison compared to 1 in 626 white men. More recent figures from the Bureau of Justice (2003) show that in 2001 1 in 6 black men were current or former prisoners, compared to 1 in 13 Latinos and 1 in 38 whites (Chaddock 2003). See also Chapter 8.

38  See Lipietz (1987), Jessop (1991) and McDowell (1991).

# Part I

# MEASURING AND THEORIZING INEQUALITY AND UNEVEN DEVELOPMENT

The contemporary world is characterized by extreme levels of wealth and poverty and despite increasing numbers of conferences, strategies, resolutions and accords to eliminate poverty, absolute poverty is increasing in some nations. So far there seems to have been little progress towards realizing the humanitarian goals set by world leaders at the start of the new millennium. These goals were to eliminate hunger, achieve universal primary education, promote gender equality and empower women, reduce child poverty, improve material wealth, combat HIV/Aids, ensure environmental sustainability and develop a global partnership for development. In particular the specific target of freeing the world from poverty by 2015 seems remote.

In 1999, just over 1.2 billion people lived on less than $1 a day, 30,000 children died every day of preventable diseases, one woman died every minute in pregnancy or childbirth and 800 million people suffered from malnutrition. Overall the poor are becoming poorer. During the 1990s 54 countries became poorer, 20 in sub-Saharan Africa, 17 from Eastern Europe and the Commonwealth of Independent States (CIS), 6 from Latin America, 6 from East Asia and the Pacific and 5 from Arab states. About half of the countries in Latin America experienced either a decline or stagnation in their income in the 1990s. Furthermore, 21 countries went backwards in their development measured on the human development index (HDI), a composite measure based on life expectancy, education and income in the 1990s. By contrast in the 1980s only four of the countries tracked by the UN experienced reversals on these measures. More specifically a larger proportion of the population went hungry in these countries than in the 1980s, in 14 countries infant mortality increased and in 34 life expectancy fell (UNDP 2003).

The 1980s was however dubbed the lost decade owing to the low levels of economic growth compared to the 1960s and 1970s, especially in Latin America, following the introduction of structural adjustment policies. These policies have been applied more widely under the prevailing neo-liberal orthodoxy, especially for countries seeking assistance from the supra national institutions, and resulted in the 1990s being a decade of despair or another lost decade for many people in many countries. Perhaps what is more dramatic about the 1990s is that this decade also saw some very high levels of growth in the high human development countries and spectacular rates of growth in urban China, especially in the coastal regions, and in parts of India. At the same time rising inequalities within countries, including an absolute decline in the real incomes of the poorest decile in the United States and the UK, means that the world has become an even more divided place, between the rich and poor countries but also between the rich and poor within countries.

This dismal account of the divided nature of the contemporary world is not however universally agreed and it is also correct to argue that overall world wealth has increased, illiteracy has fallen and life expectancy has increased. Moreover, some of the declines in the development measures can be attributed to deaths associated with HIV/Aids which has especially affected states in Africa, though the comparatively higher death rates from this virus can also be attributed to their poverty as the cost and the pricing policies of global corporations have until recently prevented these countries from obtaining the anti-viral treatments. So despite the adverse outcomes, many advisers stick to the neo-liberal agenda that others would argue has contributed to and exacerbated world poverty. Thus it is very important to be clear about the foundations for these claims and to understand the measures of income, inequality and development that are used. Furthermore it is also important to understand the processes underlying the statistics.

Part I therefore begins in Chapter 2 by reviewing some of the changes in the distribution of income on a world scale and the various measures of inequality and poverty. Different measures indicate different degrees of poverty and inequality and likewise whether they have increased or decreased, widened or narrowed. A further issue is that national GDP per capita figures disguise inequalities within countries; for example, the measure would increase even if only the incomes of the rich had expanded. Moreover income is only one measure of well being so the measures used by the United Nations Human Development Project, the HDI and its elaborations in the Gender Development Index and the Human Poverty Index, are each discussed and illustrated. While these measures encapsulate some of the wider issues affecting human choice, dignity and well being, they can also be criticized for being very summary. Indeed it would be unrealistic to expect the whole of human experience to be encapsulated by a few statistics. Nevertheless they can be used as benchmarks to identify change and to assess the performance of government, as bargaining tools, and moreover as means of generating interesting questions and hypotheses for further research. One of the points raised in the chapter and is a theme throughout the book is that countries with equivalent measures of wealth can have quite different measures of development which suggests that despite globalization nation states can still use their resources to affect human welfare in different ways.

Interesting and informative as these measures are in conveying aspects of the contemporary world in an efficient way, they are purely descriptive. They encapsulate certain trends but in themselves cannot explain those trends. Thus in Chapter 3 a number of theories that explain some of the underlying processes are discussed. Contemporary societies are characterized by globalization, which in this book is used descriptively to refer to the increasingly interconnected nature of the contemporary world and the corresponding need to contextualize analyses of people and places within this wider and more complex context. While a north–south divide is still very evident there are important divisions within the north and south, giving rise to an intricate pattern of uneven development. Moreover the focus of the book is concerned with increasing social and spatial divisions or the divided nature of the contemporary world. Correspondingly the theories discussed in Chapter 3 are ones which foreground the material connections between people and places across the globe. In particular theories of the international or global division of labour and global value chain analysis are highlighted, though reference is also made to a neo-liberal perspective and to the development of clusters.

Without some measures of inequality and some understanding of the processes generating them none of the declarations, intentions and commitments made by world leaders can ever be assessed. Understanding these processes however makes it clear that fundamental changes in the prevailing system of economic and social regulation, that go further than simply redistribution within the system, will be necessary if the inequalities are to be redressed. The purpose of Part I is therefore to document and try to explain contemporary social and spatial divisions at a general level. These issues are then explored in Parts II and III by referring to more detailed studies of people in particular places, at different spatial scales. Part IV deals with some of the processes contributing to the inequality and ways of responding to this inequality by nation states and by more radical protest movements.

# 2

# UNEVEN GEOGRAPHICAL
# DEVELOPMENT WITHIN
# THE GLOBAL ECONOMY

Globalization has increased economic and social interaction between countries but social divisions are widening and while the proportion of people living in dire poverty has fallen in recent years the absolute number has risen with the increase in world population. Currently 2.8 billion people (44 per cent) of the world's population live on less than $2 a day, and around 1.2 billion on less than $1 a day, while the richest 1 per cent receive as much income as the poorest 57 per cent (UNDP 2002).

As the world has become more homogeneous in some ways, inequalities between and within countries, regions and cities have been widening, and a significant number of countries in sub-Saharan Africa and the Commonwealth of Independent States (CIS)[1] have experienced declines in income. This chapter documents patterns of economic and social inequality between countries and begins by exploring different concepts and measures of economic and social well being.

## CONCEPTS AND MEASURES
## OF DEVELOPMENT

Income is crucial in contemporary capitalist societies and is often used to measure individual and national welfare and to compare living standards between people and between countries. Income is usually measured by gross domestic product or GDP, defined as the market value of all goods and services produced within a country; increases in GDP mark economic growth of the country as a whole, and increases in GDP per capita reflect increases in average individual incomes. Figure 2.1 illustrates the distribution of income across the world in 1993 and indicates the stark inequalities, with the 14 per cent of the world's population living in Western Europe, North America and Oceania having 45 per cent of the world's income. Overall, the richest 1 per cent of people (50 million households) received as much income as the bottom 57 per cent (2.7 billion people). More explicitly the 'total income of the richest 25 million Americans was equal to the total income of almost 2 billion poor people'

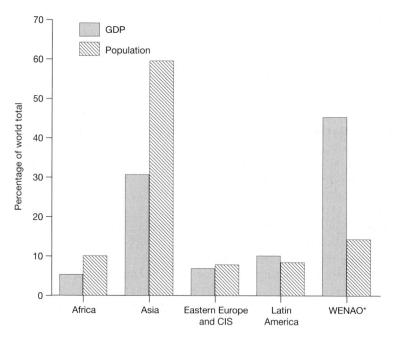

*Figure 2.1* World population and income: unequal shares.

*Source*: Calculated from Milanovic and Yitzhaki (2001).

*Note*:* WENAO = Western Europe, North America and Oceania.

(Milanovic 2002: 89).[2] Income is also unevenly distributed within countries further widening the extent of inequality; for example the assets of the 200 richest people, some of whom can be found in every continent, are greater than the combined income of 41 per cent of the world's people.[3] Moreover in the last few decades, while there have been significant increases in overall income, the distribution across the globe is generally thought to have become more uneven.

Overall there has been an expansion in world GDP, from just over five to eight thousand dollars per capita between 1975 and 2000 but this increase has not been experienced equally.[4] Rapid growth has taken place in the high income OECD countries, East Asia and the Pacific, especially the urban areas of China, and in urban India since the 1980s, but declines have occurred in sub-Saharan Africa and in the CIS countries. Here internal inequality has simultaneously widened, meaning that some people have experienced very severe declines in their income. Similarly in much of Latin America and the Caribbean, since the 1980s, increases in GDP per capita have been small while inequality has increased, again leaving many people with lower real standards of living.[5]

Comparing trends in economic growth and measures of inequality across countries is complicated as there are different ways of measuring GDP and different measures of inequality. Writers and politicians use measures selectively to support their arguments: advocates of neo-liberalism select measures that highlight growing prosperity and critics highlight those that point to rising inequality. However statistics themselves do not lie, rather they measure different aspects of reality. Understanding the different measures is therefore an important part of understanding the debates.[6]

Table 2.1 provides details of two frequently used statistics of income inequality, the Gini coefficient and the percentile ratio, and provides values for three countries. The countries chosen are two with high human development in Western Europe but with different welfare regimes and one with medium human development in South America.[7] With the exception of one measure for Denmark, inequality has increased, though the levels of inequality are lower in Denmark than in the UK, both of which have measures considerably below those of Mexico. These statistics illustrate how countries can still organize their internal allocation of resources in different ways, while remaining within a framework that tends towards inequality. Table 2.2 provides details of two different measures of income used to calculate these statistics, purchasing power parity (PPP) which measures real living standards and market exchange rate (MER) which reflects the comparative economic standing of countries. On a world scale World Bank data show that inequality has been increasing, though to a greater or lesser extent depending on which of these measures is used (see Table 2.3 and Table 2.4).[8]

World trends are the outcome of diverging tendencies at the nation state level, which in turn can disguise important widening divisions between regions and cities within countries. The OECD countries, East Asia and the Pacific, especially the urban areas of China, have experienced high and largely sustained increases in GDP per capita measured in purchasing power parity (PPP) since the mid-1970s, with some dips in the late 1990s associated with the Asian financial crisis. China has moved from having one-twenty-first of the OECD average in 1975 to one-sixth in 2002 and India from one-fourteenth in 1980 to one-tenth today (UNDP 2002). As indicated in Table 2.2, however, the PPP measure does not fully reflect the relative standing of countries and in an increasingly interconnected world potentially misrepresents well being, as possession of PPP dollars does not necessarily empower people to buy the globally marketed goods of which they are increasingly aware owing to satellite TV and the Internet. Furthermore as Robert Wade (2001: 9) points out:

> The reason why many of the poorest countries are hardly represented in negotiations whose outcomes profoundly affect them is that the costs of hotels and offices and salaries in New York and Geneva must be paid in US dollars or Swiss Francs, not in PPP dollars.

Thus although people may experience increases in their real living standards internally, they can be frustrated by their inability to share in the increasingly pervasive global consumer culture and correspondingly improvements in

*Table 2.1* Measures of income inequality

| *Measure* | *Meaning* | *Values and illustrations* | | |
|---|---|---|---|---|
| Gini coefficient | Measures the deviation from total equality. The Gini coefficient measures the area under the Lorenz curve, which plots the cumulative % of households on the *x* axis and the % of income they receive on the *y* axis. In a perfectly equal society *n*% of the population would have *n*% of the income, $x = y$. Thus a 45 degree line represents full equality. In a perfectly unequal society one person would have all of the income thus $y = 0$ for all $x < 100$, and $y = 100$ when $x = 100$ and so the shape of the curve would be the same as the axes, a left facing right angle. Usually the Lorenz curve has a new moon or loop shape of varying widths depending on the extent of inequality, and the Gini coefficient measures the area of this moon type shape. 0 = full equality; 1 = full inequality; see Figure 2.2. | Denmark | 1987 | 1997 |
| | | | 0.254 | 0.257 |
| | | UK | 1979 | 1999 |
| | | | 0.270 | 0.345 |
| | | Mexico | 1984 | 1998 |
| | | | 0.448 | 0.494 |
| Percentile ratios | A measure of the ratio between any two deciles in the distribution. Frequently used measures are: 90/10 which would measure the ratio between the extreme ends of the distribution specifically of the income received by the 90th decile or the top tenth of the distribution to the 10th decile, the lowest tenth in the distribution; 90/50 which gives the ratio of the top tenth to the median; 80/20 which gives the second highest to the second lowest decile; and 50/20, the median to the second lowest decile. These latter two measures exclude the extremes in the distribution. | Denmark | 1987 | 1997 |
| | | (90/10) | 3.22 | 3.15 |
| | | (80/20) | 2.12 | 2.18 |
| | | UK | 1979 | 1999 |
| | | (90/10) | 3.23 | 4.58 |
| | | (80/20) | 2.17 | 2.85 |
| | | Mexico | 1984 | 1998 |
| | | (90/10) | 8.69 | 11.55 |
| | | (80/20) | 4.03 | 4.60 |

*Source*: Luxembourg Income Study: http://www.lisproject.org/keyfigures/ineqtable.htm.

*Table 2.2* Concepts and measures of income inequality

| Statistic measured by: | Meaning |
| --- | --- |
| Purchasing power parity (PPP) | Real living standards. Measures income in terms of its purchasing power or ability to buy a comparable bundle of goods. This measure potentially allows for the lower cost of non-traded goods and services, such as housing or haircuts in poorer countries which may increase people's sense of well being beyond what would be expected from their money incomes. |
| Market exchange rate (MER) | Comparative living standards. Measures the country's income in terms of the dollar equivalent of its GDP measured in its own currency. This measure provides an indication of what the country or its people could buy on world markets and explains why people from richer countries with stronger currencies can buy more in poorer countries. Thus it is possible for a young British traveller to subsist in Thailand for a whole day on little more than the cost of a cappuccino and muffin in the UK, while a Thai person on average income would have to work at least a whole day in Thailand in order to pay for the same. |
| Countries of equal value | Takes each country as a single observation in measures of inequality. So a reduction in inequality in Australia (pop. approx 18M) would contribute equally to the overall measure of inequality as the same reduction in inequality in India (pop. approx 846M), even though there are 400+ Indians for every Australian. |
| Weighted by country's population | Takes the population size of different countries into account so a change in inequality in India would contribute 400* as much to the overall value as a similar change in inequality in Australia. |

*Note*: See Wade (2001) for a more detailed discussion of these distinctions.

PPP dollars do not necessarily end the desire to migrate to areas where marketable dollars can be acquired. The market exchange rate method provides a better indication of a country's relative economic standing and on this measure increases in world inequality are greater (see Table 2.3).[9] At the same time, what may appear to be high monetary incomes to people living on one or two dollars a day sometimes provide only a minimum level of welfare to

*Table 2.3* World income distribution trends

|  | *Purchasing power parity* | *Market exchange rates* |
|---|---|---|
| Countries treated equally | More unequal | Very much more unequal |
| Countries weighted by their population | Little change/more unequal | Much more unequal |

*Source*: Wade (2001).

*Table 2.4* Trends in world income distribution on different measures of inequality

| *Inequality measure* | *1988* | *1993* | *Change (%)* |
|---|---|---|---|
| Gini coefficient | 63.1 | 66.9 | 6.0 |
| Poorest decile share of income (%) | 0.88 | 0.64 | −27.3 |
| Richest decile share of income (%) | 48 | 52 | 8.3 |
| Median/lowest (50/10) ratio | 3.27 | 3.59 | 9.8 |
| Richest decile to median (90/50) ratio | 7.28 | 8.98 | 23.4 |

*Source*: Wade (2001) based on Milanovic (2002).

low income earners in wealthy countries. Three separate investigations by journalists provide experiential accounts of trying to survive on the minimum wage in different cities in the United States and the UK[10] and as Fran Abrams comments:

> So is it possible to get by like this in London, on the minimum wage? By the end of my spell here I'm quite sure it isn't. I just about manage to break even on my budget, but only after living for the best part of a week on a single bag of pasta. Then my pay slip arrives, and I find I haven't been paid for most of the scheduled extra hours I spent doing offices or for my overtime. . . . No I'm sure you can't live on these wages in London. And yet somehow, by staying with relatives or living in hopelessly overcrowded housing, by always walking or catching the bus, by juggling two jobs or even three as well as studying, tens of thousands of people in London do just that.
> (Abrams 2002: 61[11]; see also Box 2.1)

Further controversy arises over whether countries should be treated as individual cases or whether they should be weighted according to their population. Treating countries equally is appropriate if the object is to assess the effectiveness of different policies and theories of development, while weighting them by their populations is appropriate if the objective is to reflect world inequality overall, recognizing that the results will be highly influenced by

---

*Box 2.1*  Living on the minimum wage in London

| Income received (£/month) | | Expenditure (£/month) | |
|---|---|---|---|
| Pre-tax pay | 418.00 | Bed-sit | 260.00 |
| Illicit admin. charge | 10.00 | Transport | 76.90 |
| Hourly pre-tax pay | 3.43 | Food | 137.24 |
| Take home pay | 363.24 | | |
| Housing benefit | 89.83 | | |
| Total income | 453.07 | Total spending | 474.14 |

*Source*: Data from Abrams (2002: 61).

*Note*: The accommodation was very basic – 'the bed-sit is next to the toilet (shared), about eight foot by eight. There's a sagging single bed, a dilapidated fridge and cooker and a battered 1950s cylindrical water heater. The job was night cleaning with an agency for a major hotel. The hours offered and paid for were less than those promised, and most of the other employees were non-UK citizens, who worked at night and studied during the daytime.' Hours worked = 118.75.

---

trends in countries with large populations. For example if urban China, where inequality has been narrowing, is left out of the calculations then measures of world inequality increase by far more than when it is included. As Table 2.4 indicates, if countries are weighted by population and income measured by PPP then there is little change in the trend towards equality, while on all other measures the trend to inequality has increased.[12] What is clear however is that inequality remains stark and there is little evidence of any significant narrowing.

Some countries have experienced significant increases in GDP but others, especially in sub-Saharan Africa, Eastern Europe and the CIS, have experienced declines in their relative positions even on the PPP measure. The case of Eastern Europe and the CIS is particularly important as the declines occurred as soon as they followed the advice of international institutions, embraced capitalism and liberalized their economies between 1989 and 1991. Only in the very last years of the twentieth century have the initially more affluent countries of Poland and Slovenia exceeded, and the Czech Republic and Hungary almost restored, their welfare to the pre-transition level. In the CIS, especially in Russia, GDP has yet to recover and these countries have experienced the return of illnesses associated with poverty, especially tuberculosis, increases in infant mortality and the death rate and dramatic falls in life expectancy. In the case of Russia between 1991 and 2001 deaths exceeded births by 6.7 million.[13] Despite these outcomes the logic of liberalization has scarcely been questioned, rather the failure has been attributed to poor governance,[14] something strongly advocated to improve well being in sub-Saharan Africa, endorsed by the New Partnership for Africa's Development (NEPAD) and recommended more widely in the UNDP (2002) report.[15] This recommendation rests more on belief, and perhaps intellectual fashion, than on evidence. The UNDP (2002) report could find no direct association between democracy and development or between democracy and equality[16] and moreover severe constraints exist on the options open to elected governments, as will be discussed further in Chapter 8. One reason why the liberal market model is not questioned is that there have been successes and spectacular winners reflecting its inherent dynamism but its similarly inherent inequality inevitably brings losers too, something not often mentioned or planned for when liberalization is promoted.

Contemporary development in the global economy is therefore deeply asymmetric and although there is a simple north–south divide, reflected in Figure 2.1, some less developed countries, and some areas within nearly all less developed countries, have been growing rapidly, notably in China and other parts of East Asia and the Pacific while others have been stagnating, especially in sub-Saharan Africa, the CIS and Latin America. Thus there are several centres and several peripheries as well as deep divisions within the centres and within the peripheries (Castells 1996). Indeed the late Chinese leader, Deng Xiaping, was well aware of this possibility several decades ago and even promoted the slogan 'let some get rich first, the others will follow', which implies that all areas and all people will benefit from liberalization one day, but some will do so sooner than others. In China growth rates have indeed been consistently high and absolute poverty has declined, nevertheless the old social guarantees have been broken and significant numbers of people displaced from the old, now uncompetitive factories find themselves living in shanty towns in the otherwise glittering modern cities.[17] In rural areas, poverty and low incomes remain. Having said this however the Chinese government did not relinquish all controls over development and followed a rather slower transition than was attempted by Russia, and although inequalities have increased in China, they do not seem to be quite as extreme as in Russia where the benefits of transition at least initially have been confined to the richest decile. Figure 2.2 plots the Lorenz curves for 1988 and 1993 for Russia and illustrates widening inequality; whereas in 1988 the top decile received approximately 20 per cent of the income, in 1993 they received just over 40 per cent. By contrast the share received by the lowest decile declined from just over 4 per cent to just under 2 per cent in the same period. The Gini coefficient is measured by the size of the area between the Lorenz curve and the 45° line of equality (see Figure 2.2) and more recent figures for this statistic are presented in Table 2.5, along with other transitional countries in Eastern and Central Europe. In all of the countries levels of inequality on this measure have increased with a very marginal fall in Russia in the latest period, which in 1998 only had 57 per cent of its 1989 GDP. The countries that have restored or exceeded their 1989 GDP level were geographically closer to Western Europe, initially more affluent than the CIS states and maintained lower levels of inequality. Of the CIS states only Belarus, which took a rather slower transitional route, is anywhere near its 1989 GDP level and has lower levels of inequality, which is perhaps both a cause and a consequence.

Different ways of measuring GDP portray different aspects of reality and so they need to be looked at separately for an informed view of inequality. Moreover

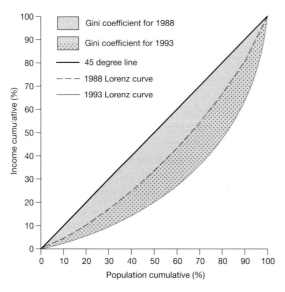

*Figure 2.2* Growing income inequality in Russia.
*Source*: Data from Milanovic (2002).

*Table 2.5* Changing GDP and widening inequality in transitional countries

| Country | Gini coefficient 1987–90 | Gini coefficient 1993–4 | Gini coefficient 1996–8 | Index of GDP in 1999 (1989 = 100) |
|---|---|---|---|---|
| Poland | 0.28 | 0.28 | 0.33 | 122 |
| Slovenia | 0.22 | 0.25 | 0.30 | 109 |
| Hungary | 0.21 | 0.23 | 0.25 | 99 |
| Czech Republic | 0.19 | 0.23 | 0.25 | 95 |
| Belarus | 0.23 | 0.28 | 0.26 | 80 |
| Estonia | 0.24 | 0.35 | 0.37 | 77 |
| Turkmenistan | 0.28 | 0.36 | 0.45 | 64 |
| Russia | 0.26 | 0.48 | 0.47 | 57 |
| Ukraine | 0.24 | | 0.47 | 36 |
| Georgia | 0.29 | | 0.43 | 34 |

*Source*: Calculated from Dabrowski and Gortat (2002).

the veracity of the measures depends on accurate recordings of output in a whole range of locations and some societies depend far more than others on informal exchanges of goods and services which never enter the calculations. Some of these transactions are deliberately hidden and form part of an illegal sector; others simply reflect different customs and practices. Thus in a society such as the UK where childcare is frequently provided by unpaid relatives, the GDP would be lower than in a country such as Sweden where there is a higher level of collective childcare in the formal sector, other things being equal. Likewise some aspects of GDP are not unambiguously positive: some goods are polluting, some activities may arise simply to counter pollution that in another country may have been organized in more environmentally friendly ways in the first instance, and individual valuations of goods also vary – a magazine or music valued by some people would be obscene or offensive to others. Furthermore GDP could derive from production of armaments or from education, which would have quite different effects on welfare. GDP is therefore a crude measure but providing care is taken to be precise about its strengths and limitations it provides some indication of relative well being and what is unambiguously clear is that on most measures the scale of world inequality is 'grotesque' (UNDP 2002: 19); see Figure 2.1.

To overcome some of the limitations of GDP various writers have proposed the broader concept of development, though this too has been criticized. Amartya Sen (2000) defines 'Development as Freedom' which he believes is denied to vast numbers of people, perhaps even the majority. His conception of freedom has economic, social and political dimensions and in his view the denial of freedom in these different spheres limits people's capacities to provide for themselves and to shape their own lives. Thus for Sen (2000: 10), 'freedoms are not only the primary ends of development, they are also among its principal means'. More specifically he argues that:

> Sometimes the lack of substantive freedoms relates directly to economic poverty, which robs people of the freedom to satisfy hunger, or to achieve sufficient nutrition, or to obtain remedies for treatable illnesses, or the opportunity to be adequately clothed or sheltered, or to enjoy clean water or sanitary facilities. In other cases, the unfreedom links closely to the lack of public facilities and social care such as the absence of epidemiological programs, or of organised arrangements for health care or educational facilities, or of effective institutions for the maintenance of local peace and order. In still other cases, the violation of freedom results directly from a denial of political and civil liberties by authoritarian regimes and from imposed restrictions on the freedom to participate in the social, political and economic life of the community.
>
> (Sen 2000: 3–4)

He then argues that an expansion of freedom in these spheres would contribute to development:

> Political freedoms (in the form of free speech and elections) help to promote economic security. Social opportunities (in the form of education and health facilities) facilitate economic participation. Economic facilities (in the form of opportunities for participation

in trade and production) can help to generate personal abundance as well as public resources for social facilities. . . . With adequate social opportunities, individuals can effectively shape their own destiny and help each other. They need not be seen primarily as passive recipients of cunning development programs.

(Sen 2000: 11)

To explain the origins and processes generating the unfreedoms is complex and measuring the different dimensions of development difficult but the United Nations Human Development Project (UNDP) has approximated some of Sen's ideas. The UNDP (2001: 9) argues that 'Development is about expanding the choices people have to lead lives they value', which can be enhanced by income, knowledge and the possession of good health. Correspondingly they designed the Human Development Index (HDI) in 1990 to provide a measure of economic and social well being that goes beyond output or income but still enables countries to measure their own progress and their performance relative to other states and that can be placed alongside more conventional measures such as GDP. Over time the HDI has become more sophisticated[18] and new measures have been introduced including the Poverty Human Development Index, the Gender-related Development Index and the Gender Empowerment Measure (GEM), and the UNDP (2002) includes a measure for the quality of governance and extent of democracy. However as the measures become more complex the number of countries for which they are calculated declines; even the basic HDI is not calculated for 18 countries.

The basic Human Development Index is designed to measure the capacity for 'living a long and healthy life, being educated and having a decent standard of living' (UNDP 2002: 34). It is a composite measure based on three criteria:

1   knowledge (measured by the adult literacy rate and a combined enrolment ratio in primary, secondary and tertiary education);
2   life expectancy at birth; and
3   standard of living as measured by adjusted income per capita in PPP.

These three elements are often related in that life expectancy and knowledge are higher in countries with high GDP per capita, but there is not a simple one to one correspondence between these measures, indicating that development can take place without high income, which alone cannot guarantee development. The UNDP classifies countries on their HDI values into high, medium and low levels of development and there are disparities between the rankings on HDI and GDP per capita; see Table 2.6. Luxembourg ranks highest on GDP per capita but only sixteenth on the HDI and the second ranked country, the United States, is sixth on the HDI, while Norway is ranked first on the HDI but third on GPD per capita.

Luxembourg has a population of just over one million people and is the administrative centre for the European Union. Not all the wealth registered

*Table 2.6* Development and economic growth: some differences

| Country | HDI rank | GDP per capita rank in PPP |
|---|---|---|
| *High human development* | | |
| Norway | 1 | 3 |
| United States | 6 | 2 |
| UK | 13 | 20 |
| Luxembourg | 16 | 1 |
| Ireland | 18 | 4 |
| Greece | 24 | 34 |
| Czech Republic | 33 | 39 |
| Costa Rica | 43 | 57 |
| Qatar | 51 | 26 |
| *Medium human development* | | |
| Mexico | 54 | 55 |
| Cuba | 55 | 90 |
| Russian Federation | 60 | 58 |
| Saudi Arabia | 71 | 45 |
| Brazil | 73 | 60 |
| China | 96 | 96 |
| South Africa | 107 | 51 |
| India | 124 | 123 |
| Botswana | 126 | 64 |
| *Low human development* | | |
| Pakistan | 138 | 131 |
| Tanzania | 151 | 172 |
| Angola | 161 | 125 |
| Central African Republic | 165 | 150 |
| Mozambique | 170 | 160 |
| Sierra Leone | 173 | 173 |

*Source*: Data from UNDP (2002).

*Note*: A rank of 1 on either the GDP or the HDI measures indicates the highest level of GDP per capita or the highest level of human development; higher ranks indicate lower levels of GDP or HDI.

there is likely to remain within the country, so the population as a whole does not necessarily benefit from this apparent prosperity. Ireland too has a large disparity between its GDP per capita ranking (4) and HDI ranking (18); again the country has a small population of around 3.89 million and in the 1990s experienced very high levels of economic growth, stimulated both by EU membership and a large amount of foreign direct investment. However, some of the profits from inward investment are likely to be repatriated, again meaning that not all of the income generated within the country remains there to enhance development. The difference between the United States and Norway probably reflects the different political priorities or welfare regimes, which influences the internal allocation of resources. Norway is a more social democratic country and correspondingly a more equal society, something that is picked up by the differences between the two countries on the poverty measure; see Table 2.7. Looking at the other imbalances, the difference in Greece probably reflects the high priority given to education as well as the presence of an informal sector which lowers the registered GDP. There are stark differences in Costa Rica and Qatar, with the former having higher levels of human development as measured by the HDI, being ranked 43rd in the world, compared to its position as only the 57th richest country on the GDP per capita measure. Qatar, by contrast, is the 53rd richest country on the GDP measure but only 90th in terms of its human development. This probably reflects the uneven distribution of Qatar's oil wealth, while in Costa Rica the more limited resources are devoted more directly to expanding welfare. The most striking differences in the medium human development countries are between Cuba and South Africa. The latter country is extremely wealthy but has high levels of inequality reflected in the low literacy and enrolment ratios and low life expectancy and correspondingly a low HDI. Cuba, by contrast, despite having limited resources has prioritized health and education for all the population. Botswana now has a very low life expectancy, only 40 years for men and women, owing largely to HIV/Aids, which has affected one-third of the adult population.[19] African countries are overrepresented among countries with low levels of human development and only Tanzania indicates a much better HDI rank than its level of GDP would suppose. The reverse situation exists in Angola and is probably linked to the high level of mineral deposits and the high death rates linked to the effects and aftermath of the long civil war. The HDI is more broadly based than GDP and so provides a broader conceptualization of well being, but it is still a very summary calculation, and subject to measurement errors. Nevertheless at a minimum, comparing countries, relative positions on these measures provides ideas for further research.

The HDI provides a better sense of well being than GDP but to provide an even clearer indication of the proportion of the population that benefits from development, the UNDP devised the Human Poverty Index. The HPI-1 is a measure of the proportion of the population living in absolute poverty and is calculated for developing countries, while the HPI-2 is calculated for selected OECD countries and includes a relative measure of income, to reflect how someone who may be affluent in global terms may still experience poverty if they do not share in the overall well being of the country where they live. Table 2.7 provides details of these measures and some illustrations of countries with similar rankings on the HDI but different levels of poverty. Looking at both measures, Cuba has the smallest proportion of the population living in poverty, though the average Cuban would be far poorer in material terms than the average citizen of the United States, for example. The internal inequality in the United States results in the average African American having a lower absolute

47

*Table 2.7* The Human Poverty Index

| Country | GDP per capita rank (1–173) | HDI rank (1–173) | HPI rank | Poor (%) | Living on less than $1 a day (%) |
|---|---|---|---|---|---|
| *Medium and low human development (HPI-1)* | | | HPI-1 (1–88)* | | |
| Mexico | 55 | 54 | 11 | 9.4 | 15.9 |
| Cuba | 90 | 55 | 4 | 4.1 | — |
| Thailand | 70 | 70 | 21 | 14.0 | 2.0 |
| Brazil | 86 | 73 | 17 | 12.2 | 11.6 |
| Uganda | 149 | 150 | 67 | 40.8 | — |
| Tanzania | 172 | 151 | 54 | 32.7 | 19.9 |
| Niger | 168 | 172 | 88 | 62.5 | 61.4 |
| *Selected OECD countries (HPI-2)* | | | HPI-2 (1–17)* | | |
| Norway | 3 | 1 | 2 | 7.5 | 0 |
| Sweden | 17 | 2 | 1 | 6.7 | 0 |
| Canada | 7 | 3 | 12 | 12.3 | 0 |
| Belgium | 9 | 4 | 13 | 12.6 | 0 |
| Australia | 12 | 5 | 14 | 12.9 | 0 |
| United States | 2 | 6 | 17 | 15.8 | 0 |
| United Kingdom | 20 | 13 | 15 | 15.1 | 0 |
| Ireland | 4 | 18 | 16 | 15.3 | 0 |

*Source:* Data from UNDP (2002).

*Notes:* * The HPI-1 is calculated for 88 and HPI-2 for 17 countries owing to the lack of data, thus some difference between the HDI rank and the HPI rank will be due to the smaller number of countries involved but it is still possible to compare countries with similar HDI ranks and different HPI ranks.

HDI-1 is a composite measure based on three dimensions: (1) Longevity: the probability at birth of not surviving until 40; (2) Knowledge: the adult literacy rate; (3) Economic provisioning: measured by the % of the population not using improved water sources and the % of children under 5 who are underweight.

HDI-2 is a composite measure of four dimensions: (1) Longevity: the probability of not surviving until 60; (2) Knowledge: % of adults lacking functional literacy skills; (3) Standard of living: % of the population living below the income poverty line (50% of the median disposable household income); and (4) Social exclusion: % long-term unemployment.

A rank of 1 on either the GDP or the HDI measures indicates the highest level of GDP per capita or the highest level of human development; higher ranks indicate lower levels of GDP or HDI.

life expectancy than the average Chinese or Sri Lankan.[20] Ireland ranks 4 on the GDP per capita measure but has 15.3 per cent of its population living in poverty, which casts some doubt on the efficacy of the policies that have stimulated the spectacular growth in the so called Celtic Tiger to trickle down and benefit the population as a whole.[21]

The comparison between Uganda, with 41 per cent in poverty and Tanzania with 33 per cent, also suggests that a more even distribution of resources can contribute to a higher level of development, at least as measured by the HDI, even when GDP per capita is substantially lower.

Gender inequality is also significant and universal. As a consequence of uneven access to resources, compounded by selective abortions and infanticide, it has been estimated that more than one hundred million women are missing from the world, especially in societies where resources are absolutely scarce and where there is a cultural preference for boy children.[22] Since 1995 gender inequality has had a higher profile in world politics, most notably with the Fourth UN Conference on Women in Beijing, which specified a platform for action to which many countries subscribed. The UNDP has also recorded two measures of gender inequality since 1995: the Gender Development Index (GDI) and the Gender Empowerment Measure (GEM); see Table 2.8. The GDI compares the relative capacity for 'living a long and healthy life, being educated and having a decent standard of living' between women and men, and in no country do women outperform men on these measures, thus in this sense gender inequality is universal. The GEM measures women's role in economic and political decision making, by measuring the proportion of women in parliament and high level managerial and professional jobs as well as gender disparities in earned income to reflect economic independence.

There is no one to one correspondence between economic growth measured by GDP per capita and the position of women on either of these measures but some correspondence between the HDI and the GDI. Life expectancy at birth is higher for women everywhere except for six countries and even there the differences are less than a year. However, women are widely believed to have a biological advantage over men of an average of five years, if given equal treatment, so gaps of less than five years are taken as representing social inequality.[23] Everywhere, women's estimated earned income is substantially lower than men's and women have lower literacy rates and enrolment ratios. Generally these latter differences are quite small except in the Yemen where there is a 42.3 per cent difference, Mozambique 31.4 per cent, Pakistan 29.65 per cent and India 23 per cent respectively, where either men are more able to exert their claims when resources are scarce or because women's potential claims are invalidated for religious or social reasons. In Lesotho, by contrast, 93.6 per cent of women but only 72.5 per cent of men are literate.

The Gender Empowerment Measure (GEM) is only calculated for 66 countries and as Table 2.8 demonstrates, there are considerable differences in the ranking of countries on the HDI and GEM.[24] Norway and other Nordic countries are high on the GEM while the United States falls from sixth position

*Table 2.8* Gender and development

| GDP per capita (rank) | HDI (rank) | Country | GDI (rank) | GEM (rank) |
|---|---|---|---|---|
| 3 | 1 | Norway | 3 | 1 |
| 12 | 5 | Australia | 1 | 10 |
| 2 | 6 | United States | 6 | 11 |
| 11 | 9 | Japan | 11 | 32 |
| 18 | 12 | France | 12 | — |
| 20 | 13 | UK | 10 | 16 |
| 4 | 18 | Ireland | 17 | 17 |
| 34 | 24 | Greece | 25 | 41 |
| 28 | 27 | Rep. of Korea | 29 | 61 |
| 53 | 37 | Poland | 36 | 30 |
| 57 | 43 | Costa Rica | 41 | 26 |

*Source*: data from UNDP (2002).

*Notes*:

(i) Data for a selected number of high human development countries is shown because data on the GEM is only available for 66 countries and too many gaps exist in the data after Costa Rica to make a rank comparison valid.

(ii) GDI – the HDI is calculated separately for women and men and then expressed as a f/m ratio.

(iii) GEM is calculated on:

- Economic participation and decision making – a combined measure of women's and men's percentage shares of administrative and managerial positions and professional and technical jobs.
- Political participation and decision making – women's and men's share of parliamentary seats.
- Power over economic resources – income measure based on the female to male earned income ratio.

These components are weighted by the population to derive an 'equally distributed equivalent percentage' (EDEP) measure. For example if the population was made up of 52% women and 48% men then for equality it would be expected that the distribution of parliamentary seats would also be 52% women and 48% men. Each variable is then indexed by dividing the EDEP by 50%. For technical details see UNDP 2002: 257.

(iv) A rank of 1 on either the GDP or the HDI measures indicates the highest level of GDP per capita or the highest level of human development; higher ranks indicate lower levels of GDP or HDI.

on the HDI to eleventh on the GEM, largely because of the relatively small proportion of seats held by women in parliament. Japan and Greece experience similar falls for similar reasons. Thus outside of the Nordic states, and to a lesser degree Eastern Europe, while women have achieved similar basic standards of living to men in terms of life expectancy and knowledge, there seems to be a barrier to women reaching decision making roles in society. International agencies have now mainstreamed gender and international aid has become conditional upon gender impact assessments so many less developed countries at least formally take gender issues more seriously perhaps than some OECD countries. Similarly, gender budgets are slowly being developed to assess the gender differentiated impact of state expenditure and fiscal policies but even in countries where equal opportunities policies have been in place for several decades, progress towards gender equality is slow, suggesting that gender remains a key structural feature that profoundly influences individuals' life chances.[25]

These statistical measures provide valuable summary descriptions that can be used to compare countries' economic and social performance with each other and over time.[26] As with all measures there are a host of problems, for example the GDI measure has been criticized for overstating the significance of earned income relative to the other two components, which are of equal significance especially to poorer countries, and the GEM for overstating the significance of formal systems of representation and the formal economy,[27] but if nothing else they can be used to generate questions about why countries perform differently on the different measures, and these questions can then stimulate more sophisticated empirical analysis and theory. In addition to these measures more qualitative investigative studies which examine the experience of life and work of people in different places are also valuable and perhaps counter some of the national stereotypes at either end of the income distribution, by making it clear that there are poor people living in Britain and America, not only among recent migrants, just as there are some very wealthy Africans, Brazilians, Chinese, Indians and Russians.[28] Likewise some migrants from less developed countries are now among the elite in their countries of destination. Gayatri Chakravorty Spivak (1999) writes about how the great grand niece of Bhubaneswari, a fighter for the national liberation of India and one of the characters in her 'Can the subaltern speak?', migrated to the United States and now has an executive position in a transnational corporation. Spivak refers to her as working for the New Empire and comments that: 'When the news of this young woman's promotion was broadcast in the family amidst general jubilation I could not help remarking to the eldest surviving female member: "Bhubaneswari hanged herself in vain", but not too loudly' (Spivak 1999: 311).

In today's global world, there is a prevailing economic and social system, which generates these inequalities albeit to different degrees in different places, depending on the nature of integration in the global system and the extent to which nation states seek and are able to control both the nature of the integration and the internal distribution of resources.

## CONCLUSION: FROM MEASUREMENT TO UNDERSTANDING

High levels of poverty and inequality lead to widespread concern, social protests and a search for remedial strategies from NGOs and local, national and supra national institutions. Different motivations underlie the desire to redress poverty and inequality: a sense of social justice and outrage that the world's resources are increasingly appropriated by an already rich minority; a sense of compassion for people struggling to survive and forced to watch their children die from curable diseases; or a sense of self-preservation among the rich who do not want to live in fear of terrorist attacks or be imprisoned by walls protecting their wealth. Whatever the reason, if the problems are to be tackled effectively then a clear understanding of the processes generating and sustaining the inequalities is necessary.

This chapter has reviewed the positions of different countries on different measures of inequality, GDP per capita and on the various indices developed by the UNDP. It has shown that inequalities are wide and in general increasing, but are more complex than a simple north–south divide, as there are high levels of inequality in both rich and poor countries and correspondingly rich people in poor countries and poor people in rich countries. These measures of income and well being are by no means perfect and cannot encapsulate the real experiences of life in different parts of the globe. Nevertheless they are a useful starting point, and raise questions for further research and analysis. In particular the differences that exist between countries with similar levels of GDP per capita but differences on the various HDI measures indicate the continued significance of state policies at the national level, despite increasing global economic integration.

Underlying these descriptive measures there are two broad sets of issues to explain: first, the relative performance and positioning of countries within the global economy and second, how countries allocate and distribute resources internally. Theories of uneven development address the first, and are discussed in Chapter 3. The welfare regime perspective linked with a regulatory approach already discussed in Chapter 1 but elaborated further in Chapter 8 and referred to in the case studies throughout the book address the second.

## Further reading

A. Sen (2000) *Development as Freedom*, New York: Anchor Books.
UNDP *Human Development Reports* (a new report is produced every year with updated statistics and a new theme).

## Website

UNDP (United Nations Development Programme): http://www.undp.org/.

## Notes

1 CIS includes: Azerbaijan, Armenia, Belarus, Georgia, Kazakhstan, Kyrgyzstan, Moldova, Russia, Tajikistan, Turkmenistan, Uzbekistan and Ukraine.

2  See Milanovic (2002) for a comprehensive discussion of measures of inequality. He uses household survey data and a measure of income that allows for inequalities within countries.

3  These 200 people exist throughout the world: North America 65, Europe 55, other industrialized countries 13, Eastern Europe and CIS 3, Asia and the Pacific 30, Arab states 16, Latin America and the Caribbean 17, sub-Saharan Africa 1. Between 1994 and 1998 their income was growing at a rate of $500 per second (UNDP 1999: 38).

4  See UNDP (2002).

5  See Krugman (2002b).

6  See Milanovic (2002).

7  The terms high and medium human development follow the UNDP (2002) classification.

8  See also Milanovic (2002) and Wade (2001).

9  See also Wade (2001).

10  See Ehrenrich (2001) on the US and Abrams (2002) and Toynbee (2003) for the UK.

11  Abrams (2002) worked in three different kinds of employment and three locations, night cleaning in London, a bottling factory near Doncaster and a care home near Aberdeen. Only in Doncaster was she able to break even and that was largely because she lived in a low cost damp caravan. Mistakes in the pay received were regular and always on the side of the employer.

12  See Wade (2001).

13  See Dunford (2002).

14  See Dabrowski and Gortat (2002).

15  See Chapter 8 for more details.

16  Dabrowski and Gortat (2002) find an association between economic performance and good governance, but economic performance is measured in terms of the extent of liberalization and percentage of small firms, etc. rather than economic well being.

17  The changes have generated a 'floating population' of 200 million peasants and laid-off factory workers who are now seeking work in China's cities. Stanley (2001) estimates that there are now '3 million such migrants among Shanghai's 16 million people. Scorned as seedy outsiders by locals, they dig the city's ditches and earn its lowest pay.'

18  This means that it is not possible to compare measures of the indices from the different reports over time but the 2002 report has produced a consistent time series going back to 1975.

19  See UNDP (2002).

20  See Sen (2000).

21  A term used as an analogy to the four South East Asian countries, which experienced similarly spectacular growth levels in the 1980s until the mid-1990s.

22  See Sen (1992) and (2000).

23  The countries with lower female life expectancy are Botswana, Zimbabwe, Lesotho, Pakistan, Zambia and Malawi. But the differences elsewhere in some countries with low levels of human development are not as high as might be expected – hence Sen's missing millions.

24  Because GEM is calculated for only a small proportion of countries it does not make sense to discuss differences in ranks for countries lower than 43 on the HDI.

25  See Budlender et al. (2002).

26  Care has to be exercised in comparing the values of the indicators from different reports as the precise nature of the measures has changed over time, but a consistent series of data, calculated backwards to 1975, is provided in UNDP 2002.

27 See Bardhan and Klasen (1999) for a detailed discussion of the statistical calculations and their limitations.
28 Roman Abramovich, the new Russian owner of Chelsea Football Club in the UK, (who is also the Governor of the Chukot Autonomous Area, Russia's easternmost region) spent an estimated $300 million on buying the club and signing new players during the summer of 2003 (Moscow Times.com: http://www.themoscow times.com/stories/2003/07/29/044.html (last accessed August 2003)).

# 3

# THEORIZING UNEVEN DEVELOPMENT

Spatial inequalities are stark, uneven development is persistent and abject poverty remains in a world of growing opulence. For a variety of reasons many people would like to change this situation but as Michael Zinzun argues:

> Theory is necessary to figure out what's really going on. People always want to be a saviour for the community. It's like they see a baby coming down the river and want to jump in and save it. We need to stop being so reactive to situations that confront us. Saving babies is FINE for them . . . but WE want to know who's throwing the XXX babies in the water in the first place.
> (Sandercock 1998: 88 citing Zinzun)[1]

Theories are necessary for understanding and explaining social change. Theories relate to general processes, not to each and every happening, so they are not substitutes for detailed explanation but rather help contextualize specific changes. For every country and for every moment there are particular sequences of events that lead to specific outcomes and these can rarely be predicted in advance, but there are also broader processes, especially in the context of a global economy, that shape or define particular social systems within which some outcomes are more likely than others and it is these broader processes that theories address and by so doing contribute to understanding.

Theories of development can be divided broadly into three categories: those proposing that market processes lead to convergence; those arguing that capitalism is inherently uneven; and those that are critical of the whole concept of 'development' considering it to be ethnocentric and value laden. Without discounting the multifaceted nature of 'development as freedom' (Sen 2000) this chapter focuses on theories of uneven development because they are consistent with the empirical trends described in Chapter 2, because access to material resources is a necessary though not sufficient condition for overall well being, and because understanding the foundations of material inequalities is a prerequisite for change.

There are a whole range of theories of uneven development: world systems theory,[2] theories of uneven exchange,[3] the development of underdevelopment,[4] and postcolonial perspectives.[5] These theories place particular emphasis on relations between countries as single entities, emphasizing the imperialist, neo-colonial and unequal power relationships between the so called more and less developed worlds and how these have developed historically. In this chapter, and indeed in the book as a whole, I have chosen instead to foreground materialist theories of change, which highlight the increasing complexity of the connections that are simultaneously within and between places and between people and that thereby contribute more directly to explaining the varied and multifaceted spatial and social divisions in the new global economy.

More specifically, this chapter highlights two theories that foreground geographical and material connections between people in different places: the new international (global) division of labour and global value or commodity chain analysis. These provide the background for Parts II and III, which consider how people have been affected by and have responded to the restructuring of the global economy in different countries, regions and cities of the world. First though, some attention is given to the neo-liberal convergence model that has dominated world economic thinking since the 1980s, despite a growing volume of counter-evidence, and to the clusters perspective that has become prominent in new economic geography and among policy makers.

## NEO-LIBERAL THEORIES

The essence of neo-liberalism is that development and modernization are enhanced by open markets and free trade. Free markets are said to allow the factors of production, labour and capital to flow to where they are most efficient, eventually leading to an equalization of factor returns and convergence between regions or to a balanced pattern of development. Labour and capital are predicted to move from areas of surplus to areas of deficit, stimulated by higher returns – wages or profits respectively. Thus, labour should move from poor to rich regions and capital should move in the opposite direction until wages and profits are equalized across regions resulting in an efficient and balanced pattern of development. Technology should also move from capital rich to capital poor regions as diminishing returns set in, allowing the less developed regions to catch up. Within this perspective the role of the state should be confined to providing a stable framework within which free markets and private capital can flourish; it should not therefore regulate prices or wages, which would distort factor flows, and neither should it be involved in productive activities, which should be privatized if not already in the private sector. This is the model that the IMF and World Bank insisted developing countries follow in order to qualify for loans or assistance, first with the structural adjustment policies in the 1980s in response to the debt crises, then in the strategies urged on the transitional economies in the 1990s. The model persists even now, despite growing criticisms, as part of economic reform programmes, albeit coupled with poverty reduction strategies, improved participation and better governance (see Box 3.1 and Chapter 9).

*Box 3.1* Structural adjustment and economic reform programmes

*What are they?*
Structural adjustment programmes (SAPs), now often termed economic reforms, consist of an array of policies that the IMF and World Bank require countries to follow in return for financial assistance, usually to resolve a debt problem.

*Origins of the debt crisis*
In 1973 and again in 1979 OPEC (oil producing and exporting countries) formed a cartel and significantly increased the price of oil, which contributed to recession in oil dependent countries. Some of the 'oil dollars' flowed into international banks, making investable funds plentiful at a time when demand for investment was low, so interest rates or the cost of borrowing fell. Less developed countries borrowed large sums of money to assist their development. Not all the funds were invested productively and over time as the developed countries began to recover from the recession, competition for funds increased and interest rates rose. Some of the borrowing countries lacked the necessary hard currency to repay their debts, leading to the debt crisis. Countries either defaulted or had to borrow more to finance repayments. As interest rates continued to rise, repayments increased and the scale of debt worsened in a vicious cycle. At various times the IMF or the World Bank has stepped in to bail out countries in debt but in return the countries have to accept SAPs.

*What do SAPs involve?*
SAPs typically involve: devaluing the currency; opening the economy to free trade and expanding exports; restricting government expenditure; ending price protection and subsidies. The underlying objective is to increase the foreign exchange earnings to facilitate debt repayment (typically made in dollars). Devaluation should encourage exports by making them relatively cheap and discourage imports by making them relatively expensive.

*Problems*
When applied to the real world, and especially to less developed countries this textbook model for recovery fails to recognize that many domestic industries have already been eliminated by global competition, that it is extremely hard to initiate new industries without any form of protection or public support and reliance on traditional primary commodities as the main source of export earnings is extremely precarious given the fluctuations in world prices and generally adverse terms of trade (i.e. the relative price of primary commodities to manufactured goods falls), so debt repayment in foreign currency continues to be difficult, even when the volume of exports increases. Cuts in public expenditure mean reduced health and education provision, which damages long-term prospects. It is often assumed that

continued . . .

women can fill the gaps but increasingly they are working in the export sectors so their workload intensifies.

*Continuing debt crisis*
Measures to reduce the debts of the most heavily indebted poor countries (HIPC), mainly in Africa, were introduced in 1996 but participating countries have to have an established track record of following IMF/World Bank economic adjustment and reform programmes, plus a poverty reduction strategy to ensure some finance for basic health and education (IMF 2002a). As a consequence debt repayments have fallen and more is being spent on education and health (IMF 2002b), but progress has been slower than anticipated partly because the incomes of these countries have fallen with the decline in world commodity prices, especially coffee.

*Critiques of SAPs*
The UNDP *Human Development Report 1997* states that: 'in many cases [SAPs] meant trying to balance the economy at the cost of unbalancing people's lives' (UNDP 1997:17) and refers to the 1980s as the lost decade.

*Alternative solutions*
Cancel more debt; reschedule loans; make debt repayable in local currencies to stimulate local endogenous rather than export growth based on primary commodities with erratic price fluctuations and long-standing adverse terms of trade; redistribution of resources to countries with debt – through a Tobin tax (tax on foreign exchange dealings) or through government bonds.

There are, however, important contradictions between the pure market model and its application by national governments and supra national institutions. The model is based on a series of restrictive assumptions about perfect competition. Where these conditions are broken, for example when there is a tendency to monopoly as with the utilities – water, electricity and transport – or where positive social externalities exist as in health, education or care then a 'market' case for state regulation can be made (see also Chapter 8). Furthermore, factor flows do not always follow the model's predictions; while capital is allowed to move freely and does flow from rich to poor countries, a much greater volume flows between the richer countries themselves leading to a cumulative concentration of development in the already rich regions rather than a spreading out of growth evenly across the globe. The European Union, the United States and Japan together dominate flows of foreign direct investment (FDI). Between 1998 and 2000 they accounted for three-quarters of global FDI inflows and 85 per cent of outflows and for 59 per cent of inward and 78 per cent of outward FDI stocks (UNCTAD 2002: 9). Further, while some labour moves from poor to rich regions many more people are held back by migration restrictions, which suggests a certain lack of faith in the model otherwise strongly advocated. In fact only something like 2.9 per cent of the world's population have migrated

(UN 2002), and even within the European Union where citizens can move freely between member states only 0.1 per cent choose to do so each year such that on average only 2 per cent of the population in the EU are citizens from other member states (Litske 2002). This suggests that given a choice many prefer to live and work at home.[6] Similarly while poorer countries are urged to open their economies to the world market many rich countries subsidize their own production such that in reality world trade is far from being free or fair.

The efficacy of the orthodox neo-liberal strategy is also challenged by the empirical evidence in Chapter 2 which illustrated rising inequalities and in the case of Central and Eastern Europe large-scale losses in GDP as well as child poverty (see also Gowan 1995; UNICEF 1997). Moreover, the nations that have succeeded in obtaining high levels of economic growth in recent decades, such as the four tigers of South East Asia, have done so with the state playing a direct role in the economy (Amsden 1993; Wade 1990), in marked contrast to countries in Latin America which have more dutifully followed IMF prescriptions. Historically during most periods of economic transition new growth trajectories have depended on significant increases in state involvement in economic management (List 1909; Freeman 1987). A director general of MITI (the ministry of international trade and industry in Japan) once said that the 'philosophy underlying the industrial policy of Japan is the principle of free competition and the market place' (Fukukawa, cited by Freeman 1987: 32). He nevertheless indicated three spheres in which the dictum did not apply: public goods and services, such as utilities and industrial infrastructure; international relations and the environment; and 'achieving optimal resource allocation from a long term dynamic viewpoint'. The first two of these spheres are characteristically found in textbook laissez faire economics, though neglected by some policy makers, but his final point suggests that, although the market is effective as a mechanism for distribution of goods and services, it fails as a mechanism for economic modernization. One reason for this is the cumulative and dynamic mechanisms associated with growth processes which have recently been recognized by endogenous growth theorists working within an otherwise neo-classical framework and by the new economic geography associated with Paul Krugman. But the idea of cumulative growth has a much longer history within a Keynesian perspective[7] and on the basis of a slightly different understanding among Marxist scholars,[8] all of whom predict uneven growth or divergence in the absence of any counter measures from the state or from the pressures of congestion within developed regions. Some of these theories see unevenness or clustering as a necessary first step that may even out later, others see unevenness and inequality as inherent within capitalism itself.

## CLUSTERS AND UNEVEN DEVELOPMENT

Uneven development is something of a paradox in the global economy, where the development of ICT and dematerialized products might suggest that geographical distance no longer matters. Yet at every spatial scale, the globe, the nation, the region, the city or locality, economic activity is clustered. Even within the neo-classical tradition, uneven spatial development or the formation of clusters has been recognized, and the stimulation of clusters has become a key tool of regional development policies (Porter 2000). There are different variants within this tradition: the abstract and formal new economic geography of Paul Krugman (1998), the business economics of Michael Porter (1998a, 1998b; 2000) and the new economic geography following the cultural or institutional turn of writers such as Ash Amin and Nigel Thrift (1994) or Michael Storper (1995) which in turn built upon the work of Italian scholars on industrial districts and the Third Italy (Bagnasco 1977; Piore and Sabel 1984).

Krugman's (1998) approach is highly abstract and analytical and focuses on the balance between centripetal and centrifugal forces (see Box 3.2), the outcome of which will determine the size and distribution of spatial concentrations. Centripetal forces, which tend towards geographical concentrations, include: market size – the larger the market the more powerful its attraction to firms; functional linkages between firms – the higher the number the greater the clustering; thick labour markets – that is the presence of a pool of labour with diverse skills; and finally pure external economies, including knowledge spillovers, similar to Alfred Marshall's ideas about the advantages that 'people following the same skilled trade get from near neighbourhood to one another. The mysteries of the trade become no mysteries; but are as it were in the air' (Marshall 1961: 271). Krugman (1998) recognizes that regional specializations can evolve accidentally but having done so economies of scale and external economies[9] will lead to cumulative growth, leading to 'lock in', or path dependency, reinforcing unevenness. The centripetal forces are however opposed by centrifugal forces, which tend towards the dispersion of activities and include immobile factors such as labour, which may be unwilling to move, land rents, which may be lower outside the existing concentrations, and pure external diseconomies such as congestion. While recognizing the importance of intangible factors, that is, the pure external economies or diseconomies, Krugman gives greater emphasis to the measurable components – economies of scale and transport costs – as his primary aim is to build mathematical models. Correspondingly he argues that the relative balance between concentration and dispersion will depend on the relative significance of economies of scale and transport costs. More specifically, the greater the economies of scale and the lower the transport costs the greater will be the tendency for spatial clusters. As transport costs, or geographical distance have become less important geographical concentration will tend to increase as firms can supply a wide range of markets from a single location.

For Krugman specialization leads to increased efficiency, comparative advantage[10] and cumulative growth within clusters because firms there experience cost savings or revenue increases as a consequence of mutual interaction within the locality, thereby generating divergence between regions or a centre–periphery pattern of economic development. This contrasts with the

---

*Box 3.2* Cumulative growth process

| Centripetal forces | Centrifugal forces |
| --- | --- |
| Market size – linkages between firms | Immobile factors – labour |
| Thick diverse labour markets | Land rents |
| Pure external economies – knowledge spillovers | Pure external diseconomies – congestion |

The tension between the centripetal and centrifugal forces determines the extent of spatial clustering. Quantitative new economic geography, for example Krugman (1998), emphasizes the relative significance of economies of scale and transport costs, specifically, the greater the economies of scale and the lower the transport costs the greater the tendency for spatial clusters. Writers following a more institutional approach such as Amin and Thrift (1994), Storper (1995) or Asheim (1996) emphasize the significance of institutions, untraded interdependencies and local learning or mechanisms of knowledge transmission.

---

neo-classical perspective which sees comparative advantage arising from natural resources and makes Krugman's ideas more attractive to policy makers because the competitive advantage arises from past humanly located economic activity rather than natural resources and correspondingly is potentially capable of being replicated.

The other two perspectives, the cultural or institutional turn in economic geography[11] and Porter's business economics, emphasize the more intangible factors, especially the role of institutions in passing on knowledge. The cultural/institutional turn in economic geography refers to institutional thickness (Amin and Thrift 1994), untraded interdependencies (Storper 1995) and learning regions (Morgan 1997), and both this perspective and Michael Porter (2000) refer to clusters. Local institutions such as chambers of commerce, state led agencies and local social networks contribute to the development of local or regional innovation systems, which are said to play a vital role in transmitting knowledge, crucial in the modern 'knowledge based era'. For Porter (2000), clusters represent concentrations of interconnected companies and supportive institutions, such as universities, state agencies and trade associations. Similarly to Krugman (1998), Porter notes that firms buy and sell from each other within clusters but he also strongly emphasizes the way that proximity enhances trust, access to information and stimulus to innovation, which in turn generate productivity increases, innovation and the development of new firms. All of these factors potentially stimulate cumulative growth. Similar to François Perroux's (1950) concept of growth poles, however, Porter's concept of cluster is rather abstract in terms of geographical space. The cluster is defined by the connections between firms and institutions rather than precise territorial boundaries and consequently the spatial dimensions of clusters can vary from being localized within a small region of a country to stretching across a continent. This 'conceptual elasticity' provides boundless policy applications, which some writers have criticized arguing that it adds more to Porter's status as a business guru than to analytical precision (Martin and Sunley 2001).

Ann Markusen likewise recognizes the significance of clusters or what she sees as 'sticky places in slippery space' (1996), but criticizes approaches linked with the cultural turn in economic geography, similarly for lacking analytical rigour, specifically arguing that they contain too many 'fuzzy concepts' (1999) to test empirically. However, Allen Scott combines elements of the two new economic geographies by arguing that 'a strictly economic logic of production will take us only so far in understanding industrial organizational processes. . . . [T]rans-actional systems are always and of necessity embedded in historically determinate social conditions' (Scott 1998: 78). Similarly to Krugman (1998), Scott (1998) recognizes the existence of locational constraints within an increasingly globalized economy but combines the measurable aspects of transport costs with spatially dependent transaction costs which include the less measurable aspects of externalities in order to develop a schematic model of different kinds of clusters, with different potentialities for expansion (see Box 3.3). Regional motors or super clusters (that is, large and dynamic clusters or, drawing upon an analogy with Quah's (1996) work on social divisions, 'superstar regions' (see Chapter 7)) will emerge where spatially dependent transaction costs are hetero-geneous, i.e. where there is a range of activities including currency transactions or software, where direct transport costs are low but costly face to face contacts important, and activities where externalities are high. When both spatially dependent transport costs and external economies of scale are high there is more likely to be a dispersed pattern of smaller clusters. In this case the existence of external economies of scale would promote clustering but the size of the cluster is limited by the transport costs. Similarly, small interconnected clusters will occur where the externalities are high but transport costs low. Scott suggests that this type of cluster may occur in the case of specialized sectors, such as film. Where both spatially dependent transaction costs and externalities are low the location pattern will be much more random.

Clearly Scott's model is schematic, or 'an idealized representation of reality in order to demonstrate some of its properties' (Ackoff cited by Haggett 1965) and real outcomes will be contingent on the precise range of activities present. Even so this model is quite useful for analysing the likely growth of clusters, and to explain why the geographical concentration of activity and uneven development continues in a world of falling transport costs, and correspondingly it provides an analytical explanation for why geography continues to matter.

Scott's (1998) approach is also useful because it differentiates clusters on the basis of the kinds of activities present. Clearly these possess different capacities for expansion and will generate different kinds of employment, so links are immediately established between the spatial distribution of activity and the potential for and nature of economic development in a richer sense than simply

*Box 3.3* General tendencies towards clustering

*General tendencies towards clustering*

| Clustering tendencies (A) | Dispersion tendencies (B) |
|---|---|
| Transaction intensity | Capital labour ratio |
| Labour intensity | Average establishment size |
| Endogenous levels of entrepreneurial activity | Ratio of production workers to total employment |

Clustering increases as A increases and B decreases.

*Clusters: A model of ideal types*

| Externalities | Spatially dependent transaction costs | | |
|---|---|---|---|
| | *uniformly low* | *heterogeneous* | *uniformly high* |
| **Low** | **No clusters**<br><br>Random distribution of economic activity | **Hybrid landscape**<br><br>Random dispersal and emerging hierarchal landscapes | **Central place hierarchical landscape**<br><br>Transport costs dominate location decisions – proximity to supplies or markets (Losch/Weber) |
| **High** | **Small interconnected clusters**<br><br>Clusters exist because of externalities but remain small because the low transport costs allow inter-firm relations to be sustained over distance | **Super cluster (Superstar regions)**<br><br>Clusters exist because of high externalities and expand because of the high transaction costs between some activities while the lower costs of others also allow the cluster to buy from and supply to world markets leading to local growth | **Small disconnected clusters**<br><br>Clusters exist because of high externalities but growth constrained by high transaction costs |

*Source*: Adapted from Scott (1998: 87) and reproduced by permission of Oxford University Press.

the presence of firms. In this way there are parallels with the global value chain approach discussed later in the chapter. Scott's model also pays some attention to the connections between the local clusters and the global context, and to the effects of clustering on development in terms of the well being of people where they are located, and on the people in the regions that have been bypassed. However, these connections are not generally drawn out in the clustering perspective making the approach rather partial in both senses of the word. First, because it only considers part of the developments taking place, as the firm focus fails to address the social divisions characteristic of the current trajectory of development. Second, it is very much concerned with promoting the development of clusters and so closer to the boosterism associated with the development agencies than to impartial academic research, though Krugman's approach is an exception because it is concerned primarily with abstract spatial modelling. More generally the clusters perspective does not really question the inequalities linked to this pattern of development and inherent within the capitalist system of social and economic regulation.

The cluster/local innovation approach has been most strongly used in relation to Western Europe and the United States where the 'ideal types' are found – Emilia Romagna, Baden Württemberg and Silicon Valley – but writers have also used this approach in Brazil (Schmitz 1999), Mexico (Rabellotti 1997) and China (Christerson and Lever-Tracy 1997). More generally both the international division of labour and global value chains perspectives pay greater attention to the connections between economic activity at local and global levels and there is greater space within these perspectives for analysing the impact on working people even though the focus remains on corporate strategies.

## GLOBAL/INTERNATIONAL DIVISION OF LABOUR

Economic relations between states have passed through a number of forms: the internationalization of commodities, the internationalization of capital, the internationalization of production[12] and now the globalization of all of these plus the development of global finance.

Commodity exchanges between countries have taken place for hundreds of years and initially led to only minor changes in the internal structures of the participating countries. The expansion and deepening of capitalist relations of production in the nineteenth century, however, together with colonialism and uneven power relations between states, culminated in:

> a new and international division of labour . . . one suited to the requirements of the main industrial countries [and it] converts one part of the globe into a chiefly agricultural field of production for supplying the other part which remains a pre-eminently industrial field.
>
> (Marx 1976: 579–80)

Marx is referring to the way that industrialization in some countries not only increases the demand for primary commodities from other countries but also undermines traditional manufacturing elsewhere, partly through colonial trade restrictions but also because of the competitive superiority of goods produced in already industrialized countries.

Later in the nineteenth and early twentieth centuries capital was exported from the industrialized countries, partly to increase the production and export of primary commodities and partly to develop new markets, sometimes in response to the import substitution strategies that some countries, for example in Latin America and also Ireland, introduced to facilitate the development of their own industries especially between 1930 and 1945 when the world economy was in depression or disrupted by war. By the 1960s however a new form of foreign investment was underway, in which the internal division of labour within the firm, deepened by the introduction of scientific manage-ment and Fordism,[13] was extended over national boundaries resulting in the internationalization of production. The new more fragmented labour processes, which enabled industrial work to be carried out with only a few weeks' formal training, especially in garments where traditional skills could be built upon (Elson and Pearson 1981; see Chapter 4), allowed the simultaneous concen-tration of control functions and the decentralization of the manual execution of tasks. Such decentralization was profitable because of the drastically lower labour costs, feasible because of improved ICTs, and desirable because it enabled companies to escape from the industrial militancy characteristic of much of Western Europe and North America in the late 1960s, especially in tradi-tional industrial regions. These developments further incorporated countries and regions not only into a world economy but into the growing number of transnational corporations, so generating another 'new international division of labour' in which world trade increasingly consisted of flows of components between plants of the same company made by people in different countries (Fröbel *et al.* 1980).

At the time many of these theorists made gloomy predictions about the effects of this integration on the development of the regions where branch plants were located, either on an international scale[14] or for less developed regions within countries.[15] Stephen Hymer (1975) for example argued that:

> A regime of North Atlantic Multinational Corporations would tend to produce a hierarchical division of labour between geographical regions corresponding to the vertical division of labour within the firm. It would tend to centralize high-level decision making occupations in a few key cities in the advanced countries surrounded by a number of regional sub-capitals, and confine the rest of the world to lower levels of activity and income, i.e. to the status of towns and villages in a new Imperial system. Income, status, authority and consumption patterns would radiate out from these centres along a declining curve, and the existing pattern of inequality and

dependency would be perpetuated. The pattern would be complex just as the structure of the corporation is complex, but the basic relation between different countries would be one of superior and subordinate, head office and branch plant.

(Hymer 1975: 38)

This pattern is outlined in Box 3.4, which illustrates how the vertical division of labour within the firm generates a hierarchical division across space. Specifically, while the highly paid control and strategic functions remained in major cities generally in already industrialized countries the day to day operations were located near to raw materials, labour supplies or markets in accordance with traditional location theory and correspondingly widely dispersed across the globe. This pattern tended to perpetuate uneven development as what are perceived to be routine activities provide only low wage employment, the profits from which would often be repatriated to the already industrialized countries where ownership remained, thus the value created within the region would not necessarily remain there. Transfer pricing[16] was and continues to be used to maximize profits in low tax locations. Writers at this time and subsequently tended to accept and even naturalize these pay differentials, yet as pointed out in Chapter 1 these inequalities build upon and reinforce existing differentials which are probably linked far more to social class, race and gender as well as to the prevailing norms of consumption for these

---

*Box 3.4* The hierarchical division of labour within the firm and uneven territorial development

| Level | Function | Local requirements | Locations |
|---|---|---|---|
| I | Strategy | Close to capital markets, media and government | World's major cities: New York, London, Paris, Bonn, Tokyo, Moscow, Beijing |
| II | Coordination | Close to sources of white collar labour, communications and information | Large cities, regional capitals |
| III | Day to day operations | Responsive to the pull of labour, markets and materials | Widely dispersed |

*Source*: Based on Hymer (1975).

different groups in different places than to the nature of the work. Following her experience of working in six jobs paying the minimum wage in the United States, Barbara Ehrenrich (2001) concludes that no job is truly unskilled and as illustrated in Box 3.5 some of the work she was required to do involved a wide range of complex skills.

---

*Box 3.5*  Work and pay: does pay reflect skill?

You might think that unskilled jobs would be a snap for someone who holds a PhD and whose normal line of work requires learning entirely new things every couple of weeks. Not so. The first thing I discovered is that no job, no matter how lowly, is truly 'unskilled'. Every one of the six jobs I entered into in the course of this project required concentration, and most demanded that I master new terms, new tools, and new skills – from placing orders on restaurant computers to wielding the backpack vacuum cleaner. None of these things came as easily to me as I would have liked; no one ever said, 'Wow, you're fast!' or 'Can you believe she just started?' Whatever my accomplishments in the rest of my life, in the low-wage work world I was a person of average ability – capable of learning the job and also capable of screwing up.

I did have my moments of glory. There were days at The Maids when I got my own tasks finished fast enough that I was able to lighten the loads of others, and I feel good about that. There was the breakthrough at Wal-Mart, where I truly believe that, if I'd been able to keep my mouth shut, I would have progressed in a year or two to a wage of $7.50 or more an hour. And I'll bask for the rest of my life in the memory of that day at Woodcrest when I fed the locked Alzheimer's ward all by myself, cleaned up afterward, and even managed to extract a few smiles from the vacant faces of my charges in the process.

It's not just that the work has to be learned in each situation. Each job presents a self-contained social world, with its own personalities, hierarchies, customs and standards. Sometimes I was given scraps of sociological data to work with, such as 'Watch out for so-and-so, he's a real asshole'. More generally it was left to me to figure out such essentials as who was in charge, who was good to work with, who could take a joke. Here years of travel probably stood me in good stead, although in my normal life I usually enter new situations in some respected, even attention-getting role like 'guest lecturer' or 'workshop leader'. It's a lot harder, I found, to sort out a human micro system when you're looking at it from the bottom up, and, of course, a lot more necessary to do so.

*Source*: Ehrenrich 2001:193–4. Barbara Ehrenrich is a journalist who tried to live on the wages provided by low paid work in three different cities in the United States between 1998 and 2000.

Hymer (1975) points out that there was nothing inevitable about this hierarchical division of labour. Having solved the basic problems of mass production in the early part of the twentieth century there was a choice over the direction of growth, between capital widening, that is extending the existing systems of mass production and products over a wider area while maintaining the existing capital labour ratio, or capital deepening, that is substituting capital for labour and broadening the range of products produced for a smaller but richer market. As Hymer argues:

> One possibility was to expand mass production systems very widely and to make basic consumer goods available on a broad basis throughout the world. The other possibility was to concentrate on continuous innovation for a small number of people and on the introduction of new consumer goods even before the old ones had fully spread. The latter course was in fact chosen, and we now have the paradox that 500 million people can receive a live TV broadcast from the moon while there is still a shortage of telephones in many advanced countries, to say nothing of the fact that so many people suffer from inadequate food and lack of simple medical help.
>
> (Hymer 1975: 44–5)

This imbalance in consumption patterns on a global scale continues and has become even more extreme as ever more products come on to the market while basic needs continue to be unmet. At the same time, however, while uneven development continues some of the countries that had been recipients of foreign direct investment (FDI) in the 1960s and 1970s subsequently have become generators of FDI, for example the East Asian NIEs (newly industrializing economies) such as South Korea and Taiwan. Further, despite the harsh conditions in many of the multinational branch plants many countries receiving FDI experienced increases in their overall living standards. South Korea now has many brands that are both exported to and produced in North America, Europe and Japan including Hyundai cars and Samsung electronics.

In the case of clothing one stimulus to these developments was the Multi Fibre Agreement, which already industrialized countries introduced to protect their own industries by setting quotas on the share of their markets that could be supplied from any particular country. Thus, having exhausted their own quotas firms in some of the exporting countries went international themselves to use the quotas and low cost labour in surrounding countries; for example Hong Kong manufacturers expanded into Singapore, Taiwan and Macao in the 1960s and Malaysia, the Philippines and Mauritius in the 1970s and then to China from 1978 with the opening of the Chinese economy. Similarly South Korean clothing companies have established plants in Latin America, the Philippines, Bangladesh and Sri Lanka[17] and there are small South Korean firms in Los Angeles.[18] These countries have also moved from simple assembly work to more complex forms of manufacturing. Moreover some advanced manufacturing and service sector work has moved to what were then less developed countries. Call centres which are becoming global to take advantage of different time zones and lower wage costs are growing in India which is also an increasingly important focus for software production, not simply electronic assembly but program writing and so on.[19] Furthermore, even in the case of clothing, where working hours are long and pay is low the feminization of paid employment has to some degree challenged patriarchal structures and empowered women.[20] Thus the capacity for less developed countries to change, modify and in some cases adapt FDI to suit their own interests and in some ways benefit from international firms was not fully appreciated by some uneven development theorists. On the other hand even where some upgrading has taken place, the hierarchical spatial division of labour remains both globally and within countries so social and spatial divisions continue but take a more complex form. That is, the one to one mapping between the hierarchical division of labour within the firm and its horizontal division across space becomes more intricate geographically as the hierarchical division of labour within the firm is replicated within and between countries in complex ways, especially as the recipients of FDI in the 1970s have become generators of FDI in the 1990s and beyond.

In this respect it is interesting to note that Japan now imports more televisions and video recorders from its offshore factories, largely in ASEAN countries, than it produces within Japan, a situation that has been developing since the mid-1980s when the high value of the Japanese yen, in part caused by its high trade surplus with the rest of the world, meant that the relative price of Japan's exports increased. At the same time Japanese exports of high value electronic components and devices increased, thus rather than becoming deindustrialized Japan simply advanced technologically and produces and exports even higher value products, on which the rest of the ASEAN electronics industry depends. Thus while Malaysian electronic exports have increased, including some to Japan, it has simultaneously increased its trade deficit with Japan in electronic goods from $0.5 billion in 1985 to $4.5 billion in 1995 and in terms of the machinery used to manufacture its electronics products the deficit has increased from $1.5 billion to $8.9 billion. Despite technological advances in Japanese plants in Malaysia it remains in a subordinate position in the hierarchical division of labour with respect to Japan, and although some research and development (R&D) workers are employed they are confined to modifying manufacturing designs with the main product and process innovations taking place in the plants in Japan (Wilkinson *et al.* 2001).

Foreign direct investment remains 'the main force in international economic integration' (UNCTAD 2002: 9). The value of sales by foreign affiliates is twice as high as the value of world exports of goods and services and there are now 60,000 transnational corporations who together own more than 820,000 affiliates abroad, with some 45.5 million people being employed in comparison to just under 17.5 million in 1982 (UNCTAD 2002). As some countries upgraded in the first rounds of decentralization, the transnationals have moved further afield geographically to even lower cost locations. China has received large quantities of FDI in recent years, including funds from the Chinese expatriates throughout the world, and is predicted to outstrip the US as the major recipient.[21] Some of the investment, especially in specially designated export processing zones, fits very closely to the original specifications of Hymer's model, which continues to provide a useful way of understanding the spatial distribution of activity and constraints on subsequent development unless strongly challenged by local development strategies. At the same time China has experienced very rapid rates of economic growth and has a range of high level strategic functions to manage its own economy and society as well as dealings with the global economy, reflected in the modern cities, while rural areas remain largely unchanged leading to growing internal divisions.[22]

Manufacturing employment as a share of total employment continues to decline in the richer countries, partly as a consequence of the internationalization of production. Mass clothing manufacture, for example, has declined quite dramatically, one region particularly affected being north east England, discussed in Chapter 5. However, not all manufacturing goes offshore. For example, Los Angeles, often thought to be the heart of the US new economy, has the highest number of manufacturing jobs in the US – 663,400 in 1997, which is nearly 6000 more than Chicago and 200,000 more than the third largest manufacturing city – Detroit (Bonacich and Appelbaum 2000: 28). These jobs are found in both the new high tech manufacturing industries but also in the old style garment industry where over 100,000 workers, many recent migrants, work in the growing number of sweatshops. Similarly in the 1980s changes in US immigration laws brought capital and labour to New York, and Chinatown, with around 250,000 people and close to the financial district, became a base for the garment and textile factories and restaurant chains. The ethnic composition of the area changed to include people from Hong Kong, Taiwan, Vietnam and Malaysia as well as China.[23] There are also sweatshops close to the financial district in London. Indeed one of the ironies of globalization is that while many women workers in Bangladesh are working outside the home for the first time in new clothing factories, their compatriots who migrated to London are more likely to find themselves working at home or in small workshops with poor conditions, working illegallly for less than the minimum wage yet producing products for leading retailers.[24]

One reason why sweatshops have re-emerged is because of the increased powers of brand marketers and retailers in the last few years of the twentieth century and in the new millennium. With increasing globalization and the development of information and computing technologies the international division of labour has been taken further as key firms have specialized in branding and marketing, and decentralized the ownership of production. Thus some of the leading companies no longer play any direct role in producing the goods they sell. Nike for example does not produce any sportswear and Dell does not manufacture computers. Similarly many of the well known clothing labels do not manufacture their clothes because they can make more profit from branding and marketing products.[25] Production correspondingly takes place in a range of companies in both rich and poor countries, which are variously owned and often connected through complex subcontracting arrangements. Each firm seeks to minimise risk in the highly competitive and volatile global market. In the case of clothing their ability to subcontract production to offshore producers and small-scale local producers enables them to maximize flexibility and minimize their own risk which is especially important owing to the transient nature of fashion. It also means that they can absolve themselves of responsibility for the working conditions or at least try to do so.[26] Working conditions can be particularly desperate in the fashion industry because of global competition and because of the uncertain and transient nature of demand, culminating in long working hours, especially in the rush to meet deadlines, and low pay both offshore and in the sweatshops. Firms like to draw upon a proportion of local producers to increase the security of supply in just in time systems.

Ownership of production plants can be a liability especially in volatile goods such as fashion where product lifecycles are very short. Thus rather than having a chain of branch plants distributed across the globe, many firms simply buy the inputs they need from other, sometimes locally owned firms. By organizing production in this way the risks are transferred to the producers and the marketing firms can simply adjust their purchases to meet their specific demand requirements. This trend has been referred to as 'hollowing out' and has given rise to complex systems of subcontracting or to commodity chains. In the case of textiles and clothing, manufacturing jobs in Hong Kong declined by over 50 per cent as jobs were transferred to China, and Hong Kong specialized in trading instead (Gereffi 1999). Loss of production does not necessarily mean loss of control and neither does it necessarily lead to an improvement in working conditions as competition can be even more intense for the remaining firms. Moreover, as the length of the chain increases then conditions at the far end are likely to worsen as monitoring declines. A further reason for these developments is that large firms have become sensitive to the critiques of 'anti-globalization' protesters and sought to distance themselves from the direct production of the goods they sell. These more complex patterns of manufacturing were partly responsible for the development of the global value chains perspective, discussed below.

## GLOBAL VALUE CHAINS OR COMMODITY CHAINS

Global value chain analysis provides another way of explaining development by tracing the amount of value added produced in the different stages of a commodity's life from direct production through to final sale.[27] By focusing on a specific chain it analyses a significant, but still manageable slice of the world economy (Sturgeon 2001). Moreover, monitoring where value is produced and appropriated highlights patterns of inequality and the potential contribution of the activities to the development of the region where they are located.

Global value chain analysis developed in response to the continuing dominance of large-scale producers in already industrialized countries and the increasing complexity of their relations with suppliers, but the analysis can also be applied to retailing and primary commodities. Supermarkets and major manufacturers increasingly distance themselves from direct production. Nevertheless they closely organize their supply chains to ensure reliability and quality, together with compliance with the various ethical codes of conduct. This degree of control has become possible owing to developments in ICT, which allow close tracking of products across the globe. Thus many transactions take place in an organized way and cannot be described as trade on the open market, even though it may be taking place between independent economic actors.

A typical value chain involves the production of primary commodities in poorer countries and their final sale in richer countries. Coffee is an excellent example of such a chain; it is produced in at least 60 countries with tropical and subtropical climates, and although drunk all over the world, with an estimated consumer market value of 50 billion US dollars, the main consumers are found in the US, Japan and Europe. About 100 million people are estimated to make their livelihood from coffee (ICO 2001).

Coffee passes through a number of links between the grower and the final consumer and different amounts of value are appropriated at each stage. Despite increasing retail prices associated with coffee's growing popularity, and 'up-grading' at the consumer end, with specialist coffee bars selling cappuccinos, espressos, lattes etc., there has been a long-term relative decline in the price of raw coffee beans as the terms of trade have moved against the primary producers. The coffee trade illustrates how people can be integrated into the global economy but remain poor, the key reasons being natural and market uncertainties and a low position in the value chain. According to Robert Fitter and Raphael Kaplinsky (2001) nearly 60 per cent of the value is appropriated in the consuming countries leaving only 40 per cent for the producing countries, many of whom are heavily dependent on coffee. The situation for coffee growers has deteriorated from the mid-1980s with the growth of neo-liberal policies, the increasing control over the market by transnational corporations which could maintain retail prices while lowering prices paid to the producers, and the corresponding collapse of the International Coffee Agreement that had protected the prices paid to growers. The structure of the value chain is presented in Box 3.6.

Producers of primary commodities face many dilemmas and insecurities linked to natural hazards, which exacerbate endemic market uncertainties. Prices can fluctuate wildly from year to year and any shocks tend to a spiralling away from rather than towards any equilibrium price, a well known textbook problem (the divergent cobweb model) with market economics. If a bad frost destroys the crop in one region prices rise encouraging new entrants to the market which in the following year/s lead to an equally dramatic fall in prices owing to excess supply. More generally prices can be affected by new entrants on to the market. Vietnam, for example, under the encouragement of the World

---

*Box 3.6* The distribution of value in the coffee chain

**Producing country     (11– 40% of final retail price\*)**

**Farm: coffee growers**
On the farm the coffee is grown, picked and preliminarily processed – either a dry process or a wet process (washing the beans which adds more value). The coffee is then sent to a factory for further processing. Most production of coffee takes place on relatively small farms.

**Factory**: At the factory the beans from a wide range of producers are processed and the beans sent to an exporter.

**Exporters**: will collect beans from a number of factories and organize shipping.

**Consuming countries     (35–80% of the final price\*)**

**Importers**: import coffee and send to roasting houses.

**Roasting houses**: process coffee and create specific brands.

**Final sales in**: supermarkets, retail stores, specialist coffee shops and cafés. Some coffee shops and cafés roast their own coffee.

*Sources*: Talbot (1997); Fitter and Kaplinsky (2001).

*Note*: \* Estimates of the proportion of the final price received by producing and consuming countries vary with the variation in the price of raw coffee which fluctuates considerably, but the long-term trend over the last decade has been in a downward direction. Other estimates are that 40% of the value is retained within the producing countries and 60% in the consuming countries.

---

Bank, has become an important producer in recent years. It has increased its share of output from 4.6 per cent of the world total in 1995 to 10.5 per cent in 2000. These fluctuations in supply are clearly a major problem to countries that are heavily dependent on coffee, such as Burundi which only produces 0.3 per cent of world output but this accounts for 76 per cent of its total export earnings; by contrast Brazil, the major producer (33 per cent of world output) is far less dependent, with coffee only accounting for 5 per cent of its export earnings.[28] In the past stocks were held by the importers or exporters but now there is a greater tendency towards just in time systems to reduce the costs higher up the chain so the costs of the natural hazards are borne by the direct producers, i.e. by those who can least afford to do so, while the buyers and retailers avoid risk by having flexible contracts with their suppliers (see Chapter 4 for a discussion of the fruit chain).

Value chain analysts consider how individuals, firms and regions might upgrade within the value chain. There are different forms of upgrading: process upgrading – introducing new more efficient ways of producing given products; product upgrading – branching out into more complex products within or beyond the sector, or by expanding the number of buyers to allow greater economies of scale and reduce the risk of dependency on a single buyer; and functional upgrading – moving into new spheres such as design and marketing.[29]

In the case of coffee it may be possible to carry out some of the roasting, grinding and final packaging in producing countries, and while designer coffee shops exist in major cities worldwide, effective demand is depressed by the generally low incomes in many producing countries. Some NGOs provide guaranteed prices to a small proportion of producers and encourage upgrading into organic and fair trade brands and even Starbucks has agreed to make fair traded coffee available in its coffee shops, though so far this remains below the industry minimum standard of 5 per cent.[30] Such ethical upgrading can lead to new niche markets and increase rather than reduce sales, though their overall scale is comparatively small.[31] Given the uncertainty, it might be better to diversify production, though another way of redressing the imbalance in welfare between countries would be to revalue value, especially the prices paid to producers.

Garment producers in East Asia and the Pacific moved from simply assembling foreign inputs to selling and branding their own merchandise in internal and external markets (see Box 3.7). In this case there seems to have been a very easy transition or a 'benign escalator'. More generally while upgrading from stages 1 to 2, that is from simple assembly to using local materials, is quite frequent, the move to 3 and 4, that is into design and branding own production is much less certain and the key firms in the chain can use their power to block this kind of diversification.[32] Thus the capacity for upgrading depends not only on what is being produced but the nature of the relationships between the firms in the chain as well as the relations between the states involved, for example the greater geographical and cultural proximity of the US to Mexico will perhaps result in US firms playing a more dominant role in the North American clothing chain in contrast to their more distant role in the East Asian newly industrialized economies.[33]

---

*Box 3.7* Upgrading in the global garment value chain in East Asia

1. Assembly of imported inputs.
2. More domestically integrated, local sourcing and higher value added local production (original equipment manufacturing).
3. Designing products sold under the brands of other firms.
4. Integrating their manufacturing expertise with designing own brand name manufacturing in internal and external markets (original brand manufacturing).

*Sources*: Humphrey and Schmitz (2002 citing Gereffi 1999) and Gereffi (1999).

---

John Humphrey and Hubert Schmitz (2002) distinguish four types of relationships (see Table 3.1), which provide different degrees of scope for the suppliers to upgrade their activities. The degree of connection and involvement between the firms depends largely on the level of expertise within the supplying firm/country and the complexity of the product being produced. If the product is simple and the country/firm has a tradition of supply to foreign firms then the foreign firm may simply buy from the market. By contrast, if the country or firm is new to supplying foreign buyers then the foreign firm may initially take closer control to close the gap between domestic traditions and foreign expectations. As the complexity of the input required increases then the degree of involvement is likely to increase. Firms/producers that simply supply to the market are neither constrained nor assisted by larger firms and so the outcome is very contingent on the specific conditions of the firms themselves and their national context. In quasi chains, large firms can be quite helpful to firms seeking to upgrade, as it is often in their interest for the firms to do so. The extent of upgrading will, however, be conditioned by the requirements of the large firm, which is more likely to resist functional upgrading, which could lead to the development of a competitor rather than a more competent supplier. The best opportunities for upgrading are found within networked firms, that is where there are horizontal or reciprocal relations between firms who coordinate their requirements through sharing information, but these conditions are less likely to be found in newly industrializing countries.

*Table 3.1* Global value chains; types of relationships and capacity for upgrading

| *Governance structure* | *Capacity for upgrading* |
| --- | --- |
| Arm's length market relations | Many potential suppliers have the capacity to produce the desired products to the required standards. Upgrading depends on the firm's own capacity. |
| Networks – firms linked by complementary competences | Horizontal or reciprocal relations between firms who coordinate their requirements through sharing information. Upgrading more likely as local firms already sophisticated. |
| Quasi hierarchy – asymmetry of power in favour of lead firm | Lead firm exercises control through the supply chain in order to ensure product standards and delivery performance. Local firm may be given assistance by lead firm to meet targets but lead firm may impede functional upgrading. |
| Hierarchy – vertical integration | Lead firm owns some operations in the chain. Upgrading largely determined by preferences of lead firm. |

*Source*: Adapted from Humphrey and Schmitz (2002).

Firms in value chains should not be seen as passive victims but agents in their own right who can use their knowledge and resources to break out of dependent relationships. Whether or not they can do so depends on a whole variety of contingent factors, including the local and national state context and systems of innovation. Some national, regional or local states are supportive and provide resources for innovation, support good labour and environmental conditions, provide education and encourage locally based firms to adapt foreign technologies, and so on. This more proactive form of aid is more likely to contribute to the development of the region than financial assistance or tax breaks to firms.

Whether upgrading takes place in this way is highly contingent, but it is not impossible and it is consistent with the spread of capitalism, which needs consumers as well as producers. The impact of the upgraded firms on the regions in terms of wages and therefore markets for other products is also likely to be positive and could lead to cumulative growth. Even so it is still capitalism, the object is to make profit and there will be marked income differences between owners and producers. Product upgrading does not necessarily lead to higher incomes for the direct producers and workers may also be displaced. When the transition from clothing to finance in the case of Hong Kong was discussed, for example, there was an implicit assumption that this was a progressive development and little attention was given to the displaced clothing workers in Hong Kong, yet it seems unlikely that they would quickly transform themselves into traders.

## CONCLUSION

This chapter has argued that theories are necessary to understand economic and social change. The point of theory is not to account for each and every instance, but to identify key processes and tendencies that are shaping change. When used in empirical investigations theory helps to illuminate not only what has happened but why it has happened.

There are many theories that address uneven development and change. This chapter has focused on three theories, clustering, the new international division of labour, and the value chain approach. These theories were chosen because they emphasize material processes, which in principle directly link people and places. Even so, they tend to foreground the firm, rather than situating these developments within the wider social and political context. In the case of the new international division of labour and the value chain approach this is a problem of application rather than being intrinsic to the theories themselves. These approaches could be extended further to take these wider issues into account (see Chapter 4).

The global value chain literature considers the proportion of value retained within the region and points out that in the case of agribusiness only a small proportion accrues to the direct producers, the major share being taken by marketers or branders, who are much more likely to be within developed countries. Similarly, the international division of labour perspective refers to the hierarchy of functions within the firm and their uneven distribution across space. In both cases therefore these theories provide useful insights into why, despite industrialization in less developed countries, inequalities between rich and poor countries remain and in some cases even widen.

The possibility of upgrading within the value chain, or within the hier-archical spatial division of labour, or developing sufficient local externalities within a cluster also partly explain, together with the hollowing out of owner-ship, why the pattern of uneven development has become more complex. Manuel Castells (1996) pointed out that development in the world is 'deeply asymmetric' but no longer structured along a simple north–south divide. David Held *et al.* take the point further when they argue that:

> the old North–South hierarchy has given way to a more complex geometry of global economic and productive power relations – a new global division of labour. North and South are increasingly becoming meaningless categories: under conditions of globalization distributional patterns of power and wealth no longer accord with a simple core and periphery division of the world, as in the early twentieth century, but reflect a new geography of power and privi-lege which transcends political borders and regions, reconfiguring established international and transnational hierarchies of social power and wealth.
>
> (Held *et al.* 1999: 429)

A simple north–south hierarchy has perhaps disappeared, as overall the global division of labour has become more complex – not simply one of manufacturing in the developed countries and agriculture in the less developed countries though a much higher proportion of the working population in less developed countries derive their income and livelihood from primary products. Neither is it simply a case of the more developed countries moving on from manufacture to services in the latter part of the twentieth century, though undoubtedly a higher proportion of the population there derive their income from the service sector than ever before. What is clear from the discussion of growing inequality in Chapter 2 is that uneven development on an international scale remains pervasive. Thus an international or global division of labour continues to exist; it is hierarchical and results in people in different regions having different opportunities and profoundly influences the patterns of uneven development, if not in quite the simple spatial hierarchical way between north and south, when all the multinational firms were based in the richer countries of the world.

This increasing complexity however is not surprising given that the underlying rationale of capitalist production is profit which can arise in two broad ways. It can either arise from increasing the amount of profit from a given volume of production or market, which was one of the reasons why firms sought to move to low cost locations. Or alternatively profit can increase by extending the size of the market from which profits can be made. As Marx pointed out there is always a contradiction between the interests of an individual capitalist as a producer, which is for low cost supplies and low cost labour, and that same producer's requirements as a marketer – that is lots of income to provide a market for their products. Thus a conflict exists between the interests of individual capitalists and the interests of capital as a whole.[34] Thus while the development of branch plants and firms in low wage regions may have been driven by the search for low cost labour, capital as a whole also seeks to extend markets and thus the wages from these workers present a new source of potential demand to which these and others including locally based firms, can respond. In reality therefore capital widening and capital deepening can and do take place simultaneously. Indeed, in Marx's own words, 'the need of a constantly expanding market for its products chases the bourgeoisie over the whole surface of the globe. It must nestle everywhere, settle everywhere, establish connections everywhere' (Marx and Engels 1983: 18).

One limitation of the clusters, the global value chain and to a lesser extent the division of labour literature is that firm and corporate strategies tend to become the key objects of analysis rather than the relations between firms and their locations or even more, relations between people and places. The focus on the firm limits discussion of the national context, especially the system of macroeconomic regulation and the prevailing welfare regime, both of which profoundly shape the way similar processes of restructuring at the global level take different forms in geographical space.[35] These omissions tend to reinforce the hegemony of the dominant model of economic and social regulation, i.e. global capitalism, which is taken to be inevitable and therefore little thought is given to formulating alternative scenarios or even identifying ways in which state institutions can and do mediate 'global processes' differently to produce more egalitarian and inclusive outcomes, albeit within a capitalist framework. They also tend to privilege a western centric approach to understanding by foregrounding changes in firms primarily based in the richer countries, leading to the neglect of locally based change and processes that shape the way these firms are embedded within local economies. Furthermore, developments within firms, especially new working practices, are rather left outside the analysis. Yet, these labour processes profoundly shape people's incomes and opportunities and hence overall material well being in particular regions and localities. Thus the firm, a central plank of analysis, is like an empty box or in Marx's terms a 'chaotic conception' (Marx 1973a: 100 and see Chapter 1), and some of the processes central to the development of the region and uneven development between regions are rather neglected. Finally, questions of reproduction in the immediate sense of the social division of labour between different kinds of paid work and between paid work and caring, and in the wider sense of the sustainable regional development are rarely discussed within these perspectives. Yet, these dimensions are central to understanding the well being of people within regions and so to regional or spatial development as a whole. Thus while these theories may account for uneven growth across the globe they do not explicitly address questions such as why regions with similar levels of GDP per capita have quite different levels of development measured in terms of the HDI. To answer these questions it is necessary to link the theorizations of uneven development based around the firm and corporate strategies with theories of the state or welfare regimes in specific spatial settings, issues at least partially addressed in the remainder of the book.

## Further reading

G. Gereffi and R. Kaplinsky (eds) (2001) *Value of Value Chains*, University of Sussex *IDS Bulletin* 32 (3).
M. Power (2003) *Rethinking Development Geographies*, London: Routledge.
A. Scott (1998) *Regions and the World Economy*, Oxford: Oxford University Press.

## Websites

ICO (International Coffee Organization): http://www.ico.org/.
Insight 21: http://www.id21.org (accessed December 2003).

## Notes

1 Michael Zinzun was a former Black Panther activist.
2 See Wallerstein (1991).
3 See Emmanuel (1972).
4 See Frank (1971).
5 See Escobar (1995).
6 The EU has a positive net migration balance with other countries of about 2 per 1000 or around three-quarters of a million people. The figure peaked in the early 1990s with a high level of asylum seekers and others from Eastern and Central Europe, the vast majority of whom have settled in Germany, without which the population in the EU would have declined (Litske 2002). Indeed, migration has been considered as a solution to the problem of ageing and population decline and like other countries the EU adapts its migration policies according to the state of the labour market.
7 See for example Myrdal (1963), Hirschman (1958) and Kaldor (1970).
8 See for example Amin (1974), Gunder Frank (1971), Hymer (1975), Smith (1984) and Lipietz (1987).
9 Economies of scale exist where unit costs fall as output increases. External economies are savings that can accrue to an individual firm from the activities or location of other firms or services.
10 This contrasts with the neo-classical perspective which sees comparative advantage arising from natural resources leading to specialization, whereas Krugman sees external economies as the foundation of costs savings and further specialization, i.e. the advantages are created by human economic activity rather than naturally given.
11 Elsewhere (Perrons 2001) I refer to Krugman's approach as New Economic Geography I and the cultural turn in economic geography as New Economic Geography II.

12 See Palloix (1976).
13 These terms are discussed in more detail in Chapter 5.
14 See for example Hymer (1975) and Fröbel *et al.* (1980).
15 See for example Massey (1978) and Perrons (1981).
16 Transfer pricing occurs when transnational organizations exchanging components between themselves structure their component prices to ensure that profits are highest where taxation on profits is the lowest. Some states may resist this strategy by insisting that a certain proportion of value added is made within their countries.
17 Gereffi (1999) refers to the Triangle system where North American buyers would place orders with South East Asian NIEs, who would subcontract to other countries – China, Indonesia and Vietnam – who would then export directly to the US using their own quotas.
18 See Bonacich and Appelbaum (2000).
19 See Mir *et al.* (2000).
20 See Chapter 4 and Kabeer (2000).
21 See UNCTAD (2002).
22 See Zhang and Zhang (2003) and Wu (2003).
23 See Anderson (1998).
24 See Kabeer (2000) and Fletcher (2002).
25 See Klein (1999).
26 See Chapter 9 for a discussion of ethical codes of conduct.
27 Initially two different kinds of chain were identified: the producer driven commodity chain which was organized by firms which managed the production of a product requiring lots of inputs in a quasi vertically integrated way – car manufacturing would be a key example; and second the buyer driven commodity chain that would be coordinated by a major retailer such as a clothing seller like Gap or a major supermarket (see Gereffi and Korzeniewicz 1994). However as these divisions are not definitive they are now generally referred to as value chains.
28 See ICO (2001) and Fitter and Kaplinsky (2001).
29 See Humphrey and Schmitz (2002).
30 See James (2001).
31 See Chapter 9 and Dolan and Tewari (2001) for a similar case of ethical upgrading in the apparel sector in India.
32 See Humphrey and Schmitz (2002).
33 See Gereffi (1999).
34 See Chapter 8 for a fuller discussion.
35 See Lovering (1999) and Esping-Andersen (1990).

# Part II

# ECONOMIC INTEGRATION, NEW DIVISIONS OF LABOUR AND GENDER RELATIONS

The deepening of market relations across the globe affects the lives of almost everyone, albeit in different ways. With the increasing convergence of economic, social and political systems, and their integration through financial markets, multinational firms, supra national institutions and commodity chains, events in one location can have widespread repercussions and the fortunes of people in different areas become inextricably interconnected. Financial crises in one part of the globe quickly spread to others and impact on their real economies; for example the financial crisis in South East Asia in the 1990s led to job losses in north east England. Likewise, in the Philippines the ensuing public expenditure cuts, in addition to those associated with earlier rounds of structural adjustment, created an excess supply of highly qualified people, some of whom migrated to more developed countries to become care and domestic workers and so fill gaps created by the feminization of employment there.

More generally capitalist social relations of production have become increasingly prevalent with the expansion of international trade and more and more women throughout the globe are engaged in paid employment. The feminization of employment has expanded first with the international division of labour in manufacturing, principally in electronics and garments, and more recently with the development of agribusiness and the internationalization of services, especially call centres and data entry. These events have been linked with broader social changes especially in gender relations, leading to rather extravagant suggestions that the contemporary era is witnessing not only the increasing empowerment of women and increasing gender equality but 'the end of patriarchy' (Castells 1997). Part II explores this idea by drawing on in depth studies of female employment in agribusiness, electronics, care and sex work, as well as studies of the decline of male employment in traditional industrial

81

sectors. This discussion refers to the restructuring of economies, especially of the north east region of the UK in Chapter 5, and the development of new forms of employment, but very much focuses on the human side of globalization in the context of an increasingly interconnected world.

The feminization of paid employment in recent decades is undeniable.[1] Between 1975 and 1995, 74 per cent of developing countries and 70 per cent of developed countries had increases in women's activity rates, while during the same period, male activity rates declined in 66 per cent of developing and 95 per cent of developed countries (Standing 1999). Table II.1 depicts the labour force participation rate of women over a longer period and while there are considerable differences between the different regions of the world, a similar trend can be observed. Everywhere women's labour force participation rate is lower than men's, but with the exception of West Africa and South Asia it has been increasing.

Employment feminization was especially rapid during the 1980s and early 1990s (Lim 2002) and has increased the 'share of women in wage employment in the non-agricultural sector', one of the indicators used to calculate progress towards Millennium Development Goal 3 to 'Promote Gender Equality and Empower Women'[2] (see Figure II.1). On this measure, women have achieved parity (a figure of between 45 and 55 per cent) primarily in Western Europe and other developed economies, Eastern Europe and Latin America. This measure is designed to reflect the more empowering effect of paid work in the non-agricultural sector where women are more likely to receive wages directly, and

*Table II.1* Labour force participation rates for world regions for people aged 20-59

| Region | Earliest census* | | Latest census** | |
|---|---|---|---|---|
| | Male | Female | Male | Female |
| East and southern Africa | 92.3 | 35.3 | 89.7 | 44.8 |
| West Africa | 91.5 | 60.7 | 89.7 | 57.1 |
| East Asia and Pacific | 92.6 | 35.4 | 88.9 | 50.9 |
| South Asia | 94.4 | 39.0 | 91.3 | 29.2 |
| East and central Europe | 93.5 | 61.9 | 90.4 | 76.0 |
| Rest of Europe | 95.0 | 36.6 | 90.4 | 53.5 |
| Middle East | 93.3 | 11.7 | 91.5 | 22.8 |
| North Africa | 92.5 | Na | 91.1 | 14.1 |
| Americas | 94.1 | 32.7 | 89.9 | 46.8 |
| Total | 93.7 | 35.9 | 90.1 | 47.9 |

*Source*: ILO (2003a).

*Notes*: * Earliest census = 1950s–1960s ** Latest census = late 1990s.
The labour force participation rate is a measure of the extent of an economy's working age population that is economically active. Specifically it measures the proportion of the population who are in the labour force expressed as a ratio of the population as a whole, for the same age category. See also Tzannatos (1999) for slightly earlier figures.

less likely to be regarded as family helpers. However, it fails to take into account the nature of contemporary employment, in particular women's overrepresentation in the informal sector in manufacturing and services.[3]

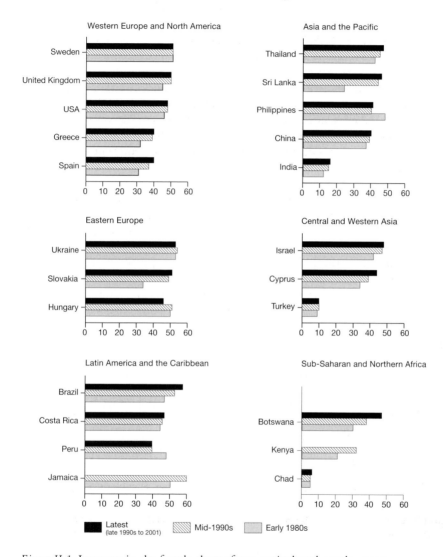

*Figure II.1* Increases in the female share of non-agricultural employment.

*Source*: Data from UNIFEM (2002).

*Note*: The data reflect the female percentage of non-agricultural employment in three different periods. The countries chosen reflect high, low and medium values for the countries in the broad regions of the world. The figure for Turkey is low because it is based on social insurance records and many women are not eligible. In most countries and regions there is an upward trend, except where parity has been reached or where there are specific circumstances such as the effects of the financial crisis in the Philippines.

Some of the changes referred to above and illustrated in Table II.1 and Figure II.1 are dramatic but whether these and more detailed statistics represent real material changes or whether they are simply statistical artefacts, due to changes in recording criteria, is less certain. In Pakistan, for example, the participation rate of rural women changed from 13.9 per cent to 45.9 per cent as a consequence of a revised Labour Force Survey definition in 1991–2.[4] International definitions are increasingly adopted and what was regarded as family labour is now more likely to be formally recorded as employment as increases in female employment are perceived positively – gender monitoring and gender impact assessments are important in national and supra national policy making and mandatory in many development programmes. The extent of feminization can also be misrepresented if employment is counted in new factories that have displaced previously unrecorded work in smaller workplaces and, conversely, if the increase in subcontracting leads to work such as home working being overlooked.

While statistics have to be viewed with caution, there is a statistical relationship between increasing participation in paid employment and falling levels of fertility for some countries (Lim 2002).[5] Given the unequal distribution of caring roles fertility decline increases women's capacity to engage in paid employment in a cumulative way, especially as employers rarely recognize or accommodate the specificity of women's caring responsibilities or issues of reproductive health. Moreover, a study across a range of countries – Brazil, Egypt, Malaysia, Mexico, Nigeria, the Philippines and the United States – found that involvement in paid employment enhanced women's power within households, giving them greater control over the use of income and resources as well as increasing their autonomy over issues such as marriage and child-bearing, contraception and sexuality and increasing their ability to resist domestic violence (UNIFEM 2000), though changes in employment that disturb traditional gender roles have also been associated with increases in violence (Pineda 2001).

Statistical trends of female employment rates are invaluable in setting the overall context, but there are dangers in moving from aggregate trends and tendencies to making inferences about the impact on social change which in reality is likely to be much more complex and varied. In all countries the rate of female labour force participation is lower than the male rate, but there are variations between countries, between regions within countries and between ethnic and cultural groups within regions and cities, depending on state policies, especially in relation to employment regulation and childcare provision, levels of development, cultural traditions and how these collectively impinge on more specific relations within the household.

At a world scale low levels of participation are found in North Africa, western Asia and parts of southern Asia, which are characterized by 'extreme forms of patriarchy', to the extent that they even give rise to gender differentiated survival rates (Kabeer 2003). Within Europe, where overall participation rates are high, they are comparatively low in Greece, Spain and southern Italy. Within regions and cities ethnic differences between women are found, for example in Latin America indigenous and Afro-Caribbean groups have higher levels of participation rates than Hispanics, especially the upper classes, and differences at this scale intensify as globalization increases ethnic diversity. So for example, within London, black Caribbean women have higher activity rates than other black and minority ethnic groups.[6] Moreover labour markets continue to be characterized by horizontal and vertical segregation to the disadvantage of women, even where there has been a long history of equal opportunities policies, so even when female participation rates are similar to men's gender inequality remains.

In the European Union, for example, women are overrepresented in activities relating to nurturing, care, clerical work and sales, while men are overrepresented in sectors and occupations involving money, management and machinery. Women hold 66 per cent of clerical and sales jobs, while men hold 80 per cent or more of jobs in the armed services, craft and related trades, plant and machine operators, and 66 per cent of skilled agricultural and fishery jobs. While women have been gaining entry into professional jobs, segregation persists, with men being overrepresented in mathematical and engineering careers and women in health and education professions. These statistics are for the EU as a whole, and while the degree of segregation varies between member states, all share this general pattern.[7] Segregation in employment builds upon and probably reinforces gender stereotypes, with women being seen as naturally suited to caring work but out of place in work dealing with complex machinery, involving physical labour, or in managerial positions, especially as the managers and supervisors of men. Again with reference to the European Union, only 21 per cent of the workforce have a woman as their immediate superior while 63 per cent have a man, the remainder having no immediate supervisor. Women managers and supervisors are much more likely to be supervising other women, and less than 10 per cent of employed men have a woman manager.[8] Likewise in Latin America women have been moving into professional work, but the overall volume of jobs available is proportionately fewer than in the EU and patterns of segregation similarly remain. More generally it seems as though men perhaps even universally find it difficult to work under female supervision or direction. A study of a micro-credit scheme to assist women in Colombia found that while men were prepared to work with their wives, if additional male employees were hired then the husband would give the orders, otherwise the masculinity of both the husband and the male employees would be threatened (Pineda 2001).

Even where female participation rates are high barriers to women's progress remain and it will be interesting to see how the increasing number of policies for gender equality at national and supra national levels challenge these patterns in the coming years, especially as just at the moment when policies for equality have been expanding and increasing numbers of women are entering paid employment, the terms and conditions on which it is offered are becoming increasingly flexible and insecure.[9] There are elite, professional and regular well paid jobs in the main urban centres throughout the globe, and women are gaining access to them, but the scale of this higher quality work varies positively with the level of development. Many jobs are part time or temporary and self-employment has increased, not because of a burgeoning of entrepreneurship, though some takes this form, but because for some people there is no alternative and for others it is because firms subcontract their responsibilities as employers in order to increase flexibility, leading to expansion of the informal sector.[10] Nevertheless although limited, the changes that have taken place have had an unsettling effect on men and led to a questioning of male identity, which becomes evident from detailed studies at the level of the household.

Households are generally understood as units of shared residence and or shared resources and through which daily and intergenerational reproduction takes place. While household structures are permeable and household relations vary cross-culturally, men have traditionally played the provider role, while women have specialized in caring, more or less universally. These differentiated roles have never been absolute and the boundaries between them are fuzzy but the feminization of employment has challenged the male provider role, in two ways: because women are acquiring incomes in their own right and because the male ability to act as sole provider has been challenged by contemporary 'feminized' forms of employment that offer different kinds of work, lower incomes and a lower degree of security.

In Europe, the displacement of heavy industry by lighter assembly work has affected men's sense of masculinity in a similar way as the displacement of cattle keeping in Kenya by a more settled form of farming, 'Shamba work'.[11] In both cases the new work was associated with women and considered demeaning. Thus work has embedded gender ascriptions, which though neither permanent nor uniform across cultures and countries, affect self-image or identity. More generally, the earnings associated with new forms of work are often lower in relative terms, for example in the maquiladoras earnings are insufficient to provide a family wage, unlike the work provided in factories associated with the previous import substitution policy, so households have to rely on two incomes (Cravey 1997). In other cases changes have resulted in a loss of male employment as in the industrial regions of northern England, in both cases leading women to become economic providers.

There is little evidence that men have been playing a greater role in child-care and domestic work as women's role outside the home has expanded, though formal time-use statistics have only recently been compiled and it is possible that formal recordings overstate the significance of regular activities in comparison to more occasional but time consuming ones such as household maintenance, more likely to be performed by men (Jackson 2001) and that men understate their domestic role as it conflicts with their self-image and prevailing conceptions of masculinity. Existing cross-cultural evidence suggests that there has been little rearrangement of domestic work and childcare, and in the context of state cutbacks in education and health provision, this increases the intensity of women's working lives and in some cases leads to younger girl children being denied education, so they can act as substitutes for their mothers, such that existing gender inequalities are passed on to the next generation.[12]

Despite arduous conditions in paid work and an increase in their total work burden, cross-cultural studies suggest that the overall impact of paid labour is positive for women, increasing their empowerment and self-esteem. Correspondingly men's reactions to changes in the perceived gender balance of power have been more negative. Some of the responses by women and men are discussed in the following two chapters, in the context of a discussion of changing forms of work across the globe as economies become increasingly integrated. In Chapter 4 therefore attention is given to more detailed studies of the feminization of employment in three key areas: the 'cool chain', that is agribusiness, the electronic circuit and the care chain in order to consider in more detail the way that paid employment has affected women's lives. Chapter 5 considers the reverse side of the coin, that is how men in declining industrial regions have responded to the loss of paid employment and how this has affected concepts of masculinity and gender relations there.

Changing gender relations is not necessarily a zero-sum game in which men are inevitably the losers. While transitional generations of men may have a sense of loss and women a sense of gain, some of the illustrations discussed in Chapters 4 and 5 indicate that these changes often become normalized and traditional gender relations can be reborn inside new employment forms and indeed men can reinvent their sense of self-worth and masculinity even within the domestic sphere. Moreover, in some instances women also collude in sustaining the more traditional masculine self-image so that unequal gender relations in some ways endure quite significant changes in material conditions. This is not to say that important changes have not taken place. Many women have been empowered by receiving direct incomes, but the illustrations of the implications of new employment forms associated with globalization through the lived experiences of women and men in Chapters 4 and 5 suggest that so far it would be premature to suggest that the end of patriarchy was on the immediate horizon.

## Notes

1 Interestingly Diane Elson (1999) suggests that it might be more appropriate to refer to these changes as masculinization because a greater proportion of the population now has employee status, once confined to men.

2 This is Millennium Indicator 11. The other three indicators under Millennium Goal 3 are the ratio of girls to boys in primary sector and tertiary education; the ratio of literate women to men 15–24 years old; and the proportion of seats held by women in parliament.

3 See UNIFEM (2002).

4 See Elson (1999) and Tzannatos (1999).

5 In the developed countries where labour force participation of women between the ages of 25 and 54 was between 60 and 85 per cent, rising through the 1990s, total fertility rates declined and were below replacement level. Elsewhere the

relationship was varied. In sub-Saharan Africa female labour force participation rates and total fertility rates are high, but in North Africa and Middle Eastern states female labour force participation remains low but total fertility rates have been falling (Lim 2002).

6 See GLA (2002) and Cabinet Office (2003).
7 See European Foundation (2002).
8 See Fagan and Burchell (2002).
9 See Standing (2002).
10 See Benería (2001).
11 See Jackson (2001).
12 See Elson (1995) and Pearson (2001).

# 4

# THE GLOBAL DIVISION OF LABOUR AND THE FEMINIZATION OF EMPLOYMENT

## Women's role in cool chains, the integrated circuit and care chains

Globalization reflects the deepening of market relations across the globe. The new international division of labour that came into existence in the 1970s as corporations decentralized parts of manufacturing has now been extended to high and low paid services and perishable agricultural commodities, to take advantage of different time and climate zones as well as lower cost labour. These developments have been facilitated by continued improvements in ICT and have contributed to the feminization of employment.

This global division of labour leads to factories, call centres and packing plants being created in poorer countries and regions but further links consisting of people moving in search of work and a better life have emerged on an unprecedented scale. The neo-liberal agenda, in particular structural adjustment policies, has reduced the proportion of regular paid employment, especially for male workers, while the feminization of employment in both rich and poor countries has generated gaps in childcare and domestic work, which are generally filled by people from poorer regions and countries creating a global care chain (Hochschild 2000). Another movement involving people, rather than goods, is trafficking, again in general from poorer to richer countries, as well as sex tourism, in which movements take place in the reverse direction, as tourists, predominantly men, travel with the implicit or explicit understanding that they will be able to buy sex as part of the holiday. There is also some 'body shopping' of men in the reverse direction as IT specialists, especially from India, are recruited to work in richer countries, especially the US (Mir, et al. 2000). This chapter illustrates some aspects of the global division of labour by considering the 'cool chain' or the fruit chain, the integrated circuit, or electronics, and the 'care chain', drawing on a range of examples throughout the world. Reference is also made to sex tourism. These vignettes of people and places in the new economy are designed to illuminate the human side of globalization

89

and how the growing interconnection between people and places shapes human lives in different locations throughout the world.[1] As the majority of the workforce in these new sectors are women there have been changes in gender roles, responsibilities and relations but whether this feminization of employment will lead to greater gender equality and the 'end of patriarchy' as has been suggested by Manuel Castells (1997) is much more questionable as the extent of change is limited by the new more precarious forms of work associated with the neo-liberal economic agenda and the continuing, profoundly gendered social context within which it takes place.

## THE COOL CHAIN

Global commodity chains form a characteristic feature of globalization and mark out a global geography consisting of the places or nodes connected by flows of commodities. Local production districts typically have only forward linkages and may be part of a single chain linked with centres of consumption elsewhere. At the opposite extreme global cities emerge where a multiplicity of chains with linkages stretching out in all directions coalesce. The global commodity chain literature is extremely valuable in highlighting the uneven production, appropriation and flow of value between nodes in the chain (see Chapter 3) but their social embedding in particular regions has only more recently been addressed.[2] This section addresses social embedding by focusing on the nature of work and life in two local links within one cool chain – the production of fresh fruit in Chile and its sale in UK supermarkets – in order to provide an understanding of the geography of the people and places within the chain and more specifically the impact of women's involvement in this sort of work on gender relations (Barrientos and Perrons 1999).[3]

Agricultural products have been exchanged between countries over the centuries but in the last few decades trade in fresh, temperate and tropical fruit (Barrientos 1996), flowers (Meier 1999) and vegetables (Barrett *et al*. 1999) has rapidly expanded. Thus, grapes are exported from Chile, roses from Israel and India and green beans from Kenya, to major sellers, especially supermarkets in countries such as the US and UK who buy from different countries and regions at different times of the year to ensure that they are continuously stocked with a full range of produce, irrespective of the season. These commodity chains are buyer driven, as they are initiated and controlled by supermarkets, whose bargaining power with respect to suppliers has increased with the expansion in their share of final sales relative to traditional wholesalers and retailers.

Chile, in the southern hemisphere, has a window in the global supply chain between January and April each year, when it supplies fresh produce including apples, pears, nectarines, peaches and grapes to northern markets in the USA and Europe, with about 5 per cent of its output going to the UK.[4] Chile first became involved in the global fruit chain in the early 1980s. The military government led by General Pinochet initiated a repressive agrarian counter-reform policy through which private commercial farming displaced poorer peasants from the land in the central regions of the country and then pursued an export led growth strategy, which was in turn intensified by the implementation of structural adjustment policies. Given the favourable climate and water supply from the Andes, in addition to the well developed infrastructure, including roads, containerized port facilities, electrification and communications, the area became characterized by mono-cultivation of high quality fruit, which has now become Chile's third largest export industry, after mining and forestry.[5]

Most of the production takes place on medium-sized farms owned by formally independent Chilean entrepreneurial farmers, formally independent because effectively they are often locked into a dependent relation with large export firms who coordinate transport and shipping and own the packing houses, where the fruit is selected, prepared and packaged. The fruit then passes from these export companies to import firms in the international points of destination. From here, fruit traditionally went on to the wholesale market, the greengrocer and then the consumer. But UK supermarkets typically circumvent the last stages of the chain and establish direct supply relations with the import companies and sometimes the exporters, and so indirectly become more closely involved with the producers. Supermarkets consider this degree of control necessary to ensure product quality and reliable delivery. Thus production is strongly characterized by agribusiness and is quite different from the peasant subsistence farming that it displaced.

The supermarkets effectively organize the chain in a just in time way, almost knowing from which field in which country their shelves will be filled on any particular day so the fruit has to be picked and packed at exactly the right moment and within the allotted time frame. Precise scheduling of production and distribution plus rapid retail turnover is crucial to profitability, far more so than for other commodities as fruit ripens and decays and potentially loses all value, thus advanced technology is applied to all stages in the chain. In production, computerized drip irrigation systems are used together with fertilizers, pesticides, hormones and new seed varieties. In distribution, advanced packing and storage facilities have been developed together with containerized transport, shipping and airfreight, and the fruit is chemically treated and maintained in computer monitored temperature and atmospheric conditions to control ripening and ensure that it arrives in 'perfect' condition. Finally, in sale, the supermarkets use high technology storage, monitoring and communication systems to coordinate the supply of fruit to match the levels sold.

Despite the technology, however, natural and market hazards remain, leading to large price fluctuations, risk and uncertainty, which each player tries to offset by developing flexible contracts with others in the chain. That is, to be sure of meeting demand, yet avoiding waste, the supermarkets have flexible contracts with a range of suppliers, who in turn develop flexible contracts with the producers, who respond by having flexible arrangements with their employees, especially the pickers and packers at the end of the chain, the majority of whom are women and highly dependent on this temporary fruit employment for survival.

In Chile, there is a clear gender division of labour in fruit production. Some women are employed in pruning, to ensure that the fruit meets the required shape, but the majority are employed in selecting, cleaning and packing, where women's 'natural' dexterity is considered important, while men work mainly in the fields and service the packing process (providing the boxes and transporting the fruit). Women packers are paid on piece rates, thus their pay depends on the volume of fruit available for packing and their individual productivity. At the height of the season, a period of perhaps three to four months, work is extremely intense; if the fruit is not picked, packed and sent off for export it is wasted. Correspondingly, large numbers of people, 400,000 or more, are mobilized into temporary fruit employment from the rural shanty towns, established when the peasants were displaced from the land, or else bused in from as far away as Santiago. Working hours for the packers can be extremely long, beginning around midday as the harvest starts to come in from the fields, and ending perhaps 12 hours later when the last of the day's output has been packed. Allowing for journey times this can mean anything up to a 16 hour working day, for six days a week. Added to this is domestic work, as the women generally prepare the day's meals prior to leaving for the fields. Interestingly, so far, Sunday remains sacrosanct, in contrast to the secular UK, where the supermarkets are now open for at least five or six hours. For this brief period, the most productive women packers can earn higher weekly incomes than the men, who are on daily rates in the fields. But overall women tend to work for shorter periods than men (as work in the fields extends beyond the harvesting season), and women are less likely to be able to find alternative employment than men out of season (Barrientos 1996); thus their attachment to the labour market is partial.

Traditionally in rural Chile women either worked on the peasant holding, caring for livestock and the kitchen garden, or if they worked on the hacienda, it was usually under their husband's contract. Thus although they worked extremely hard, it was rarely for an independent wage. Their role has therefore been transformed, from providing subsistence through domestic cultivation to earning money wages from agricultural labour on commercial fruit farms. Although the conditions are harsh, with little sanitation or even drinking water, exposure to dangerous chemicals and long hours, this is the only time that women are able to earn anywhere near a reasonable income, and can make a significant contribution to the household budget. It is however only a seasonal income and there are few alternative sources of employment open to them even at other times of the year. Moreover within the season, employment is varied and uncertain and there is no year on year guarantee that they will be employed. Indeed, regulations are few and can even be resisted by the workers, who prioritize short-term money income. For this brief period, however, they may be earning more than men, and also feel a sense of empowerment and enhanced sense of self-worth. Their earnings allow them to contribute to house maintenance, clothing and education for children as well as larger household goods, such as TVs, which sometimes have to be returned when the season ends and they can no longer afford the payments. As their earnings are precarious and insufficient to finance an independent livelihood it is questionable whether they will bring about major changes in gender relations. Even while at work there is little change in the household division of labour; children are sometimes taken to the fields with the women, indeed the state provides crèches near to the packing houses, or else cared for by other relatives, with men generally playing only a marginal role in household affairs if they return from work earlier than their wives.

Turning to the opposite end of the chain, supermarkets in the UK have captured an ever-increasing share of the market, that is:

> During the last three decades, the United Kingdom has been transformed from what Napoleon described as a 'nation of shop-keepers', with innumerable small businesses, to a supermarket culture dominated by a handful of large retailers. Their formula for success is simple – they operate efficiently, they provide a one-stop shop and they enjoy consumer confidence. Today they wield immense influence over the way we grow, buy and eat our food. They are shaping our landscape, our health and the way we interact socially, and these changes are going unchallenged because of our fast food lifestyles; consumers want quick access to a wide choice of goods at low prices.
>
> (Corporate Watch 2002)

More formally supermarkets now account for 80 per cent of all food retailing, and this share is forecast to continue increasing to 83 per cent by 2004.[6] The concentration of sales in supermarkets has been strengthened by their ability to manage global commodity chains and by the changing patterns of shopping, referred to in the quotation above, especially the once weekly 'one stop shopping', often outside 'normal working hours', that they have partly helped to create. This pattern is also associated with rising household incomes and consumption standards, including car ownership, fridges and freezers that allow bulk shopping, and even more recently with home computers, that allow shopping through the Internet, and with the feminization of employment that has contributed to the pressured lifestyles that make supermarket shopping and virtual shopping more attractive.[7]

Flexible working patterns also underpin and sustain this new pattern of trading, as extended opening hours, including Sundays, mean that labour needs simply cannot be met from a full time workforce working 9–5. Employers use ICT to match employment to customer demands as they vary throughout the day, week and year, just as they monitor and match their supply of produce to sales. Moreover ICT has also made it easier to match employment requirements with employee working time preferences, which potentially allows complementarity between employer needs and employee preferences for flexible hours. Certainly when labour markets are tight, imaginative employers devise ways of meeting employee preferences, although at other times and perhaps more generally it is the employers who retain control over the parameters of flexibility, as the following human resources manager explained: 'What we have tried to do is introduce unique working patterns so that individuals truly do come in when the function and operation requires them to.' This particular company had about one-fifth of its workforce on a 'flexi contract' which committed employees to three hours a week between January and March, nine hours a week between April and October and then 25 hours a week between November and December to match the seasonality of sales.[8] In contrast to the Chilean workers, however, these employees were permanent and would qualify for pensions and staff discounts, even though it would be very difficult to plan an independent survival strategy on this kind of contract. Overall in 2002 there were just over half a million employees in the big four supermarkets[9] with a further 2 million in retail as a whole; approximately two-thirds of the total are female, and in turn two-thirds of the women work part time with a very wide range and distribution of hours.

Nevertheless even full time workers' earnings are rarely sufficient for them to be independent or to pay for childcare. Indeed the low childcare support in the UK, despite the National Childcare Strategy introduced in 1998, and some financial support through tax credits, makes the varied hours offered by supermarkets, including evening, night and weekend work, attractive to people with caring responsibilities.[10] A pattern of shift or serial childcare is widely practised among partnered employees, with one parent returning from work as the other leaves, giving rise to complex juggling arrangements in which the adults in the family spend little time together. Similar to Chile, there can be a marginal change in gender roles, as men have to take on more domestic work and childcare when women are absent from the home, but similarly the scale on which this is done is unlikely to lead to any long lasting changes in gender relations.

Despite vast differences in national location, culture, history and religion between the two regions, unequal gender relations are pervasive. The effectiveness of the global chain depends on the ability to draw on flexible female workers at both ends. In Chile women move in and out of paid work on a seasonal basis while in the UK they are more likely to be permanently employed, but on low pay and with varied and sometimes limited hours, either during the school day or during 'unsocial times' when partners or other relatives can care for children.[11] This flexibility can only be understood by reference to gender relations in the two localities that have been built upon and shaped rather than transformed by their involvement in paid work, in both cases casting doubt on the idea that the feminization of employment will by itself transform gender relations. This limited involvement also constrains the extent to which women are willing to engage in industrial action to improve their conditions, although in the larger

UK supermarkets, unionization is high relative to the rest of retailing. In Chile, by contrast, it has been argued that the economic miracle to which fruit growing has contributed (though the benefits have hardly trickled down to the fruit workers) has made it difficult to organize workers, as they become locked into consumer credit chains:

> People are unwilling to take action or go on strike because they have such big credit card bills to pay at the end of the month. . . . They [the credit card companies] are right there beside the orchards and vineyards ready to sign people up as they leave for the day.
> (Angela Alvarez Cerda an activist for the organization Mujer y Trabajo (Women and Work), cited by Swift 1998)

Looking at this more positively, at least new workers are able to organize openly in contrast to the situation when Pinochet first embarked on the export led growth strategy. Thus while feminization of employment has led to some changes within the household and has increased a sense of empowerment, the wider context which shapes decision making remains profoundly gendered, in both cases women's partial involvement in the labour market is accepted because it is assumed that there is a male provider to fall back on, but this assumption is losing credibility, owing to falls in male employment, declining male real wages, especially in lower income jobs, and the increase in divorce and separation.[12]

## WOMEN IN MANUFACTURING: THE INTEGRATED CIRCUIT IN MALAYSIA

During the late 1960s and 1970s, in response to the crisis of Fordism, large firms in developed countries decentralized labour intensive production activities to lower wage regions and countries throughout the world, but the higher level, strategic and organizational activities remained, creating the new international division of labour, discussed in Chapter 3. Clothing and electronics were particularly central to this development and the literature highlighted what were then seen as novel and rather perverse patterns of organization whereby jeans 'are cut out in Germany, then flown in air containers to Tunisia, where they are sewn together, packed and flown back for sale in Germany' (Elson and Pearson 1981: 92),[13] patterns which have subsequently become more complex and commonplace. Similarly, highly labour intensive parts of the electronics production, for example integrated circuits, were decentralized, especially to South East Asia. States in less developed countries and less developed regions within developed countries offered lucrative incentive packages to electronics plants to locate in their regions because they were modern, high tech and thought to promote development. A Malaysian investment brochure for example advertised their suitability in the following way:

The manual dexterity of the oriental female is famous the world over. Her hands are small and she works fast with extreme care. Who, therefore, could be better qualified *by nature and inheritance* to contribute to the efficiency of a bench-assembly production line than the oriental girl.

(Emphasis by authors Elson and Pearson 1981: 91)

Women were and continue to be overrepresented in textiles and electronics plants and undoubtedly the way that the skills documented above could be combined with low pay made these locations and their workers attractive to the multinationals. As one manager of an electronics plant in Penang, Malaysia explained: 'One worker working one hour produces enough to pay for the wages of 10 workers working one shift plus all the cost of materials and transport' (Grossman 1979:7).[14] Wages paid therefore bear no relation to the value of output, the gap arising partly because the women's skills are acquired through years of learning within the household which enable them to be quickly trained up to do intricate work on the lines, but as these are assumed to be natural talents rather than material competencies, they are correspondingly underpaid (Elson and Pearson 1981). Other reasons relate to the allegedly low aspirational wages of women and the imbalance between the prevailing norms of consumption where the products are produced and those in the markets where they are sold. More generally, and in contrast to neo-classical models of wage determination, 'social wages' are paid. Thus in India lower wages are paid to people belonging to lower social castes (Rammohan and Sundaresan 2003), and in Saudi Arabia the wages of migrant labour are linked to the country of origin, having little to do with the work performed or the prevailing wage rate in the local labour market (Mahdi and Barrientos 2003). A further reason why employers favour women is because they are thought to be more disposable arising from their family roles, which makes it easier for management to adjust the size of the workforce to the fluctuations in demand.[15] Even so the work and pay was attractive to many Malaysian families who broke traditional customs by allowing their daughters to work outside of the home. Factories held visit days to demonstrate that the working and living conditions were physically and morally safe. Somewhat contradictorily, within the factories westernized, feminine fashions and lifestyles were promoted to encourage workers to identify with the company, and the electronics workers could be identified by their tight jeans, make-up and high heeled shoes. Moreover Rachel Grossman found that beauty contests were held to foster a 'passive and ornamental femininity'. Further competitions between individuals, between groups of workers within the plant, and between Asian subsidiaries were held to enhance performance, the star prize being a trip to the plant in California (Grossman 1979).[16] These mechanisms formed part of a sophisticated human resources strategy, designed to boost output while forestalling any sense of collective solidarity. Unions were discouraged and the only resistance was 'hysteria' which broke out from time to time as individual women, under intense pressure, cried out because they saw spirits down their microscopes, and unless they were quickly removed these visions could spread rapidly and bring the whole factory to a standstill. Rachel Grossman (1979) argues that these human resources strategies were particularly necessary in Malaysia rather than Indonesia, the Philippines and Thailand because the Malaysian labour market was tighter and people were more educated, indeed these characteristics also meant that Malaysian plants did the quality control work for others in South East Asia.

Despite the modernity of the factories and the high tech image conveyed by the special dust proof clothing, working hours were long and the work was detailed – connecting microchips to circuit boards with wires as fine as human hair. Little consideration was given to health and safety, a factor that continues to be neglected (Pearson 1997a). After three or four years of such intense work many women had irreparable eye damage and were called 'grannies' because they had to use glasses even though they were only 25 years old. Some were forced to leave the industry altogether and experienced considerable hardship, as they were left in an in-between world, never really being accepted back into traditional ways of life because of their westernization and yet unable to continue working in the factory, leaving them only bars and restaurants, with some ending up in the sex trade (Grossman 1979). Thus while the planning agencies considered the work high tech and therefore a desirable stepping-stone to development, which in Malaysia has been partially realized (Mitter and Rowbotham 1997), the position of women is more questionable even for those who have remained within the industry.

Twenty years on and contrary to some expectations, electronics factories are still in Malaysia and the country experienced very rapid growth, at least up until the Asian financial crisis of 1997, from which it is now recovering. The electronics industry that was responsible for about half of its exports in 1995 (Wilkinson et al. 2001) has contributed to this growth. Further, some of the factories have been upgraded through automation and many women workers are now considered skilled, because they work with sophisticated computer-aided machines. Again contrary to some of the earlier critical expectations some women have become long-term employees. For example half of the employees in an American factory established in 1972, studied by Cecilia Ng and Maznah Mohamad (1997), had been with the factory for at least 11 years and 19 per cent between 16 and 20 years. Women could also apply to become technicians, though these only constituted 3 per cent of the workforce, so promotion, from their initial positions as production operators, was rare in practice. Thus there have been some quite radical changes in the organization of work, but less dramatic changes in the relative position of women.

There have been changes as well as continuity in the human resources management strategies. The direct day to day supervision by supervisors has dramatically diminished and supervisors and managers now consult (rather than insult) the workers through the three levels of participatory involvement identified in Table 4.1. Participation is effectively compulsory and workers risk being shamed if they do not make at least two suggestions a month for solving

NEW GLOBAL DIVISIONS OF LABOUR

*Table 4.1* Empowerment through participation: levels of participation in an American owned electronics factory in Malaysia

| Name of programme | Programme content | Workers supportive or highly supportive of programme (%) |
|---|---|---|
| I recommend | Workers required to make 2 suggestions to improve the line. For good suggestions workers receive RM 2 (0.5 euro) and if taken up RM 5. Failure to make any suggestions results in names being posted on the bulletin board. Prizes for the top 200 suggestions – bed sheets, umbrellas or clocks. | 90 |
| Participative problem solving | Bi-monthly meetings of groups of workers, led by empowerment leader to pick a production problem and solve it together. Rewards for best team: extra wages, prize of a holiday. | 85 |
| Participative management process | Operators and managers brought together to discuss workplace problems | 91 |

*Source*: Information taken from Ng and Mohamad (1997).

problems. Thus whether the strategies are genuinely empowering or just a new form of control is questionable and despite the positive response rate towards participation reported in Table 4.1, some of the comments by employees are either highly sceptical of or at least give a new meaning to empowerment. For example one employee commenting on the changes stated that:

> With automation work is less tiring and teamwork becomes impor-
> tant: Now there is empowerment. When there are less materials
> to work on and when another line leader asks for extra hands . . .
> 2 or 3 girls are asked to go to that department. The line leaders
> cooperate with each other in work.
>
> (Ng and Mohamad 1997: 189)

This strategy clearly increases the intensity of work, even if some control has been devolved to the line leaders. Moreover, it encourages the workers to work hard and ensure quality by sharing the company's objectives rather than through strict supervision and thus 'the workers are the ones who monitor each other now' (respondent in Ng and Mohamad 1997: 191). Furthermore workers are persuaded that they do not need unions as there is an open door or 'Speak Out' policy in which they can voice their concerns to their immediate team leaders and if they remain unsatisfied, through successive layers of management, until their concerns are resolved. However, if their suggestions are contrary to management policy, for example asking for a union, for promotion to follow training or even that they be allowed to sit down during their shift, they are generally persuaded that these are not necessary:

> When I wrote in the Speak Out about an increase in annual leave I was called in for a personal interview. They brainwashed me and told me it was not necessary. The problem is that we ordinary workers do not know how to argue our request.
>
> (Ng and Mohamad 1997: 198)

Nevertheless, old style practices remain and workers are still encouraged to compete with one another for prizes, such as rice cookers and mini radios, in the participatory strategies referred to in Table 4.1 and to remain committed by rewards for regular attendance. In addition there are family sports activities and an annual dinner as well as subsidized transport and food. In this respect the contemporary terms and conditions of employment have echoes of Fordist employment relations and would seem to be better than those reported in studies of work in the maquiladoras, that is foreign owned assembly plants, in Mexico.[17] The outcome of the human resources policies and automation are that the number of defects has been drastically reduced from 50,000 parts per million to 3 parts per million. From the employee perspective the factories provide regular employment at wages above the local average, but for women there are few possibilities for promotion and few options outside the firm as their skills are firm specific.

## Export processing zones

In relation to manufacturing more generally, the scale of the international, now global, division of labour and the range of countries involved has expanded dramatically. Many firms are in export processing zones or EPZs, which are industrial zones with special incentives to attract foreign investors, and in which imported materials undergo some processing before being exported again.

The International Labour Organization (ILO 1998) estimated that there were some 2000 such zones throughout the world employing 27 million people; the majority were in North America (320) and Asia (225) but they are expanding throughout the world, including 47 in Africa. However, many countries have tried to integrate foreign manufacturers with local firms and so move beyond simple processing. Furthermore incentives often apply to firms within and outside the zone, especially with the expansion of free trade areas, for example NAFTA, so definitions become blurred and correspondingly the impact of EPZs is difficult to measure. In Mexico, for example, maquiladoras were first established along the United States–Mexican border in the mid-1960s, when the Mexican government set up the Border Industrialization Programme to provide employment for Mexicans returning from the US, in turn a consequence of the termination of the Bracero Programme in which predominantly male Mexican workers had been allowed into the United States to meet the labour shortages during the 1939–1945 war (Fleck 2001). There are now about 3500 maquiladoras in Mexico, with approximately 1.25 million employees or around 25 per cent of manufacturing employment (BIE 2002), and while concentrated in the border states are no longer confined to them (see Figure 9.1, Chapter 9). Contrary to the original purpose and expectation of the plans, women constitute over half (56.6 per cent in 1998) of the employees. Initially this percentage was far higher (78.2 per cent in 1979) but while in absolute terms female employment has multiplied sixfold, male employment has also increased as maquiladoras represent an increasing proportion of Mexican manufacturing and correspondingly have become more diverse sectorally, especially with the expansion of the transport and equipment industry, as well as having a wider geographical distribution (Fleck 2001). Over time the domestic manufacturing sector has been squeezed and similar to the UK (discussed in Chapter 5), young men without a history of industrial militancy are attractive to modern multinational employers. Furthermore, the supply of single females is not inexhaustible and although married women represent an increasing share of employment in the maquiladoras for the country as a whole the female employment rate is lower than the male rate, as women continue to be constrained by domestic responsibilities. With NAFTA the special terms on which the maquiladoras were established in the border areas will extend to all regions in relation to trade with the US and Canada, making the whole of Mexico attractive to North American based firms (Fleck 2001). Moreover, as in the case of South East Asia discussed above some firms have upgraded and are no longer confined to simple assembly work. They correspondingly provide a relatively wide range and comparatively stable form of employment, at least in contrast to the informal sector, though wages can be lower (Fussell 2000). Plants established in countries comparatively new to FDI and manufacturing for world markets are more likely to be simple assembly and have a high proportion of female labour. The maquiladoras recently developed in Honduras, which now account for over 27 per cent of manufacturing employment there, are all in clothing and 60 per cent or more of employees are women (ver Beek 2001). Similarly, garment production has expanded dramatically in Bangladesh. The Bangladesh Garment Manufacturing and Exporting Association (BGMEA) had 19 members in 1977 but 3581 in 2002, employing over 1.8 million people, 90 per cent of them female (BGMEA 2002; see also Kabeer 2000). Similar expansions of employment have taken place in Indonesia, Mauritius, Tunisia, Sri Lanka and the Philippines.

Women constitute the majority of people employed in these factories and the feminization of employment on this scale is a new phenomenon in many countries. Conditions in the factories also vary enormously, with some contemporary factories having far worse conditions than those described for Malaysia twenty years ago. Naomi Klein (1999) for example reports on the harsh working environment in Rosario, an EPZ in the Philippines, and the intensity of work, long working hours and poor working conditions could be replicated in EPZs elsewhere, for example Indonesia or China, and even in some of the Mexican maquiladoras (Denman 2002; see also Chapter 9) where possibly because of the large numbers of people seeking work, conditions in the factories have not really improved. In the factories directly owned by transnational companies conditions are likely to be among the best around owing to pressure from NGOs and consumer groups leading to the development of codes of conduct (discussed in Chapter 9), as well as from the workers themselves, together with the firm's concern to maintain its reputation. Referring back to garment factories in Bangladesh, where there have been concerns about the working environment, the ILO, assisted by the US Labour Department together with the BGMEA, have recently introduced a $1.5 million programme to improve health and safety and eliminate child labour (U.N. Wire 2001), partly because the Bangladeshi government recognizes that:

> 'Consumers have become very choosy,' said Commerce Minister Amir Khasru Mahmud Chowdhury. 'Now they want to know where a product is made and what were the working conditions.'
>
> (*United News of Bangladesh*, 1 October, cited by U.N. Wire 2001)

According to the BGMEA website, however, child labour had already been eliminated by 1996.[18] Moreover, contemporary competition takes place on the basis of quality as well as cost, which is more difficult to obtain from an alienated workforce. Thus some companies realize that it is in their interest to improve basic conditions, but this tends to happen either with the upgrading of plants or in plants that produce higher value products. Indeed conditions vary both within and between countries.

The ILO (1998) suggests that there are three phases of human resource management:

1   Labour supply is ample and treated as a renewable resource and working hours are long. This model is especially prevalent in labour intensive, low value added operations and is associated with high labour turnover, lateness and absenteeism and basically low levels of worker commitment such that in the end the intensification of work becomes counter-productive.

2   Managers try to motivate the workforce and enhance commitment and therefore output through performance related pay, rewards for long service and a range of services such as free breakfasts to reduce lateness, free transport to reduce fatigue, some health and dental services to reduce time off and some housing assistance and childcare facilities: '*The in plant crèches allow the mother to feed a number of times during the 10 or 12 hour shift*' (my emphasis (see below for comment)).

3   Managers involve the workers directly in problem solving and through quality control circles and work teams, which are empowered to solve their

own work problems; the supervisor layer is phased out. To be effective this strategy requires considerable training in statistics but also attitudinal change. This development is particularly prevalent in capital intensive electronic plants where the firm has to amortize its investment quickly before they are outmoded.

This last phase of this stylized model has parallels with the Malaysian electronics factory described above, where it was regarded with a degree of scepticism by the employees. What this model indicates however is the varied conditions within contemporary manufacturing plants and that the precise conditions will vary according to the nature of work in the plant, its position within the global value chain as well as the prevailing local, national and global economic conditions. This model also presumes that firms consciously plan their human relations strategies, and while the larger firms undoubtedly do so, contemporary changes in the organization of manufacturing, particularly 'hollowing out' and subcontracting, undermine these kinds of development and also make any monitoring of workplace conditions difficult (see Chapter 9). Thus there is no necessary reason why particular plants in particular locations should pass through these phases, i.e. the model could describe different plants in different locations at the same moment rather than the progression of a plant in a particular location over time.

## Hollowing out

Some of the leading companies, especially in clothing and footwear, no longer play any direct role in producing the goods they sell, as discussed in Chapter 3. Nike for example does not produce any shoes and Dell does not manufacture computers. Similarly many of the well known clothing labels do no manufacturing because they can make more profit by designing, branding and marketing products (Klein 1999). Thus they obtain products from a whole network of companies distributed throughout the globe and firms can be brought into or dropped from this network very easily. Production has thus been hollowed out and takes place in a range of variously owned companies, from large nationally owned companies in both rich and poor countries to small and medium-sized enterprises to homeworking, often connected through complex subcontracting arrangements. Similar to the fruit chain, each firm seeks to minimize risk in the highly competitive and volatile global market and these risks are frequently passed on to employees whose security depends on their employer's position in the supply chain, with home workers typically experiencing the most vulnerable conditions. Furthermore large firms can absolve themselves of responsibility for the working conditions of people in their supply chain that logistically are extremely difficult to monitor. So even where codes of conduct exist it is difficult to be sure how effective they are (see Chapter 9). What this also means is that just at the moment when women are entering into paid employment in

102

increasing numbers, the terms and conditions of employment are likely to be characterized by risk and insecurity,[19] something which also questions the simple association between the feminization of employment and greater gender equality.[20] Even so, transnational corporations, with their indirect suppliers, remain responsible for the major share of foreign direct investment and the development of export processing zones (EPZs) throughout the world.

Globalization reflects the clear and undeniable restructuring of manufacturing activity and conditions vary enormously between different plants so it is impossible to make general statements about the overall desirability of this sort of investment or the impact on gender relations. In each case it is contingent on the combination of specific conditions and also needs to be considered in the context of alternative possibilities, which in the present context seem to be declining as the increased opening of economies demanded by the structural adjustment policies and by the conditions attached to loans has undermined the competitiveness of small and medium-sized enterprises. In addition there have been reductions in public sector employment associated with cuts in public expenditure. One finding which consistently emerges from the qualitative work on women's experiences in manufacturing is that even when working conditions are extremely harsh, such work almost universally has a positive and empowering effect on women's self-esteem, though the impact on gender equality is less clear. These issues are taken up in the concluding section but this sense of empowerment in manufacturing is probably more evident than for care and sex workers discussed below.

## THE GLOBAL CARE CHAIN

One of the characteristics of the new economy is that some workers are highly paid but time starved and this generates demand for a wide range of marketized services including cleaning, catering and caring, which are low paid and in which typically women are overrepresented. For the highly paid professional people, work is rarely physically demanding, it takes place in congenial surroundings and for some may be preferable to domestic work, childcare or even leisure (Reeves 2001). Indeed some employers provide a range of concierge services or 'lifestyle fixers' to enhance the workplace and facilitate long working hours, as the human resource manager at Microsoft explains:

> 'If you want to keep staff then you have to look after them,' says Hilda Barrett, group human resources manager at Microsoft. 'That's why we try and create a campus atmosphere at our office. We have top quality, gourmet food always available and in the evenings we even run cookery classes. Oh yes, and you also get Waitrose Direct, a grocery shopping service. Who wants to waste their spare time pushing a supermarket trolley?'
>
> (Extract from 'Work Unlimited' – Anita Chaudhuri,
> *Guardian*, Wednesday 30 August 2000)

It is not clear how many employers provide services of this kind, and to what proportion of employees they are available. They are clearly a bonus or perk for employees and can easily be withdrawn if circumstances change. In south east England a new media company withdrew its breakfast service in the downturn of 2001,[21] while a major investment bank in the City of London provided all of the services identified in the quotation from Microsoft plus an emergency crèche to enable its employees to continue working if their usual arrangements failed (see Chapter 7).[22] Some employers are beginning to appreciate that they will have to be more flexible if they are to retain highly qualified women in senior positions given the contemporary unequal gender division of domestic labour. Nevertheless so far this flexibility takes place within very narrow confines. Referring to the firm that provided an emergency crèche, employees considered the firm 'understanding' because it accepted that mothers with young children might work a 10, rather than a 12 or 13, hour day in the office and allowed them to complete work at home. Rather as in the case of the maquiladora above which allowed mothers to take breaks for breast-feeding during their 10–12 hour working day, this seems to be more of a necessary concession, than an active promotion of work–life balance. In both cases there seems to have been a reconciliation of life to work rather than vice versa, and in the London case this has not been achieved by a greater sharing of tasks between women and men, but rather by an expansion of care and domestic work, most likely carried out by low paid women workers.

Demand for marketed services – domestic cleaners, childcare workers and nannies – is growing to facilitate long working hours, and more simply as a consequence of women working away from home, given the context of an unchanged domestic division of labour between women and men.[23] Individualized forms of domestic work in other people's homes have never been popular and, historically, have been abandoned as soon as alternative work becomes available. In the UK during the Industrial Revolution domestic servants quickly moved into workshops and factories, and similarly more left domestic service as opportunities became available with the development of Fordist factories in the 1920s and 1930s leading to a domestic servant crisis in both the USA and the UK. Middle class women had to carry out domestic work themselves with the aid of labour saving technologies such as washing machines and vacuum cleaners produced by these factories (see Chapter 5). Similarly, women were very reluctant to relinquish their jobs in shipbuilding and armaments and return to domestic work when men returned from the Second World War, clearly depicted in the film *Rosie the Riveter*.[24] Despite this reluctance, demand for childcare and domestic work has once again increased, reflected in the expansion of the personal service sector, itself a response to, and indication of, the continued feminization of the labour force, especially of highly qualified, higher paid mothers in dual earner professional households. Rather than leading to higher pay, however, this expansion has led to the development of global care chains, similar to commodity chains but depending directly on the movement of people, overwhelmingly women, which has changed the gender balance of contemporary migration flows.[25]

Arlie Hochschild[26] (2000) refers to the case of 'Vicky Diaz' from the Philippines, a college educated, former teacher and travel agent and mother of five who earns 400 US dollars a week working as a nanny for the 2 year old son of a wealthy Los Angeles family, while paying her own family's live-in worker in the Philippines 40 dollars a week. Vicky is a specific instance of the global care chain, which in general consists of:

> A series of personal links between people across the globe based on the paid or unpaid work of caring. Usually women make up these chains, though it's possible that some chains are made up of both women and men, or, in rare cases, made up just of men. Such care chains may be local, national or global. Global care chains . . . usually start in a poor country and end in a rich one. But some such chains start in poor countries, and move from rural to urban areas within that same poor country. Or they start in one poor country and extend to another slightly less poor country and then link one place to another within the latter country. Chains also vary in the number of links – some have one, others two or three – and each link varies in its connective strength. One common form of such a chain is: (1) an older daughter from a poor family who cares for her siblings while (2) her mother works as a nanny caring for the children of a migrating nanny who, in turn, (3) cares for the child of a family in a rich country.
>
> (Hochschild 2000: 131)

Thus rather than value being unequally appropriated at different stages in the chain, care chains take a directly hierarchical form on the basis of gender, race and generation as poorer people, generally women from poorer regions of the world, care for the children and elderly relatives of people in richer regions, that is 'mothering is passed down the race/class/nation hierarchy, as each woman becomes a provider and hires a "wife"' (Hochschild 2000: 137). Care chains are typically racialized as Caucasian workers are generally paid more than Asians and Asians more than those of African descent. In Rome, the Filipinas are paid more than people from Cape Verde (Tacoli 1999) and in Canada, Filipinas are 'housekeepers' and have to combine housework with childcare while Europeans are more likely to be 'nannies' and only required to care for children, even though the Filipina may be a university graduate and the European a qualified nursery nurse but with fewer years of education (Pratt 1999). Thus uneven development within the context of globalization, together with the undervaluation or non-recognition of qualifications from poorer countries can create a spatial dislocation by social class, as the following Filipina graduate commented:

> When my employer gave me the bucket for cleaning, I did not know where I had to start. Of course we are not so rich in the Philippines, *but we had maids*. I did not know how to start cleaning, and my

feelings were of self pity. I kept on thinking that I just came to the
United States to be a maid.

(Parreñas 2001:150, my emphasis)

Caring is sometimes a relatively easy way of gaining entry to richer countries.
People do care work while qualifying for a Green Card in the USA before moving on to other forms of employment.[27] Similarly in Canada, after two years
as a live in caregiver people can apply for full citizenship, which would enable
them to send for their families as well to take up other forms of employment.[28]
Thus while the global care chain remains in place individuals do not necessarily remain in fixed positions; care work can facilitate migration and be a
stepping-stone to other activities. Nevertheless, in the meantime care workers
can be exposed to all the hazards associated with individualized forms of
employment; in particular, indeterminate working hours, low wages and in
some instances, sexual abuse. Even in countries where they exist, employment
regulations do not always apply to live in caregivers and even when they do, may
be difficult to effect, as workers may be unaware of them, unwilling to
do anything that could prejudice their chance of permanent citizenship, or in
some cases prefer the higher short-term incomes associated with unregistered
work, for example the illegal Filipina care workers in Italy (Tacoli 1999). This
situation varies considerably between states. In the more corporatist Germany,
housework has been professionalized and although pay is low, qualifications
are required and a formalized career structure exists.[29] In France, as part of its
strategy to reduce unemployment, the government provided assistance to
people who employed childcare workers, which again provided them with
greater security or at least made it more likely that this work would be formally
recognized and employees more protected, but regulation in this form of work
is comparatively rare on a global scale.[30]

Care chains are in some ways an inevitable consequence of globalization in
the context of uneven capitalist development, unequal gender relations and
reductions in state expenditure. As women move into paid employment outside
the home, gaps arise in care and reproductive work that are filled by lower paid
women internally and when this supply is exhausted, from elsewhere. As with
other forms of economic migration it is an individual solution to a structural
problem and does little to redress the inequalities that generate the movements,
except through migrant remittances. These remittances are a vital source of
hard currency to many countries; in the Philippines in 2002 they were around
$6.5 billion per annum which is equivalent to $79 per capita while foreign aid
was equivalent to $7.4 per capita and overall they amount to over one-third
of total foreign earnings (World Bank 2003a; Sassen 2003). Indeed remittances
have been institutionalized to the extent that to facilitate transfers between
Hong Kong (where many Filipinas work as maids) and the Philippines ATMs
have been established on street corners and in 7–11 convenience stores.[31]
So while migration is an individual solution to a structural inequality, it has
become institutionalized as some states rely on remittances to resolve their

economic problems and especially debt and have developed formal labour export programmes to sustain such developments. Moreover while the migrants are considered and treated as temporary, institutional arrangements such as the banking transfers system indicate that migration itself is more permanent and in fact many migrants work as 'migrant' or 'temporary' labour over a long period.

Structural inequalities between countries can explain the broad directions of migration, and the specific links between countries can be explained by the size and nature of the development gap, geographical proximity and cultural and historical ties. Thus there are major movements of people from Sri Lanka and the Philippines to the oil rich states in the Middle East, and richer countries in the Far East, such as Japan and Singapore; from South to North America, and from Eastern to Western Europe, in addition to patterns that follow former colonial relations. However neither uneven development nor geographical or cultural proximity can explain individual migration decisions or the gendered nature of the care chain.

Individual migration decisions rest on how personal circumstances and preferences intersect with the wider economic and social context, that is on push and pull factors. In general, migrants are among the more highly skilled in their countries of origin but find themselves in the lower echelons of the societies to which they move, and this pattern exists in care work. The Filipinas who migrate to Rome, for example, are rarely the poorest, owing to the high transport costs, but people who have experienced a sudden deterioration in personal finances, arising from the death of a partner or loss of a middle class job as a consequence of public expenditure cuts (often linked to structural adjustment policies), but seek to maintain their living standard, in particular their children's education.[32] The case of Vicky Diaz referred to above is exemplary in this respect. Before working in the US, she had worked illegally in Taiwan as a housekeeper, factory worker and janitor. She was happy with the income but felt insecure owing to her illegal status so she returned to the Philippines, but only stayed three months before using her savings from Taiwan to pay $8000 to use another woman's passport to go to the US. In the last nine years she has only spent three months with her family. She is currently trying to legalize her position so that she can reunify her family in the United States. Parreñas comments that:

> Vicky claims that she works outside of the Philippines so that her family does not become destitute; it is actually more accurate to say that Vicky works in Los Angeles to sustain a comfortable middle class life for her family in the Philippines.
>
> (Parreñas 2001:88)

The relative scale of different motivations is unknown but migration also takes place to escape from low incomes or oppression at home in countries where divorce is not permitted or, in the case of young men, as a remedy for misdemeanours such as gambling, or simply for adventure and to see the world, especially where local opportunities are limited (Tacoli 1999). Individual motivations are complex and varied. However, this does not mean that the undervaluation of this form of work, the conditions experienced by care workers and indeed its hierarchical and racialized nature should be overlooked. Migrants also experience emotional trauma from leaving their own families behind and risk their own children being abused or their husbands using their remittances to form and finance new families (Parreñas 2001). In Sri Lanka, where unemployment has led to increases in female migration, especially to work as maids in the Middle East, there is also concern that husbands waste the remittances on alcohol, as they attempt to retain their status among their peers having lost the traditional means to do so through paid work (Gamburd 2003). These anxieties and tensions can sometimes be eased a little by contemporary technologies. As Carla, a South African nurse working in Brighton and Hove while leaving her children, aged 2 and 8, with her mother in the Eastern Cape, explained: 'I always leave the computer on, I have to I am a mother; there might be an email – they might need me.' This could be construed as a form of 'virtual mothering', which seems inherently contradictory but nevertheless gives a sense of closeness, especially in the case of older children, that would have been missing in the past. Carla went on to say that her own mother was more of a mother to her children even before she came to the UK as she had been a union activist and was often away from home. Indeed this pattern of grandparents playing a primary role in nurturing their grandchildren has a long history in South Africa[33] where under the apartheid system many people worked in areas where they did not have residence permits, so in this case globalization has simply extended the boundaries over which the migration takes place. Carla was fortunate in the sense that her earnings were sufficient for her children to visit at Christmas and in the summer. The examples of Carla and Vicky both indicate that it is important not to victimize people who are simply making choices to make a better life for themselves and their children, albeit within constraints outside their control. In this sense globalization has expanded the range of choices for some people, although equally may have created the circumstances that made these rather desperate choices necessary. This migration also constitutes another form of surplus extraction from poor countries, in that the emotional labour which might have been devoted to their own children is transferred to richer parents in richer countries.

While there are many different circumstances leading to emigration and likewise varying conditions within which migrants work, it is important to distinguish this kind of migration from trafficking, where people are moved either by compulsion or else totally misled about their future work, imagining work in entertainment, bars, restaurants or in care, but finding that they are required to work in the sex industry and have little choice but to comply (see below). Some countries are also far more active in protecting their migrants when they are abroad; in the case of the Filipinas in Rome the Catholic Church plays a role and helps the migrants form a community.[34] The Sri Lanka government also has a package of measures to assist and protect migrants (Banerjee 2002) as well as facilitating the flow of remittances. The diversity of individual decision making can never be fully captured in theory, but there are common trends which derive from the structured economic and gendered inequalities in which these individual choices are made.

Globalization and uneven development cannot however explain the gendered nature of the care chain, the existence of which also undermines the suggestion that the feminization of employment is associated with the end of patriarchy. Buying childcare and domestic work may liberate some women from these traditional responsibilities, but these tasks are generally only transferred to other women, thus effectively leaving the gender distribution of roles unchanged. There is a redistribution of care work between women but little is done to revalue care or to bring about a more lateral sharing of care between women and men, although this may happen to some degree as women's absence from the home forces men to play a greater role irrespective of their preferences. The extent to which men engage in domestic and caring work has only recently been measured in national time-use budget surveys and the limited quantifiable data that exist suggest that men's contribution remains low (see Table 4.2). Qualitative studies have similarly not found any major change, though it is possible that men's contribution may be understated, due to the lower frequency but greater duration of household tasks performed by men (Jackson 2001) and because men may be reluctant to reveal the extent of this work, as it conflicts with prevailing conceptions of masculinity.[35]

*Table 4.2* Total work burden by gender

| | | Total work time (minutes) | | Female work time as % of male work time | Time spent on non-market activities (%) | | Time spent on market activities (%) | |
|---|---|---|---|---|---|---|---|---|
| | | women | men | | women | men | women | men |
| 1998 | Canada | 420 | 429 | 98 | 59 | 35 | 41 | 65 |
| 1999 | New Zealand | 420 | 417 | 101 | 68 | 40 | 32 | 60 |
| 1997 | Australia | 435 | 418 | 104 | 54 | 38 | 46 | 62 |
| 1999 | France | 391 | 363 | 108 | 67 | 40 | 33 | 60 |
| 1996 | Japan | 393 | 363 | 108 | 57 | 7 | 43 | 93 |
| 1999 | Bangladesh | 545 | 496 | 110 | 65 | 30 | 35 | 70 |
| 1999 | Rep. of Korea | 431 | 373 | 116 | 57 | 12 | 43 | 88 |
| 2000 | India | 457 | 391 | 117 | 65 | 8 | 35 | 92 |
| 2000 | South Africa | 332 | 273 | 122 | 60 | 21 | 40 | 79 |

*Source*: Taken from UNDP (2003: 326).

*Note*: The data are estimates from time use surveys and from a number of sources. Only data for recent years have been included here and more countries are beginning to collect this data. Although this data should be viewed with caution, it indicates both the high total work burdens of women and their overrepresentation in non-market activities.

In market societies there is an inherent tendency for care work and personal services to be underpaid as they are highly labour intensive and in contrast to knowledge goods are intrinsically not infinitely expansible or non-rival (see Chapter 1). Care is also a composite good, having a custodial aspect, making sure that no harm comes to the individuals being cared for, and a nurturing aspect, looking after their emotional and psychological needs (Folbre and Nelson 2000). Thus measuring the quality and effects of care work is inherently difficult owing to this composite character, its individualized performance, the fact that the cared for are generally unable to effectively express their preferences and because the outcomes are associated with positive social externalities, i.e. good quality care leads to benefits for society as a whole. These factors together with the way that caring work is seen to be a natural talent of womanhood rather than a material competency or skill help to explain why wages are well below average.[36] Where care work is fragmented in individual homes, small nurseries or care homes, low pay is sustained by the difficulty of establishing collective bargaining or exercising industrial power, and, perhaps as importantly, the way that the humane empathies of care workers are relied upon to prevent them from utilizing their potential power. This is not to suggest that industrial action never takes place. In 2002 in south east England many care home workers, childcare workers and learning support assistants protested against low public sector pay and the strength of their feeling led to headlines in the national newspapers, as Paul Kelso reported:

> A mums' army marched on Brighton town hall yesterday in support of the council workers' strike. Several hundred classroom assistants, social workers, carers, librarians, housing officers and street cleaners, most of them women attending their first union rally gathered to highlight the problems they say low pay is bringing to the outwardly affluent resort.
>
> (Kelso 2002: 5)

Even though this action led to a pay increase, the new wage is equal to only 43 per cent of current average hourly earnings for all adults (40 per cent of the male, 49 per cent of the female and 65 per cent of the part time average) (NES 2002) indicating the continuing relatively low social value of care work. Moreover the pay increase applied only to those employed directly by the local authority, who are in fact more likely to have better terms and conditions of employment than those in the private sector, especially the agency workers. The former are also more likely to be local in origin, while a large proportion of people working in these sectors, especially the private sector care homes and for the multitude of agencies which supply domestic workers and cleaners to a wide range of local employers, including major national employers such as the health service, come from all over the world. Thus even amongst organized care workers, wages are barely above the legal minimum, so outside these protected environments conditions are likely to be worse. If, however, workers endure very basic living conditions and low levels of personal consumption while working, allowing some money to be remitted, the purchasing power in the countries of origin is far greater, which explains why this migration takes place, and why professional workers in poorer countries are prepared to work as domestics in the richer countries of the world. However, these movements do little to redress the structural inequalities between countries, which have been aggravated by some of the policies forced on these countries by global institutions such as the IMF. In this case countries are often forced to devalue their currencies, as part of the adjustment packages, which aggravates the wide disparity in purchasing power between foreign and local money.

In its present form the marketization of care services is likely to both reinforce and to some extent undermine gender divisions. Feminization of employment does not automatically undermine patriarchalism but rather is associated with continuing divisions between women and men over the domestic division of labour and widening differentials between women as richer households buy childcare and domestic services which are supplied largely by low paid women workers. A more radical solution would be to encourage a greater sharing of roles between women and men. As Hochschild (2000: 144) has argued, 'if fathers shared the care of children, worldwide, care would be spread laterally instead of being passed down a social class ladder'. To effect greater gender equity a redivision of responsibilities between women and men would be required, along the lines perhaps of the universal caregiver model proposed by Nancy Fraser (1997). An alternative possibility would be for the state to play a greater role in meeting the care deficit, and the level of state assistance varies between countries, being high for example in more social democratic countries such as the Nordic states. In the neo-liberal USA and UK, however, it could be argued that the state too has become a 'deadbeat dad' (Ehrenrich and Hochschild 2003: 9) and has made little or no provisions for the much higher proportion of working parents.

Obtaining equality in the workplace is unlikely while the question of providing adequate caring remains unresolved. Furthermore as long as either individual or collective care is disproportionately supplied by low paid female labour then class and ethnic divisions between women will increase and gender inequity will remain. To resolve these issues it is therefore necessary to link workplace equity and childcare provision within the broader question of over-all daily and generational social reproduction.[37] There is a wealth of evidence to suggest that neither paid nor caring work is equated solely with negative utility. Qualitative analyses of mothers in paid work, ranging from high powered managers to maquiladora assembly line workers, suggest that going to work can be a 'rest', a means of socialization and maintaining self-esteem, even where the conditions of paid work are extremely hazardous. Similarly, caring for people can sometimes be similar to leisure. Whether watching football, walking in the park or sitting on the beach with a young child is work, care or leisure, depends largely on the relationships between the people involved, rather than on what is actually taking place. This point is clearly appreciated by paid domestic workers who point out that employers often like to treat them as one of the family in some respects, and strong emotional ties can be established between carers and the children they care for, which highlights the difficult and complex nature of care work, but these emotional ties can also be used as a form of exploitation and an excuse not to pay higher wages.

More generally, the lack of clear boundaries between work, care and leisure suggests that people's time could be more evenly spread between these activities, something that might also occur if care work was given higher monetary rewards. Hochschild (2000) argues that there is a cultural embrace for the idea of sharing parenting more between parents in the United States but a lag in implementation and the same could be said of the UK, where an increasing proportion of mothers are in paid employment, but this is often part time and fitted in around caring responsibilities. Similar accommodations are simply not expected of men and so where women do play a greater role in the labour market, the caring tasks are in general combined with paid work or transferred to the market and carried out by lower paid women, some of whom may be part of the global care chain.

## SEX TOURISM, SEX WORKERS AND TRAFFICKING

Service sector employment in domestic work, cleaning and catering has also expanded as a consequence of the rapid expansion of tourism, now a leading global industry, facilitated by advances in transport and communications technologies and by growing levels of affluence in some parts of the world. Already about 10 per cent of the world's population take vacations and by 2020 the figure is expected to rise to 25 per cent or 1.6 billion people (ILO 2001a).[38] The major flows are between rich countries and Europe is by far the most popular destination, partly because of the large number of intra European flows, but poorer countries now represent a third of all destinations (WTO 2002).

Tourism is a multi million dollar industry, a major source of foreign exchange earnings and employment and is frequently encouraged by development agencies but there are also negative features: it can threaten local ecosystems especially by overuse of water and distort the local economy through 'dollarization' which diminishes the value of local currency and drives people into dollar earning jobs. Thus qualified professionals turn to taxi driving, where dollars can be earned. Further, the overall quality of employment for local people is low; adults and children work in hotels, fast food outlets and as roving beach vendors but managers and owners from major hotel and leisure chains appropriate the higher wage jobs and profits (ILO 2001a). NGOs such as Tourism Concern[39] are trying to bring about more socially responsible tourism, but one particularly insidious and growing aspect of this form of globalization is sex tourism, in which women and children are overrepresented.

Sex tourism is growing but formal statistics are rare because it is generally clandestine. Governments are ambivalent, because it contributes to foreign exchange earnings and makes a significant contribution to the economies of the countries involved. This is estimated to be between 2 and 14 per cent of GDP, when the associated range of activities from medical practitioners, cleaners, waitresses, food vendors and security guards are taken into account (Lim 1998)[40] but it also brings the risk of HIV/Aids and human rights violations. Generally sex work is illegal and sex workers are open to prosecution and fines, which perversely also contribute to government revenue. At the same time, however, official tourist agencies still promote the erotic qualities of their population and informally condone the framework within which sex tourism takes place.

The broad geographical pattern of sex tourism reflects uneven development and the gender and race hierarchy, on a regional and global basis. There are specific historical and economic reasons to explain why countries become involved and similarly diverse reasons why individuals become sex workers but the root causes are poverty, uneven development and unequal gender relations. Similarly to the global care chain, sex workers travel from poorer to richer regions, nationally and internationally, while tourists, predominantly but not exclusively men, travel in the reverse direction to destinations known for sex tourism throughout the globe, often advertised on the Internet. Within this pattern there are more regionally based movements, such that Japanese sex tourism is largely to South East Asia, including Cambodia, Thailand and the Philippines, while Americans travel to the Caribbean and Western Europeans travel to Eastern Europe, but sex tourism is global and takes place in countries as far apart as the former Soviet Union, Brazil, Thailand and, somewhat surprisingly, Cuba, where it revived in the 1990s following economic decline.[41]

Havana was renowned as a sex city in the 1950s but following the revolution in 1959 the Castro government almost eliminated the sex trade and rehabilitated sex workers. Now, however, it is 'back with a vengeance' following the economic crisis in the early 1990s, with a decline of 25 per cent in GDP in 1991 alone. This was initiated by the withdrawal of Russian financial support (following its transition to a market economy), falling primary commodity prices, and the continuing US trade blockade (Clancy 2002). The crisis has affected the whole system of reproduction and the way of life. As Ruth Pearson (1997b) argues, the state had previously provided good quality health care, education and to a lesser extent housing; basic goods were available at low prices via rationing, and higher value imported goods could be obtained via workplace and political organizations in return for exceptional work performances or contributions to citizenship by working for the sugar harvest or building programmes or else for participation in political structures. In this way not only material life but the values of the socialist society were sustained (Pearson 1997b). The crisis effectively ended this arrangement, as the state could no longer guarantee even the provision of basic goods or transport services to work. To boost food production peasants were allowed to market their produce independently and other workers could work on their own account in a range of services, as full employment could not be guaranteed. Goods intended for official distribution channels were increasingly diverted towards these private markets, making it difficult to survive on the basis of regular wages and rationed entitlements so people had to draw on all sorts of resources and the tourist sector, where dollars could be earned, was particularly attractive. The government renewed its efforts to promote tourism, already initiated in the late 1970s and 1980s in response to the debt crisis, and, while not officially condoned, sex tourism formed part of this expansion. Even though attempts had been made to equalize childcare responsibilities and domestic work, women retained primary responsibility for household tasks and so experienced the shortages most immediately. In this context, foreigners provide a means of access to hard currency for which some mainly young people are prepared to exchange social and sexual services as Ruth Pearson points out: 'Many young women in Cuba explicitly seek out the companionship of foreigners for an evening in a disco, a weekend in a tourist hotel, or an extensive relationship involving travel, possibly migration, study abroad or a steady income' (Pearson 1997b: 690). Cubans call these people *jiniteros* (literally jockeys) as they 'take foreigners for a ride' (Pearson 1997b).[42] The government has tried to curtail their actions through fines and by banning prostitutes from tourist hotels, sending them to rehabilitation centres or returning them to their home areas. For repeated offences, they could be jailed, and significantly harsher penalties exist for the organizers (Clancy 2002), but tourism is a vital source of foreign exchange and so a certain ambivalence or toleration prevails. There has been some economic recovery since the mid-1990s and as indicated in Chapter 2, Cuba has a far higher level of human development, as measured by the HDI, than its level of GDP would suppose indicating that despite the crisis, aspects of the comparatively high quality social infrastructure remain in place.

Elsewhere in the Caribbean some sex workers also work on a very casual basis and from the tourist's perspective the transaction can be ambiguous varying from 'romances' with some economic exchange such as presents and dinners to formal prostitution. The sex workers themselves, however, distinguish between clients and more steady partners, with whom they might feel some affection, which adds to the ambiguity, making the tourists sense that this is not real prostitution or certainly very different from prostitution at home (O'Connell Davidson and Sanchez Taylor 1999). In the case of the Dominican Republic, sex workers interviewed by Denise Brennan (2003) were often single mothers, who were looking for income and possibly a long-term escape from poverty through marriage with their clients, though in practice this rarely happened and when it did, was not always sustained. Brennan argues that these sex workers were better off than many as there were no pimps and the women themselves controlled the work. More generally while the tourists may be in a state of ambiguity, the sex workers themselves have a clear idea that what they are doing is work and few sex workers report enjoyment from the work itself (Lim 1998). Thus, the circumstances of sex workers are varied; some people make a conscious choice to engage in this flexible and sometimes remunerative employment, albeit within severe constraints, while for others it may be a form of debt bondage or slavery.

Sex work is generally paid better than most occupations for low and even medium skilled women, although the range of earnings varies according to the type and location of the sex work. In Indonesia call girls working in high class nightclubs can earn a higher annual equivalent income than middle level civil servants but women working from the streets or cheap brothels may receive little more than $1 for each client.[43] Similarly in Malaysia, prostitutes working for about 12 hours a week in the cheapest hotels would still earn more than an unskilled factory worker and about 70 per cent as much as a skilled manufacturing worker, working full time (Lim 1998). Sometimes it seems to be the only choice, especially for single women with caring responsibilities, as a sex worker explained:

> I earn enough to look after my two young children. It's so difficult to get someone to look after them when you work elsewhere. Here I come only when I need the money and it's easy to find a baby sitter for just one day.
>
> (Sex worker cited by Lim 1998)

For some, sex work is a conscious choice and a means of obtaining foreign currency and far higher incomes than could be obtained from any other activity; nevertheless it is the male owners of bars and restaurants that earn the higher incomes. A very small proportion of clients are women, who also seek to combine sun and sand with sex from men while on holiday.[44] For both women and men sex tourism often reflects the way that sex workers are constructed through exotic, racialized stereotypes that can be bought comparatively cheaply in the safety of other and distant locations (Kempadoo 2001).[45]

This discussion of sex work is derived from studies where sex workers were able to give their views freely and had consciously chosen this form of work, albeit within constraints. However, there has also been an expansion of trafficking, that is the forced transportation for the purpose of exploitation, generally prostitution, of women and children, to the countries known for sex tourism and to the affluent cities throughout the world. Some sex workers travel freely to the centres of sex tourism in South East Asia from the surrounding region, thus there are many Burmese women in Thailand and Nepalese women in India as well as women from Bangladesh in Pakistan and India (Banerjee 2002) so the links between poorer and richer countries extend throughout the international wealth hierarchy. Other women and children from these regions are trafficked to these areas as well as to Western Europe and the United States in addition to women and children from Eastern Europe and once there, they are effectively prisoners, as their passports and documents are taken away and they have no legal status, leaving them extremely vulnerable to exploitation.

The European Union as a source of demand for trafficked women recognizes that it has a responsibility to address this issue (see Box 4.1) and has introduced a number of programmes to assist trafficked women and proposed higher

---

*Box 4.1* The European Union's recognition and condemnation of trafficking

Ladies and gentlemen

It is shocking that we are discussing today an issue that should have been solved with the abolition of slavery: the market in human beings.

Western Europe is at the heart of a modern day slave trade. Up to half a million women and children are being trafficked into this region each year. They are bought and sold into forced prostitution; to domestic labour as servants or forced into sham 'marriages' where they are held as prisoners, raped and often forced to provide their so-called 'husbands' with children.

The bodies of several hundreds of these women are discovered each year. Europol estimates that many are never found. The trade is international, well organized and growing. One CIA report estimates that traffickers make up to a quarter of a million dollars with one woman trafficked and re-trafficked.

Even if they manage to escape from the trafficker, or report to the authorities, women can find themselves facing further trauma. The cruel reality is that trafficked persons may be treated as illegal migrants and criminals and face arrest, detention or expulsion. So the victims are further victimized. Many trafficked women fear deportation. It might seem an escape from the trafficking situation – but the reality is far more complicated. Often, the victim has borrowed money to leave her country of origin. Deportation means returning home with empty hands. Debts she will never be able to pay off. Fearing that she might be ostracized by her family or her community.

Deportation means that the trafficked person is put at the mercy of the traffickers again. Trafficked people not only have to fear reprisals from the traffickers but also harassment, arrest or detention from authorities in their own country.

(Anna Diamantopoulou, 2002 EU Commissioner responsible for employment and social affairs)

---

penalties for those caught trafficking. The EU has suggested that the women be given the right to stay and work if they testify against the organizers of trafficking but so far only Spain has implemented this intention within its national law. Elsewhere in the EU and more generally, despite UN protocols, trafficked adults and especially children have little protection and so far the emphasis seems to have been on penalties for the organizers, rather than assisting their victims to remain in Europe, should they wish to do so (see Box 4.2).

Sex tourism and trafficking both reflect uneven development and intense poverty that drives women usually unwillingly into this form of work as well as uneven gender relations which allow double standards of morality for men and women, though as illustrated in Box 4.2 women cannot be relied upon to protect other women from trafficking and neither can solidarity be assumed within ethnic groups.

---

*Box 4.2* Trafficking in the UK

*Sisters deceived hundreds into prostitution*

Two sisters were jailed yesterday for heading a multi-million-pound prostitution business in which hundreds of Thai girls and women were brought to Britain to work in a chain of brothels.

Southwark Crown Court in London heard that the women were the 'controlling minds' of a gang which tempted the women into the country by promising them cleaning jobs but sent them to a 'finishing school' for prostitutes.

The women were then sent to one of 15 brothels in London and the south-east where they had to earn almost £50,000 – by having sex with about 500 customers – to buy their freedom.

Judge Paul Dodgson said it was one of the most serious cases of its kind imaginable. He told the sisters, Bupha Savada, 45, and Monporn Hughes, 40: 'Each of the girls you controlled had to perform sexual acts with literally hundreds of men. These were girls in a foreign land totally dependent upon you and you used them as animals.'

During the hearing one of the women forced to work as a prostitute, an 18-year-old known by the alias Mary Pondee, told how at the age of 16 she was duped into leaving her job working in the rice fields of northern Thailand by the promise of cleaning work which would earn her twice as much in a month as she earned in a year.

She said she borrowed £650 to pay the gang's fixer in Thailand for a false passport, visa and plane ticket. Within days of arriving in the UK she was in a brothel in Southampton. She said she was told that her family in Thailand would suffer if she did not cooperate.

Savada was jailed for five years and Hughes for three-and-a-half. Both women pleaded guilty to controlling prostitution.

*Source*: Morris (2003).

---

The continuation and indeed the expansion of sex work on a global scale clearly questions the idea that the feminization of employment automatically enhances gender equality. Paid employment may challenge private patriarchal structures and relations within the home, but the wider public sphere which sets wage differentials for different kinds of work and sustains the double standards of morality for women and men remains profoundly unequal.

## IMPACT OF EMPLOYMENT ON GENDER RELATIONS – END OF THE FAMILY, END OF PATRIARCHY?

One of the common themes emerging from the qualitative experiential accounts of women's increased participation in paid work is that despite exploitation, it provides a sense of freedom, a space and time where they can be themselves and some enjoyment from socializing with other women. Paid employment is also found to raise self-confidence, self-esteem and respect from other people in their household, so overall women are to some degree empowered by independent incomes.[46] Fruit pickers, retail, electronic, textile and care workers and even some of the sex workers share these sentiments. Manuel Castells (1997) goes further by suggesting that the feminization of employment, together with new reproductive technologies, the feminist movement and the global culture in which ideas quickly spread, pose a challenge to patriarchalism. According to Castells (1997: 134–5):

> Patriarchalism is a founding structure of all contemporary societies. It is characterised by the institutionally enforced authority of males over females and their children in the family unit. . . . The patri-archal family, the cornerstone of patriarchalism, is being challenged in this end of the millennium by the inseparably related processes of the transformation of women's work and the transformation of women's consciousness.

He goes on to say that: 'if the patriarchal family crumbles, the whole system of patriarchalism, gradually but surely, and the whole of our lives, will be transformed. This is a scary perspective, and not only for men' (Castells 1997:136). Castells argues that although the feminization of employment places an 'unbearable burden on women's lives' as the paid work is generally added on to their other roles, it increases women's bargaining power relative to men and undermines men's role as sole or main provider. He emphasizes that he is only identifying and illustrating tendencies but a wide range of qualitative studies support his ideas, in this respect. In Argentina married women report that their husbands found their paid work and incomes challenging:

> 'My marriage started to break down when I started work. . . . I had more chances than he did.'
> 'Everything was OK till I became a worker, then I would come back at six or seven o'clock at night to find nothing had been done and the children were unfed and dirty. I would tell him to help, but he became violent. Several times he "aimed" at me. The thing he most hated was that his shirts were not ironed. Also, he resented my handing him money because it was mine. So he could not go drinking when he pleased.'
> (Textile worker in Argentina; Aceros 1997)

A manager at a textile factory in Bangladesh makes a general observation along similar lines:

> 'Girls in my factory marry several times because if there is any trouble with the husband, they think they can survive on their

own, they need not stay. They walk out and simply get married again.'

(Textile factory manager reported in Kabeer 1995)

Employment in the electronics factories in Malaysia also allowed single women greater freedom, even in the 1970s. Grossman (1979) comments that given the severe working and living conditions she was about to ask the women why they worked there but then:

After casting a sidelong glance at the men at the next table, Tuti shot the rest of us a conspiratorial smile, eyes twinkling. I stared straight into the coffee I was stirring, pulling the Malay words together in my mind to ask why they had come to work in this factory. Suddenly I laughed to myself, realizing that part of the answer was right there at this coffee stand at 11 o'clock at night. . . . They come for the money, of course but also for the freedom. They talk of freedom to go out late at night, to have a boyfriend, to wear blue jeans, high heels and make up. Implicitly they contrast this freedom with the sheltered regulated lives they would lead with their families in Malay villages and small towns. They revel in their escape from the watchful eyes of fathers and brothers.

(Grossman 1979: 13)

More recent studies have focused on men's reactions to these changes, and similar associations have been found between the expansion in the number of women receiving independent incomes and men's loss of self-esteem. In Sylvia Chant's interviews with low income men in 1997 in Guanacaste, Costa Rica, Martín, a 30 year old bricklayer, commented, 'A woman who has her own money loses affection for her husband. Many marriages have been ruined because of this' and similarly Luís, a waiter, argued that a man who cannot provide for his wife and children loses self-esteem and that his social image 'isn't worth anything' (Chant 2001: 211). Similarly in Sri Lanka husbands of migrant maids are stigmatized for being unable to fulfil traditional gender roles.[47]

Entry into paid work alone does not automatically bring about change. Many women are made to hand over their earnings to their husbands or families and men can also adjust to the changing circumstances by investing new arrangements with a revised, but nonetheless important, self-image. Javier Pineda (2001) reports on Cali, Colombia, where male employment had declined and women had been offered the chance to be involved in a micro-credit scheme financed by WWB.[48] While women owned the new activities several men explained how nonetheless there was a complementary division of tasks, as Ramón argued:

it is very easy to get adapted to the work because we have to divide the work. I took production and she does management. Why?

Because for a good worker it is very difficult to manage and produce
at the same time, because it does not pay. Then she is at the front of
the business and I am at the front of production.

(Ramón, a respondent in Pineda 2001: 81)

Other men described this same division of labour as women doing the 'light'
work while men did the 'heavy' work, thereby maintaining self-image through
retaining stereotypical gender divisions in the new context. This division might
reasonably be referred to as mental and manual labour, but if it was it would
imply a reversal of the gender hierarchy. The fact that it is not described in
this way suggests that existing unequal gender relations have a resilience or
durability that can withstand changing forms of employment, thus work is not
only a bearer of gender relations (Elson 1999) but something through which
unequal gender relations can be reborn.

More formally, these issues have been theorized in the cooperative conflict
model (Sen 1990; Kabeer 1995), which explores intra household bargaining
and the criteria for breakdown. At the micro level, despite growing individualization and the 'crisis' of the family, most societies continue to be organized
around households, which are institutions based on relatively long-term
commitments and through which individuals obtain their survival needs and
within which children are born and raised. Households take different forms in
different societies, their boundaries can be permeable, and their composition
can change over time. Within households there are generally different possible
divisions of labour between members, especially relating to the amount and
timing of paid and unpaid work in order to meet their survival needs or lifestyle
preferences. The cooperative conflict model perspective considers how households arrive at their choices.[49] Households rest on implicit contracts relating
to the claims and obligations that household members can legitimately make
on each other. When there are different possible divisions of labour, individual
household members are likely to benefit to different degrees depending on
the choices made, which is why there is conflict as well as cooperation, so final
choices are arrived at through bargaining. This can be done in a very implicit
and sometimes non-conscious way. The bargaining position or power of the
different household members is influenced by their relative fallback positions,
that is the level of utility or well being they would experience if the cooperation
broke down. Those with the stronger fallback positions, that is those less
dependent on the cooperative arrangement, have more power to affect the
outcome. The influences on the fallback positions are profoundly gendered.
Generally, and especially in the past, they have 'coalesce[d] in favour of men as
a category' (Kabeer 2000: 29) largely because market/public work, in which
men are overrepresented is perceived to be more important than home/private
work, and thus men are perceived to make more significant contributions to the
household and correspondingly merit a greater say in how household resources
are allocated.[50] Even with the growing feminization of employment, men's
market work is generally paid more than women's market work. Effectively

this model would support Castells's (1997) claims and that of the textile manager in Bangladesh quoted above, that paid work increases the fallback position of women and makes them less dependent on the survival of the household and correspondingly increases their likelihood of breaking the implicit contract. Given the continuing unequal division of domestic labour and childcare, generally women remain within the household and it is men who leave or are asked to leave, as their contributions fall, and their presence becomes less essential, giving rise to the increase in female headed households observed throughout the world.[51]

The impact of the feminization of employment can, however, be overstated. It is important to recognize that not all women are able to keep the income they earn; contemporary paid work is often flexible and insecure; vertical and horizontal segregation remains and women are concentrated in lower positions and lower paying sectors, so the relationship between paid work and the strength of the fallback position is also gendered. Moreover, as argued in the introduction to Part II, feminization of employment can sometimes be more of a statistical artefact than a reflection of material change. Further, many women are entering the paid labour force just at the moment when the terms and conditions of employment have deteriorated and when men's employment and incomes are in decline, making more than one earned income a condition for survival or to maintain customary lifestyles. In the Mexican maquiladoras for example wages paid have been found to be lower than those paid in industries arising from the previous import substitution strategy. Thus whereas in the past men would earn a family wage, allowing the traditional gender division of labour, in the new context more than one income is necessary. Women and men therefore negotiate together to organize their working times, for example offsetting night and day shifts to cover their childcare responsibilities, or draw upon other generally female younger relatives to live with them to help care for children, while continuing their own education (Cravey 1997). These circumstances, combined with the increasing trend towards having to pay for public services such as health and education, as well as the pervasive and insistent global marketing of consumer goods, means that women's incomes are generally fully absorbed even where they have control over them, making this control more of an added responsibility than an opportunity to exercise individual discretion and empowerment.[52]

Household situations vary. They exist within local, regional and national and now global settings all of which may shape social norms and the nature of bargaining within individual households. That is, in addition to patriarchal or unequal gender relations within the household the economic, social and cultural context is characterized by widening spatial inequalities as Diane Elson (1999) argues:

> There is a paradox at the heart of contemporary restructuring as far as many women are concerned. On the one hand, their bargaining power in relation to the men in their households, their communities,

their networks, and the organizations of their civil society, may often (though not, as we have argued, inevitably) be increasing as a result of their greater participation in labour markets. But at the same time their households, their communities, their networks, the organization of their civil society are more and more at the mercy of global market forces that are out of control.

(Elson 1999: 618)

In particular, the global context heightens competition and the way that corporations have responded to this through chains of subcontracting has increased the insecurity of paid employment. Furthermore, despite the evident gains from paid work for women as individuals, two key problems remain which limit women's bargaining position in the household and correspondingly the extent to which it is possible to argue that patriarchy is challenged by the feminization of employment. First of all structures remain that limit the range of paid work that women do, for example their presence in high level jobs, partly but not exclusively determined by their continuing responsibility for organizing, managing and carrying out a disproportionate share of reproductive, domestic and care work together with working hours that deny these responsibilities. Second, the kinds of jobs in which women are disproportionately represented are on average paid less than those where men are overrepresented, reflecting in part the intrinsic economic properties of caring work but also the low social value given to certain jobs simply because women do them, i.e. pay norms themselves embody gender inequality.[53] The extent and depth of this continuing gender inequality varies between countries, some of the differences being captured by the Gender Empowerment Index discussed in Chapter 2, which in turn reflects the prevailing gender order. But there are other factors relating specifically to the nature of gender relations that influence the relative position of women and men discussed further in the following chapter, that suggest that even when the income generating capacities of women and men have been reversed, that is where women are the sole earners in dual person heterosexual households, despite the loss of role as provider men are still able to retain a dominant position by redefining their masculinity.

## CONCLUSION

This chapter has examined the effect of globalization on people's lives, in particular the way that there has been a geographical expansion of capitalist social relations of production as a whole range of activities from agribusiness to care work have become integrated on a global scale. It has put some substance to the idea that people in different places are increasingly interconnected through their work and explored the idea that the feminization of employment will profoundly alter gender relations. While changes have taken place, and many women feel more empowered as a consequence of independent incomes, the impact on gender equality is more limited, partly because of the new circumstances in which paid work is carried out and partly because unequal gender relations seem to be resilient to material change and reappear in new ways with new forms of employment. This theme is continued in the next chapter in the context of the profound restructuring of economies in the transition from Fordism to post-Fordism and the new economy in the richer regions of the world, but here the emphasis is on old industrial regions, that have experienced losses of employment in predominantly male areas of work.

## Further reading

Lordes Benería (2003) *Gender, Development and Globalization: Economics as if all People Mattered*, London: Routledge.

Barbara Ehrenrich and Arlie Hochschild (2003) *Global Women*, London: Granta Books.

Naila Kabeer (2003) *Gender Mainstreaming in Poverty Eradication and the Millennium Development Goals: A Handbook for Policy-makers and Stakeholders*, Ottawa: Commonwealth Secretariat, Canadian International Development Agency.

## Websites

CEDAW: http://www.un.org/womenwatch/daw/index.html (Convention for the Elimination of All Forms of Discrimination Against Women).

UNECE: http://www.unece.org/oes/gender/Welcome.html (United Nations Economic Committee for Europe).

UNIFEM: http://www.unifem.org/ (United Nations Development Fund for Women).

## Notes

1 These flows are termed 'counter geographies' by Saskia Sassen (2003).
2 See Barrientos (2001) and Rammohan and Sundaresan (2003).
3 This section draws heavily upon research carried out with Stephanie Barrientos (see Barrientos and Perrons 1999). Stephanie carried out all of the research in Chile and has published widely on this commodity chain as well as ones in South Africa and Kenya; see for example Barrientos (1996), Barrientos and Kritzinger (2003).
4 See Gwynne (1998).
5 See *Economist* (2001).
6 See Keynote (2000).
7 Internet shopping still represents a relatively small share of the market, but has been increasing rapidly – see Chapter 6.
8 For further details of this study see Perrons (1999b).

9 Tesco, the largest company, employs 190,000 people in the UK with an overall total of 240,000 worldwide (http://www.tesco.com/careers/); Sainsburys employs between 120,000 and 130,000 (http://www.sainsburys.co.uk/recruitment); and ASDA/Walmart employs 122,000 (http://www.asda.co.uk/asda). At the time of writing, 2003, Safeway was in the process of being taken over by Morrisons and so the current employment figures are not clear; prior to the takeover Safeway employed about 90,000 (http://www.safeway.co.uk) and Morrisons 50,000 (http://www.morrisons.plc.uk/); some job losses are expected as a consequence of the merger, though the new company will displace Asda/Walmart as the third largest supermarket in the UK in terms of turnover and numbers of employees. Source: company websites, accessed in November 2003.

10 By contrast single parents, without any support from families or friends, find working outside school or nursery hours very difficult.

11 About 25 per cent of the supermarket workforce is made up of young people, especially students with no caring responsibilities, especially for night and weekend work.

12 See for example González de la Rocha (2003).

13 See also Elson and Pearson (1998).

14 Rachel Grossman's (1979) study was carried out in the mid-1970s and focuses on an electronics plant in Penang, Malaysia where 90 per cent of the 1400 strong workforce was female.

15 See Chant (2002).

16 See also Ong and Peletz (1995).

17 See for example Cravey (1997) and Fussell (2000). Maquiladoras is the term used to refer to foreign, mainly US owned, plants in Mexico, that employ Mexican labour to make components or products that are mostly exported back to the United States. The early maquiladoras were generally assembly firms in electronics, car and general machine components and textiles. They receive subsidies from the Mexican government in the form of preferential tariffs on imported materials and semi-finished products and tax relief on their profits.

18 'On 4 July 1995 Bangladesh created history by signing a MoU on elimination of child labour from the garment sector. It was the culmination of long and arduous negotiations with ILO and Unicef and concentrated effort of BGMEA. BGMEA takes pride in declaring the garment sector of Bangladesh child labour free since 1 November 1996' (http://www.bgmea.com/social.htm, last accessed December 2002).

19 See Standing (1989, 1999); Benéria (2001), González de la Rocha (2003).

20 See Elson (1999).

21 See Perrons (2003); also Chant (2002).

22 The fact that these women do not have relatives to do this work, either their parents or their own older children, can also be a reflection of affluence, in that their own parents, if not in work, could be retired and living leisured lives, while their older children (if they have any given falling fertility rates) are likely to be in education.

23 See Ehrenrich and Hochschild (2003).

24 *The Life and Times of Rosie the Riveter* directed by Connie Field. Further details can be obtained from the Clarity films website http://www.clarityfilms.org/story.html.

25 See Momsen (1999); Willis and Yeoh (2000).

26 Arlie Hochschild (2000) draws her examples from Rhacel Parreñas, now published in Parreñas (2001).

27 See Gregson and Lowe (1994).

28 See Pratt (1999).

29 See Fölster (1999).

30  See Fagnani (1998).
31  See Day (2003).
32  See Tacoli (1999).
33  Personal interview with Carla, contacted in relation to my research on the Brighton and Hove labour market. Personal communication from Stephanie Barrientos in relation to intergenerational care patterns in South Africa.
34  See Tacoli (1999).
35  See for example Cravey (1997) and Gamburd (2003).
36  See Phillips and Taylor (1980).
37  See Perrons (2000a) for a fuller discussion.
38  There was a 0.6 per cent decline in tourism between 2000 and 2001 reflecting the weakening of economies and also the effects of the September 11th terrorist attacks.
39  For more details see http://www.tourismconcern.org.uk/campaigns/ campaigns_ human_rights.htm (last accessed July 2003).
40  The study was carried out in the mid-1990s and was based on four countries – Indonesia, Thailand, Malaysia and the Philippines. It was estimated that between 0.24 and 1.5 per cent of the female population was involved.
41  See also Chant with Craske (2003).
42  See Brennan (2003) for a similar interpretation of sex tourism in the Dominican Republic.
43  See Chant and McIlwaine (1995) for a discussion of the hierarchy of sex workers in the Philippines.
44  See Blindel (2003) who reports on work carried out by Jacqueline Sanchez Taylor and Julia O'Connell Davidson on women sex tourists and their male partners or 'beach boys', in Negril in the Dominican Republic. Similarly to the women sex workers they have varied attitudes to their role; some acknowledging prostitution, others seeking an escape from the island through marriage. Likewise some of the women clients consider themselves to have found romance as well as a way of 'helping out' people in a poorer country.
45  Foreign women are also seen to possess traditional feminine qualities of caring, docility and eagerness to please, that have been lost by women, especially by 'working' women, in richer countries (see Ehrenrich and Hochschild 2003).
46  See for example Afshar (1998), Kabeer (2000), Denman (2002).
47  See Gamburd (2003).
48  WWB is Women's World Banking, a financial NGO.
49  The cooperative conflict household bargaining model was developed in part to redress the individualism associated with traditional neo-classical economics. In orthodox economic household allocation models decision making is assumed to take place on the basis of atomistically determined individual tastes and preferences and final preferences determined by a benevolent dictator who acts on the basis of maximizing overall household utility.
50  Furthermore women often express adapted preferences (Agarwal 1997) or a perceived interest response (Sen 1990), that is they behave in an altruistic way and prioritize family welfare. By so doing they attach less value to their own well being perhaps because they recognize that their fallback position has been so low that they simply did not press their claims. Bina Agarwal (1997) however suggests that the idea that women have a less sharp perception of their individual interests is debatable as women may be altruistic in relation to children but less so in relation to husbands with whom they drive hard bargains and from whom they hide resources; furthermore what might appear to be altruistic in the short term may be self-interested in the longer term.
51  For further details see Chant (1997), Duncan and Edwards (1997).
52  See Pahl (1989) and Chant (1997).

53 A top UK company which received awards for having one of the lowest gender pay gaps explained that it had 'controlled for market value', that is assumed away differences linked to the market value of occupation, thus if, for example, the difference between the earnings of software engineers and personnel managers in the company was no greater than would be expected given the market rates of pay for these occupations, then there was assumed to be no gender pay gap in the company.

# 5

# THE NEW GLOBAL DIVISION OF LABOUR AND THE OLD INDUSTRIAL REGIONS

## Uneven regional development in the UK

Times are changing. Globalization and economic restructuring are associated with profound changes in the nature of work, the spatial division of labour, the gender distribution of employment and the economic and social foundations of particular places. As new activities emerge within or are captured by particular places others go into decline, and as the new and old are rarely matched geographically some regions expand while others languish, with differing implications for the people living there.

Capitalist societies are dynamic so change is continual but there are periods of comparative growth and stability, periods of crisis and periods of more profound restructuring. During the last decades of the twentieth century there was a fundamental restructuring of economic activity from Fordism to post-Fordism in Western Europe and North America and now a 'new economy' with global dimensions has emerged. What is different about the new era is that the coherence of local and regional economies has been destabilized by the deepening global division of labour, which embeds activities within global rather than local economic space, so their fortunes are increasingly shaped by decisions taken elsewhere.

Regions have never been entirely autonomous and are shaped by their links with national and transnational organizations and national, colonial or neo-colonial powers. In the current era the interconnections have become complex and more immediate as there are now numerous centres and many peripheries and the speed and volatility of global processes can have quite sudden and dramatic impacts on regional economies. To analyse how patterns of regional development have changed in the contemporary context, this chapter develops some of the ideas of uneven development and theories of social change discussed in Chapter 3 but in a regional context. It focuses on the changing fortunes of older industrial regions that have experienced decline in both their traditional industries and subsequently in the new branch plants that came to replace them. While the specific history of any region is unique, it is important to understand some of the broader processes shaping the context within which regional futures have been forged. Correspondingly this chapter briefly outlines some ideas from the French Regulation School and then takes this analysis further by exploring what has happened subsequently in the new economy as regional futures are influenced ever more strongly by the global as well as the national context. In this way, while the chapter focuses on changes taking place in a specific old industrial region in the UK, the processes described have some resonance for understanding change elsewhere.

Economic restructuring is also associated with social change, in particular in the gender balance of employment, especially in regions where there have been absolute losses of male manufacturing jobs. In this chapter the effects of the feminization of employment on the end of patriarchy and concepts of masculinity are also addressed but, in contrast to Chapter 4, more from the perspective of men who have lost their role as the main economic provider, rather than from women who have gained new economic independence, though clearly these issues are interdependent. The first section begins by identifying some of the general processes shaping regional change; it then discusses the main ideas of regulation theory, which provides a general theoretical context for the discussion of more specific developments within the UK and in north east England, elaborated in the next two sections. The final section addresses the social aspects of economic change by discussing the effects of the feminization of employment on understandings of masculinity.

## THE DETERMINATIONS OF REGIONAL WELFARE

Regional economic well being is profoundly influenced by the range of activities present. Depending on the level of development, prevailing welfare regime and income distribution, all regions, especially in richer countries, have a certain level of economic activity associated with social reproduction, including health, education and welfare, planning, housing, utilities of various kinds, which may be public or private, private services and retailing. The scale and quality of these services will vary geographically, often following the spatial hierarchy along the lines predicted by central place theory; for example key hospitals are found in regional capitals with the most prestigious being in the national capital. These activities will offer a range of employment from highly paid professionals to low paid cleaners and care workers. Elsewhere this pattern will be repeated though in smaller cities and towns there will be a narrower range of job opportunities associated with their more limited functions and in very poor countries the scale of services outside the capital regions is generally much more limited and sometimes non-existent.

The most profound economic differences between regions for any given level of development are therefore largely determined by the presence of activities, which market their goods and services beyond the immediate region to the rest of the country and indeed to the world, that is by their role in the national, international and global divisions of labour. These activities will likewise offer a range of employment opportunities, the extent of which will depend on the existence or otherwise of high level functions. Their presence will also generate demand for more goods and services, some linked with social reproduction and provided locally, while others will be linked to the exporting firms. Activities supplying external markets will always have some multiplier effects on the local economy, rather in the way predicted by export base theory. In the contemporary global era, however, the scale of the latter will depend on whether they have any suppliers and whether their suppliers are regional, national or global. Thus overall the scale of the multiplier effects will depend on the activities' role and position in the supply chain. Primary sector activities generally have few if any suppliers, or backward linkages, unless they have upgraded and process and package their product. Manufacturing activities typically have a larger range of backward linkages, hence their attraction to regional development agencies, though whether or not these are located in the region will depend on firm specific strategies. With just in time strategies there has been a move back towards regional concentrations, but even so, being 'within the region' can be interpreted fairly broadly; in the case of Japanese foreign inward investment in the UK for example, it could include locations anywhere within the EU, or more narrowly within an hour's driving time.

Key or superstar regions (see Chapter 7) house global cities, where the management and strategic planning of large international organizations and corporations are found in addition to the most prestigious institutions in government, education, health and culture as well as a whole range of other manufacturing and service activities associated with contemporary economies. The majority of regions however have a more limited range of activities that produce goods and services for external markets. In many cases these activities are locked into supply chains that are controlled outside the region and correspondingly are vulnerable to decisions taken elsewhere.

These are the general theoretical ideas which underpin much of orthodox regional development theory, the new economic geography developed by economists, and to some extent the value chain approach discussed in Chapter 3. They are general tendencies and to varying degrees are universal, in the sense that they are valid for capitalist societies in the contemporary era but they relate largely to issues at the micro scale, such as markets, costs and factor availability, which influence firms' decision making. Thus they do not really incorporate the macro tendencies or broader processes shaping the trajectory of capitalist society within which these decisions are made. These broader processes, including uneven development over time and between states, the role played by the state, the organization of work and social reproduction, also influence the fortunes of regional and individual well being and these are explored more in the section below which begins with a discussion of the French Regulation perspective.

# ECONOMIC RESTRUCTURING AND UNEVEN REGIONAL DEVELOPMENT: A REGULATIONIST PERSPECTIVE

Within Marxist political economy, capitalist economies are inherently prone towards crisis because they are intrinsically anarchic. Thus, while managers plan their own operations in the finest detail, for example in the fruit chain supermarkets use sophisticated tracking systems to ensure a continuous supply of produce, there is no conscious organization of the economy as a whole – it is left as it were to Adam Smith's invisible hand. Referring back to Box 1.2 in Chapter 1, there is no guarantee that the commodities produced will be sold, and this is crucial, because without sale there is no profit and without profit there is no incentive for capitalists to invest in a new round of accumulation. Without reinvestment there would be an economic downturn and crisis as both Marx and Keynes pointed out.[1] Workers would be without jobs and capitalists without investment opportunities. Despite the tendency towards crisis, however, there have been long periods of growth and stability. French Regulation theory[2] developed a series of intermediate analytical concepts between the abstract categories of Marxist political economy and specific empirical forms in order to explain this apparent paradox.[3] The key concepts are: regime of accumulation, an industrial paradigm or labour process and mode of regulation.

A *regime of accumulation* signifies a period of relative stability and a balanced organization of production, income distribution and consumption. More specifically it refers to the parallel development over a long period of the conditions of production, including the productivity of labour, the degree of mechanization, and the relative importance of different branches of production on the one hand and social reproduction (that is, individual household consumption, collective consumption, investment and government spending) on the other (see Lipietz 1992). Several regimes of accumulation can be identified in the history of capitalism but the most significant one is the regime of intensive accumulation that emerged after the 1939–45 World War and lasted until the early 1970s called Fordism. During this period, western capitalist economies experienced unprecedented levels of growth, sustained by new sources of energy and new methods of production or a new *labour process* based on 'scientific management' or Taylorism. Taylorism[4] involved a conscious effort to effect a rigid division of labour between knowledge or mental skills which were to become the prerogative of management and manual skills with which workers would carry out tasks in prescribed ways. Together with the flow line principle of Henry Ford that allowed divided tasks to be recombined efficiently, tremendous increases in productivity occurred. In reality however the rigid division of labour between mental and manual tasks could only be approximated, but even so there were enormous increases in labour productivity, which enabled real wages and profits to rise simultaneously, thereby potentially offsetting the two main sources of crisis. Wages would be sufficient to buy the products and profits sufficient to encourage investment so capital accumulation could continue in an outward spiral of growth, thus suspending the tendencies towards crisis. These processes underlay the unprecedented economic boom from the 1950s to the 1970s in Western Europe and the United States.

Technology develops within specific sets of economic and social arrangements and technological possibilities are not realized automatically, that is social change does not automatically follow technology. Indeed the technology underpinning Fordism existed and was being implemented in the 1920s, but sustained accumulation did not follow. Changes were necessary in the prevailing *mode of regulation*, that is in the institutional arrangements within which individual and social reproduction takes place. Antonio Gramsci (1971) writing in 1934 argued that 'in America rationalisation of work and prohibition are undoubtedly connected' as the new working practices required more disciplined and reliable workers and a relatively stable social life. The 'new methods of work are inseparable from a specific mode of living and of thinking and feeling life' which led to 'the biggest collective effort to date to create with unprecedented speed, and with a consciousness of purpose unmatched in history, a new type of worker and a new type of man' (Gramsci 1971: 302). The novelty of the new more disciplined and intense working practices necessary to effect the productivity increases are graphically illustrated in the film *Modern Times* (1936) in which Charlie Chaplin plays a factory worker who has difficulty adjusting to the robotic nature of assembly line work and under the strain of the job, goes berserk creating mayhem in the factory. So to encourage people to accept these new working conditions Henry Ford introduced the $5 a day and this raised the effective demand necessary to sustain this national regime of accumulation by raising working class consumption norms. Thus, during the Fordist era the expansion of the national market was an important generator of growth. Gramsci argues that there was a danger that higher wages would be spent on alcohol and 'womanizing' while 'the new industrialism wants monogamy and relative stability' – hence the role of prohibition (Gramsci 1971: 304). Gramsci's account seems rather functionalist and it overlooks the way that women constituted a high proportion of the workers in the newly developing consumer goods factories (see Glucksmann 1990). Nevertheless it is one of the few accounts that explicitly links questions of production and reproduction (see also McDowell 1991) and Fordism was undoubtedly associated with the development of new more privatized lifestyles organized around increasingly suburbanized, heterosexual, nuclear families, with a male economic provider and a dependent wife whose primary role was housework and childcare. Thus the 'new man' needed a new type of woman, a homemaker, or at least this was the ideal to which the increasingly affluent worker aspired. Roger Miller (1991) illustrates how advertising not only played a role in boosting demand for the new consumer goods but also in convincing the wives of the more affluent workers that they should become house workers by using the new domestic technologies themselves, rather than manage domestic servants. Domestic workers were in any case becoming increasingly scarce owing to long distances between the areas of lower working class accommodation, still in the centres of cities, and the newly developing suburbs for the rising working class, but also because they found the jobs available in the growing number of factories producing these consumer goods far more attractive than domestic work. In addition to changes in individual lifestyles greater direct involvement by the state was necessary to manage social reproduction as a whole.

Only in the post-1945 era did the state organize an effective mode of regulation to match the new production system. In most Western European countries and to a lesser degree in North America the welfare state expanded to increase the quality and coverage of education, housing and health provision as well as widening access to social security, which sustained consumption during periods of unemployment or illness. Credit systems were introduced to enable employees to buy high cost goods such as cars and housing, which were beyond their immediate earnings and savings capacities, yet central to the regime of accumulation. Without stable accommodation for example, there would be limited demand for the new consumer goods, but the home also provided a space for rest, recuperation and indeed self-expression which was increasingly denied in the new forms of work. The state also assumed some responsibility for managing the economy through Keynesian demand management in order to ease cyclical crises, and enable capitalists to realize the profits from long-term investments. In short, the linked development of production and reproduction on an individual and social scale gave rise to a new stability or coherence in the system and it was this that underlay the unprecedented economic growth in the third quartile of the twentieth century.

These stable, indeed booming outcomes were not however achieved by careful planning but by a fortuitous combination of circumstances together with pressure from active trade unions who secured wage increases through collective bargaining and unofficial campaigns. The success of their actions varied across countries, but it was possible to obtain increases in living standards without threatening economic competitiveness or profits because of the increases in productivity. The state was also prepared to concede more to workers because of the threat posed by the Soviet world, which in the late 1940s and 1950s seemed to offer a realistic and more inclusive model of development.

During the era of Fordism there were large increases in welfare for the working class. Miriam Glucksmann (1990) recounts the experiences of women employees moving to areas of industrial expansion in the UK, mainly the Midlands and west London, and their evident delight in the higher quality housing, well heated homes with running water and indoor bathrooms. Later on the wider accessibility of goods such as cookers, washing machines, fridges and freezers eased the physical burden of domestic work. Many people experienced regular employment for the first time and even the unemployed benefited to some degree from the expanded welfare state in the fields of housing, education and health. People could plan their life strategies, as the future seemed relatively secure.[5] It was a period of rising prosperity and intergenerational upward mobility. The affluent worker became the subject of sociological research, and more specifically led to the embourgeoisement thesis, which highlighted the changing character of the working class, who were enjoying more affluent and privatized lifestyles (Goldthorpe 1968). It is important how-ever not to over-romanticize. With reference to the UK, the film *Saturday Night and Sunday Morning* (1960, dir. Reisz), based on Alan Sillitoe's novel of the same name, graphically portrays the grimy back streets from which people longed to escape and the intensity with which people had to work in order to earn sufficient wages to afford the suburban dream of a private house with a garden.[6] Moreover, the suburban dream was not always what it seemed and for some, especially women, became more of a nightmare, as this concretization of the gender division of labour left them spatially as well as socially confined to the home. These issues are graphically demonstrated as they are taken to an extreme in the film *Stepford Wives*.[7]

During Fordism, inequality overall was declining,[8] but gender and ethnic inequalities remained and even intensified. Thus while the white male factory worker could reasonably aspire to a house and garden, women working in factories could not afford an independent lifestyle.[9] Furthermore migrant workers and longer-term migrants who were recruited to fill the growing labour shortages in the expanding regions throughout the UK and the rest of Western Europe were subjected to explicit racism in work and everyday life as well as being overrepresented in the very lowest paid jobs or hardest forms of work in construction, transport, cleaning and caring (Berger and Mohr 1975; Rex and Tomlinson 1979; King 1998).

The forms in which the regime of accumulation developed also varied considerably between states (see Jessop 1991) as did the form of the welfare state: a minimalist one in the United States, a more comprehensive one in the UK and a universalistic or social democratic one in the Nordic countries, and while there was an unprecedented economic boom, the rates of growth varied between countries, being lower in the UK and the United States than in France, Germany or Japan. All regions in these countries were affected by and largely benefited from increases in national prosperity and regional inequalities narrowed in much of Western Europe (Dunford 1994). However, even though Fordism was a nationally based regime of accumulation, the processes gen-erating uneven regional development were not suspended and the regions where the Fordist industries were located grew faster than others. In the UK new industries were predominantly located in the Midlands and South East; in the United States, in what subsequently became known as the frost belt around cities such as Chicago and Detroit; in Italy they were concentrated in the north west of the country and in France in and around the Paris region. The boom lasted until the late 1960s and early 1970s when the virtuous cycle of growth based on sustained increases in productivity, profitability and rising working class consumption was broken and crisis conditions returned to the economies of Western Europe and the United States.

## Crisis of Fordism

The French Regulation School attributes the crisis of Fordism to the inability to sustain productivity increases within the technology then available. No matter how carefully production was organized and coordinated there was always a balance delay time as a result of work tasks being of different duration so it was simply impossible to increase productivity any further. Raising the intensity of work through speed ups was ineffective as it led to social resistance, sabotage and absenteeism, culminating in widespread industrial unrest throughout much of Western Europe, especially France and Germany, where students joined industrial workers in their protests in the late 1960s. The hard and intense working conditions together with the collective nature of work in large workplaces were also conducive to industrial organization, both formally through trade unions and informally by militants. Indeed union density was at its highest in many countries during this era, certainly in the UK and the United States.[10] So, the virtuous growth cycle was broken because declines in productivity meant that wages and profits could no longer rise simultaneously and the state had fewer funds with which to finance collective consumption. The costs of many state services were also rising because of their inherently technologically unprogressive and labour intensive nature, leading to a fiscal crisis of the state and subsequently to declines in welfare spending.[11] Moreover the widening of industrialization and the increasing openness of world markets meant that wage increases became more of a threat to competitiveness than an internal generator of growth through additional demand.

In response to these problems new electronic technologies with automatic feedback mechanisms were developed which reduced delay times and allowed greater productivity. They also made it possible to monitor the output of individual workers more closely and thereby limit formal or informal sabotage. Quality circles with greater worker involvement were also introduced in place of the long production lines, which sometimes made work more interesting but also made workers more accountable, in part by making them more responsible to each other. These strategies are sometimes referred to as empowerment and in some ways workers do have greater control and responsibility, but the objective is generally dictated by a desire to increase productivity and profitability.

A further solution was to decentralize production to areas of lower cost labour both to less developed regions within countries and to other countries leading to the new spatial and international divisions of labour discussed in Chapter 3. Existing processes could be established and operated profitably there, because of the lower social resistance and lower wage costs. These conditions were reinforced either by the absence of trade unions or by so called 'sweetheart agreements' made between single unions and the companies, as a consequence of the competitive bidding for plants between different locations. People in these locations were also new to and often welcomed the employment in the context of few alternative ways of making a living. The new units typically carried out a smaller range of tasks and thus allowed a far more advanced central-ization of production with a decentralization of the operating units, which in turn made it possible to break up large working class concentrations and create environments that minimized the convergence of conflicts at the point of production, that is the structural contradictions of capitalist labour processes were to some degree dispersed geographically.[12]

In addition to the growing tensions within the workplace there were also problems in the sphere of reproduction. It was increasingly difficult for the state to continue to finance welfare, as resistance to taxation grew and as profits and wages began to decline, and the family could no longer be relied upon to fill the gaps. Fordism, in some ways, assumed a dependent wife, yet increasingly women were becoming part of the workforce but with few, if any, collective facilities to substitute for childcare and domestic work, so tensions in the home grew. In Beck's (1992: 14) terms, there is a fundamental contradiction within what he terms industrial society, which is similar to the idea of Fordism with-in the French Regulation School. The contradiction arises because the dominant normative model and accepted way of life is the nuclear family, but this is 'based on ascribed "feudal" sex roles for men and women' which are incompatible with real living conditions. He argues that this contradiction is even more apparent in the subsequent era of the second modernity, the risk society or post-Fordism, and indeed the new economy, leading to problems of social sustainability (see Chapters 7 and 8).

## Post-Fordism

The analysis of Fordism has been quite powerful and persuasive to the extent that until recently the subsequent period has been called post-Fordism which is often defined negatively in relation to Fordism. There is good reason for this because economic growth rates have been much slower and income inequalities have been widening at most geographical scales – within cities, between regions and nations and between the more and less developed parts of the world – though in absolute terms overall wealth has risen. The state has also become less involved in ensuring citizen welfare, as more and more countries endorse the neo-liberal agenda and either by choice or compulsion cede some powers to supra national and local institutions of governance whose role is oriented more towards providing a 'good business climate' rather than directly promoting social welfare (see Chapter 8). On the other hand it is important not to over-romanticise the Fordist era, or overlook the fact that some of the gains enjoyed by the working classes in developed countries continued to depend on the underdevelopment of regions elsewhere in the world, on the widespread use of low cost migrant labour and continuing gender inequalities.

Post-Fordism has never quite had the same degree of analytical rigour as Fordism perhaps because different countries initially followed rather different pathways out of the crisis. These pathways correspond to different political traditions in terms of the key relations between the state, capital and labour, that is in the mode of social regulation. Danielle Leborgne and Alain Lipietz (1991) defined five such pathways, neo-Taylorism, Californian/individualist, Toyotism, West German (corporatist) and Kalmarism (social democratic), which correspond in some ways to the three welfare regimes outlined by Gösta Esping-Andersen. Links between these approaches were drawn in Chapter 1 and are discussed further in Chapter 8. Given these different pathways, however, in order to consider links between economic restructuring, uneven regional development and ways of life, the next section focuses on just one of these pathways, the neo-liberal route, consciously followed by the Conservative government in the UK after 1979, and one to which increasing numbers of countries are being drawn. To set the context for the regional discussion some of the main elements of economic restructuring and employment change in the UK are briefly outlined.

## ECONOMIC AND SOCIAL RESTRUCTURING, EMPLOYMENT CHANGE AND SHIFTING REGIONAL FORTUNES IN THE UK

Economic restructuring affects the kinds of jobs available, the rates of pay, incomes and lifestyles of people in places. In the UK the transition from Fordism to post-Fordism and beyond has taken place within a neo-liberal pathway and has been associated with rising income and earnings inequalities, an increase in poverty, especially child poverty, increasing social polarization and social exclusion.[13] The UK now has the highest proportion of the population living in poverty of any EU country and almost the highest level of child poverty, with 19.8 per cent of children living in poor households in 2000. Italy is the only country that has a higher level, at 20.5 per cent; the figure for the US is 22.45 per cent and for Mexico 26.2 per cent (UNICEF 2000). Thus the UK is in the bottom four OECD countries on this measure (see Figure 5.1). Although many of these children are in households where no one is in employment, just over half are in households of low paid workers, which have doubled since 1977.[14] Despite the government's pledge to end child poverty by 2020, the 2001 census shows that 6 million people or 20 per cent of the workforce were still in low paid jobs and 2 million children were being brought up in households where no one was in paid employment.[15]

Empirically, the transition from Fordism to post-Fordism has been associated with a marked sectoral shift from manufacturing to services and an increase in the feminization of the labour force (see Figures 5.2a and 5.2b). Figure 5.2a illustrates the relative rise of the service sector and Figure 5.2b the relative proportions of women and men in these sectors. In 1978, 7.1 million people worked in manufacturing, while in 2002 the figure was only just over half this amount (3.8 million); so while one in three male jobs and one in five female jobs were in manufacturing, today these figures are just under one in five for men

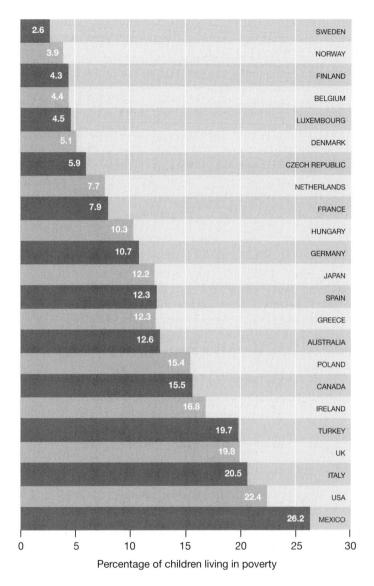

*Figure 5.1* A league table of child poverty in rich nations.

*Source*: UNICEF Innocenti Report Card Issue No.1, Florence: Innocenti Research Centre (2000).

*Note*: The diagram shows the percentage of children living in relative poverty defined as households with income below 50% of the national median.

and one in ten for women (ONS 2003).[16] Some of the increases in the service sector can be explained by a transfer of roles from the home to the state in the earlier period, and from the family to the private sector in the more recent past. Thus, in Esping-Andersen's (1999) terms, there has been a defamilialization of

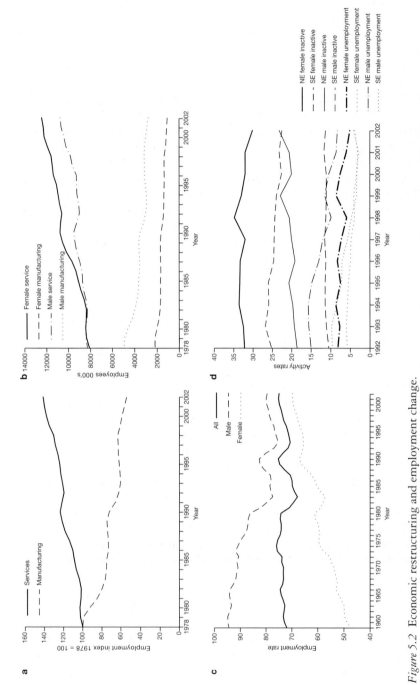

*Figure 5.2* Economic restructuring and employment change.
(a) The rise of the service sector; (b) Women and men in manufacturing and services; (c) Feminization of the labour force in the UK;
(d) Economic activity rates in north east and south east England.

*Source:* ONS (2003). Data compiled from the LFS and DfEE.

domestic work and care. Financial services have also grown very rapidly in the last twenty years and now account for about one in five jobs (ONS 2003). The service sector is characterized by a wide earnings distribution from highly paid professional workers, such as lawyers, to manual workers such as carers and cleaners. Overall there is a lower mean and greater variance in earnings than in manufacturing, which in part explains the widening earnings inequalities. However, inequalities have not increased to the same degree in every country. They are greater in the UK, USA and New Zealand, which have followed neo-liberal routes, than, for example, in Germany that followed a more corporatist route out of the crisis of Fordism, so the wider context of relations between the state, capital and labour needs to be considered in order to understand the well being of people in specific places. The section below discusses the nature of changing gender composition of employment in the UK in some detail, in order to be able to realistically assess its likely implications for gender relations.

The feminization of employment is illustrated in Figure 5.2c which plots the long run trend in the employment rate for women and men of working age between 1959 and 2002.[17] It clearly shows a trend towards convergence, although in the past five years both male and female employment have been increasing, with the male rate growing faster than the female rate, so currently the female and male rates are increasing in parallel.[18] Despite this recent increase in the male employment rate, the male economic activity rate[19] has continued to fall and the male economic inactivity rate[20] has continued to rise, despite falling male unemployment. The outcome of these different changes indicates that although a higher proportion of men seeking work have now obtained jobs, the long run increase in economic inactivity especially for older men is continuing, with 27.2 per cent of men between 50 and 64 economically inactive compared to 32.9 per cent of women aged 50–59.[21] This category conflates conflicting possibilities; it includes those taking early retirement (29 per cent of this age group), some of whom will have good pensions, but also people whose skills have been rendered redundant by economic restructuring and those who have become inactive as a consequence of ill health (55 per cent) following years of industrial work[22] (though who is and who is not considered medically fit for work tends to vary with the number of jobs available as much as absolute health).[23] Correspondingly, inactivity rates vary considerably by region and are far higher in the northern and western regions than in the south and east. Figure 5.2d illustrates both regional and gender differences in these rates. During the 1990s male economic inactivity rose as unemployment fell; female inactivity and female unemployment both fell. Generally, across the country, the female economic inactivity rate is far higher than the male rate, but in north east England in 2002 the male exceeded the female rate. By contrast both the female employment rate and the female economic activity rate have been increasing, indicating that more women wish to be and indeed are active in the labour market, although, as in the past, the absolute differences between women and men in the 25–49 age groups remain very wide indeed (with at least 20 per cent of women but under 10 per cent of men being economically inactive), reflecting

*Table 5.1* Rising male and falling female economic inactivity

| | 1992 | | 2002 | |
| --- | --- | --- | --- | --- |
| | *women* | *men* | *women* | *men* |
| All | 29.1 | 13.3 | 27.0 | 16.2 |
| 25–34 | 30.1 | 5.0 | 24.9 | 7.0 |
| 35–49 | 22.8 | 5.5 | 21.9 | 8.2 |
| 50–60/64 | 38.2 | 26.2 | 32.9 | 27.2 |

*Source*: Barham (2002).

the continuation of differential caring roles between women and men (see Table 5.1).[24]

Employment rates tend to overstate the convergence between women and men's working patterns. For example with reference to Figure 5.2c it is important to recognize that the working age for women currently ends at 59, while for men it is 64. Moreover when slightly less aggregate statistics are considered, which to some extent reflect the quality of employment measured by the extent of participation in paid employment (Figure 5.3a) and earnings differentials (Figure 5.3b), it is clear that gender inequalities in the labour market remain significant.

As the measures become more differentiated, gender differences become even more marked and these differences have profound effects on lifetime earnings and life chances. In particular, the labour force remains highly segregated by sector – in 2000, 74 per cent of manufacturing workers were male while 81 per cent of health and social workers were female – and by occupation, with 79 per cent of computer analysts and programmers being male, while 92 per cent of care assistants and 80 per cent or more of retail checkout operators, cleaners, catering assistants and primary school teachers were female (Cross and Bagilhole 2002). The kinds of occupations where women are concentrated generally receive lower earnings, which are reinforced by vertical segregation, such that in more mixed and even female dominated occupations, women are concentrated in the lowest levels. These differences have important effects on well being; the difference in lifetime earnings between a low qualified married/co-habiting woman with children and a low qualified married man with children is £0.5 million. As qualifications increase, then the difference falls, such that a high skilled mother would earn just over £0.25 million less than her husband/partner over her lifetime (Rake *et al.* 2000).

As discussed in Chapter 4 and in the Introduction to Part II, the feminization of employment has taken place at a time when jobs are less likely to provide a family wage, thus in effect what has happened is that more people have to work to maintain living standards. In the UK to a greater extent than other EU countries there has been a growing polarization between the high and low paid, which in part explains the increase in the numbers of working poor and the

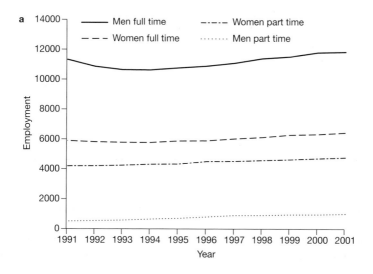

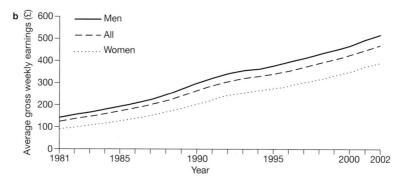

*Figure 5.3* Employment and gender (a) Differential extent of employment for women and men; (b) Earnings differentials in the UK.

*Source*: Data from NES (2002).

increase in child poverty.[25] In contrast to less developed countries and other countries in the EU much of the expansion of employment has been in the formal sector, but the character of the formal sector has changed. Privatization, deregulation and flexibility have been emphasized leading to an abundance of low paid and part time jobs, and an increase in the number of people employed through agencies, especially for low paid work in hospitals.[26] Thus while these employees are still within the formal sector their terms and conditions of employment have deteriorated significantly.[27] This polarization is also reflected in regional inequalities, which have been sustained and widened by the uneven distribution of both the scale and quality of employment. Thus while the nature of work and the composition of the workforce changes, uneven regional development continues and gender inequalities widen. To illustrate the impact of restructuring on life, work and uneven regional development the following section focuses on north east England.

## INDUSTRIAL RESTRUCTURING AND WAYS OF LIFE IN NORTH EAST ENGLAND

North east England is characterized by traditional industries. Historically it has been one of the poorer regions of the UK and this relative deprivation continues despite the establishment of some new modern industries. The region is characterized by above average male unemployment and inactivity and below average levels of GDP per capita. There is a comparatively high proportion of poor households and the region is currently losing population through both natural decrease and emigration. As a consequence a high proportion of the region qualifies for special assistance (see Figure 5.4 and Table 5.2) but despite these policies regional inequalities in the UK have been widening in the 1990s and early years of the new millennium.

During the era of Fordism regional income inequalities narrowed in the UK, similarly to other parts of Europe, but inequalities associated with the differentiated economic base remained; the production of mass consumer goods was disproportionately concentrated in the Midlands and south east of the country, leaving north east England with the more traditional sectors of mining, iron and steel, heavy engineering, chemicals and shipbuilding. In terms of industrial transformation this region was therefore a 'pale shadow' of Fordism (Hudson 2000: 133). Nevertheless Fordism was a national regime of accumulation and the policies for social reproduction in terms of health, education and welfare applied nationally; jobs were created in these areas. For people in employment, unionization was high, wages were sustained through collective bargaining and jobs were relatively stable with the concept if not the reality of a job for life and a reasonable expectation of career progression.[28] In terms of family life the division of labour was very traditional. Indeed, the male economic provider model was perhaps more intense in this region, partly because a high degree of domestic labour was necessary to renew male workers working in heavy industries. What is more, industrialists had resisted alternative kinds of employment developing in the region to minimize competition for labour.

By the early 1960s, however, unemployment levels were high relative to other regions and regional policies were strengthened to try and effect a greater balance between the booming south east and Midlands, where labour was scarce and inflationary pressures increasing, and the depressed north, in order to permit higher overall growth (McCrone 1969).[29] Similar arguments have been made in relation to the contemporary epoch but in the prevailing neo-liberal climate, rather than any form of redistribution the intention is to create greater balance between the regions through faster growth in the poorer regions, though how this is to be achieved is more of a mystery, especially as most regional theory now predicts cumulative differentiation between regions. In the 1970s however

142

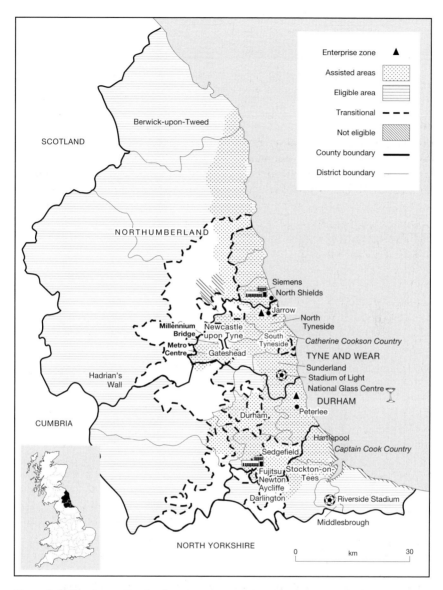

*Figure 5.4* North east England.

*Source*: Adapted from ONE North East (2002).

new activities were attracted to the north east by the policies but also by the lower cost labour. Indeed the fact that Fordism was not generalized 'meant that the old industrial regions provided a space to which decentralised plants could go' (Hudson 2000: 133). Correspondingly, this region became a 'global outpost' receiving a whole range of unconnected branch plants – from household

Table 5.2  Selected indicators of regional uneven development in England

| | | UK | North east | South east | London |
|---|---|---|---|---|---|
| GDP per capita index UK = 100 | | 100 | 77.3 | 116.4 | 130 |
| Employment rate | | 74.4 | 68.5 | 80.1 | 70.9 |
| Unemployment rate | | 5.2 | 6.9 | 4.0 | 6.6 |
| Occupational dispersion | Higher managerial (%) | 10.2 | 6.9 | 14.2 | 14.0 |
| | Routine occupation (%) | 10.2 | 12.8 | 7.8 | 6.7 |
| Net migration | International | +183,000 | 3000 | 4000 | 124000 |
| | Inter-regional | — | –4000 | 14,000 | –69,000 |
| Population | % natural change | 1.2 | –0.8 | 1.1 | 6.1 |
| Standardized mortality ratio UK = 100 | | 100 | 110 | 92 | 95 |
| Average weekly earnings (£) | men | 488.2 | 418.6 | 526.6 | 667.7 |
| | women | 365.5 | 318.4 | 381.6 | 483.1 |
| Trade union membership | Public sector (%) | 66 (m) | 73 (m) | 58 (m) | 60 (m) |
| | | 22 (w) | 65 (w) | 45 (w) | 53 (w) |
| | Private sector (%) | 56 (m) | 33 (m) | 17 (m) | 15 (m) |
| | | 14 (w) | 18 (w) | 11 (w) | 12 (w) |
| Qualifications | Pupils with 5 GCSEs | 51.0 | 43.9 | 55.5 | 48.6 |
| Rich and poor | Highest quintile (%) | 20 | 12 | 29 | 28 |
| | Lowest quintile (%) | 20 | 24 | 14 | 17 |
| Social benefits | Households (%) income support/WFTC | 16 | 21 | 10 | 16 |
| Internet access | Households (%) | 32 | 25 | 39 | 40 |
| Crime | Notifiable offences per 100,000 | 10,440 | 10,139 | 8012 | 14,474 |

Source: ONS (2003).

Notes: Data relate to the latest year available in ONS data set – all data between 1999 and 2002.
WFTC = working families' tax credit. Figures for rich and poor indicate the proportion of households in the region in the highest and lowest quintiles of household incomes nationally.

electrical equipment producers to clothing and crisps. These plants offered new types of work – routine assembly, rather than physically strenuous manual labour – and the gender composition of employment shifted towards women.[30] The new plants had parallels with the level III activities referred to in the Hymer model of the spatial division of labour discussed in Chapter 3 (Box 3.4), and were externally controlled by parent companies and subject to market forces on a global scale. Thus while the firms provided employment they rarely included high level activities and neither did they generally establish sub-supply industries within the region, so the multiplier effects were limited. Some firms were very transitory leading to the same workforce in the same plant with the same management producing different products for different owners as one firm succeeded another.[31] In other words the firms that came to the region were largely the dependent or cost sensitive plants rather than developmental or performance plants or ones with a world product mandate, and correspondingly the multiplier effects were limited.[32]

There was also a decentralization of government offices including tax and social security, a continuing expansion of health, education and welfare and later on a whole range of private sector services connected with care homes (Beynon *et al.* 1993). With the exception of professional work in the public sector, the majority of new jobs were at the lower end of the employment hierarchy, so while regional inequality narrowed on conventional measures of regional distress such as unemployment and out migration, new forms of inequality in the form of individual job opportunities and in the region's capacity for autonomous growth were emerging.[33] Moreover, many of the firms that came in the late 1960s and 1970s had disappeared by early 1980s, leaving the region in a relatively weak position on both the traditional and new measures of regional inequality.

With the further development of post-Fordism and the new economy from the 1980s until the present this pattern has largely continued, though with the development of new technologies and increasing globalization the composition of the incoming firms has changed and with deregulation the terms and conditions of employment have become more flexible. Mining has now all but disappeared from the north east region. While manufacturing has declined, there is still some steel production, chemicals and pharmaceuticals, but work is less secure, with many employees on flexible contracts.[34] In Hartlepool 6000 employees worked in a fully integrated steel works in the 1970s but only about 600 remained by 2002, employed in a specialist pipe mill; numbers have varied over time, owing to widespread use of flexible contracts, as a respondent in Lydia Morris's study of Hartlepool in 1989 explains:

> They kept about 50 men as core workers, to keep the place ticking over when there were no orders in. When an order came in they'd bring the old pipe men in under a short term contract. The mugs then have to go back with no protection, no holiday pay, no sick scheme and no pension fund. They work for so many weeks then

they're laid off again. The union's hands are tied now – they've no muscle. There are so many on the dole.

(Morris 1995: 6)

Thus even within the traditional industries the terms and conditions on which employment is offered are now quite different from those prevailing in the Fordist era. There have also been significant declines in the clothing industry where women are overrepresented. At least 3000 jobs were lost in the 1990s, many being due to the decision by Marks & Spencer to reverse its strategy of only buying from British producers, in an attempt to redress its falling profits. Outward processing of textiles has been renewed with the opening up of Eastern Europe, which has the double advantage of comparatively low cost labour and much shorter lead times (two weeks)[35] compared to South East Asia (nine weeks) which are crucial in the fashion sector. The ensuing loss of female jobs has in general received less press coverage, less response by politicians and less academic interest than job losses in the heavy industries.

Foreign direct investment has continued and there have been spectacular successes and failures. One of the key successes was the attraction of the Nissan car factory to Sunderland in 1984 which currently employs around 4500–5000 people, produces 330,000 cars per year and is one of the most productive plants in Europe. In 2001 it secured production of the Micra, induced by a government grant of £40 million, which is expected to guarantee its future for the next five to ten years.[36] Another successful inward investment is the South Korean firm, Samsung, established at Wynard near Sunderland in 1987, which produces electronic consumer goods and employs around 3000 people, though its future is less certain.[37] However, two of what were at the time seen as flagship developments, Fujitsu in Newton Aycliffe and Siemens in North Shields, both closed in 1998, and raise doubts about the viability of the branch plant economy (Pike 1999). These companies manufactured memory chips and while Fujitsu had been operating successfully for over seven years Siemens had only been open for a year.[38] Both the opening and closing of these firms demonstrate how the fortunes of regional economies are shaped by their integration into the global economy. In the case of Siemens, a German based multinational, the initial location decision reflects the firm's desire to expand output in the rapidly growing global market by taking advantage of lower labour costs and government grants in a relatively deprived region. The subsequent closure was a response to the global price fall, in turn a consequence of advances in chip technology, which doubled their capacity, leading almost immediately to excess supply worldwide. Thus the global nature of the market intensifies volatility. As pointed out earlier, one of the characteristics of capitalism is that there is intense internal planning within firms but little or no planning between firms. Thus firms all over the globe expanded their capacity in response to the high prices, but as the technology changed, the global market was quickly flooded, leaving many firms with excess capacity. In this case, the Asian financial crisis (see Chapter 8) exacerbated these market changes because South Korean producers, in particular, began to price extremely aggressively to increase foreign earnings. The significant outcome for the north east was that the price of chips fell from $50 when Siemens was being planned to less than $1 when it opened, making it no longer viable.[39] This example also illustrates the vulnerability of a branch plant economy; as Ulrich Schumacher the president of Siemens Semiconductor Group commented, 'this decision is not a reflection on the plant's employees or any other local domestic economic factors' (BBC Online 1998). The freer liberal market economy in the UK, however, makes disinvestment far easier than in the more regulated German economy.[40]

The closures had widespread reverberations on the local service economy, for example 85 taxi drivers were chasing the work of 50, which lowered their earnings, even though they worked longer hours. As a 23 year old taxi driver called Lee commented:

> Siemens' people were going to the airport and university umpteen times a day and the odds were you'd get a fare on the way back. Of course we're feeling the pinch. Some of us bring in sandwiches instead of going to the chippie.

He goes on to say that he had to curtail his all day sessions at the pub on his day off: 'when I used to have maybe half a dozen pints, six trebles and then God-knows what. Now I'll go later and drink less' (young taxi driver, cited by Carroll 1998).

Lee was not alone in reducing his consumption, so there was a significant fall in the takings of the local bar. This example also illustrates one of the differences between more and less developed countries, in that while globalization leads to similar processes being played out in different locations their impact varies according to local conditions. In this case, the impact of the closure has led to reductions in income and a decline in what some may consider unhealthy and excessive consumption. Even where people were made redundant there would be some redundancy or unemployment pay, so although people would become poor relative to the population as a whole, the absolute circumstances were far less desperate than the far greater job losses in, for example, the Philippines, where the economy was devastated by the financial crisis, as were many people's lives; they became destitute, having no welfare state to fall back on.

More generally in north east England, hotels, restaurants and cleaning firms were also affected by the closure, but the impact on local manufacturing was less obvious as local supply networks had not been developed. Indeed the closures highlight the dilemma of inward investment for local development planners, in that if firms do become embedded, something which has been strongly advocated in some of the regional development literature, especially on clusters and learning regions, and adopted by policy makers to counter the transience of branch plant economies of the 1980s, then the costs of closure, should it happen, are wider and deeper. Indeed there is an imbalance between the advice to continue to attract and embed multinational corporations, which are now thought to give more autonomy to local managers to develop relationships with local suppliers, and the reality of their more limited impact on local economies, arising from their low degree of local sourcing, limited range of employment and knowledge transfer, and overall confining of the regional impact largely to the employment offered, rather similar to the call centres discussed below. In the case of the north east however there is little empirical evidence to suggest that the institutional efforts to effect embeddedness have been successful, which as it turns out and in contrast to the prevailing advice, may have been to the region's advantage.[41] Subsequently, a call centre (see below) temporarily occupied part of the Siemens' site, and in 2000, a Californian semiconductor company moved into the plant, but the scale of employment does not yet match the level Siemens promised. Some high technology firms have remained in the region, for example 150 IT and software firms, including Sage, a firm originating in Newcastle, that has become a major European producer of accountancy software, as well as 400 new media firms (ONE North East 2002). In addition there are firms in pharmaceuticals and engineering.

## The cultural sector and call centres

In character with the new economy, there has been a shift not only in the nature of manufacturing employment but also from manufacturing to services and some of the fastest growing sectors are now cultural industries and call centres, both of which are potentially export oriented so in principle capable of stimulating regional growth. The cultural sector includes activities as diverse as opera, crafts, football and clubbing as well as tourism. The north east has many rural areas of outstanding natural beauty and a heritage coastline, which have been marketed to attract tourism. There is Catherine Cookson Country in South Tyneside, Captain Cook Country along the coastline from Hartlepool to North Yorkshire and two world heritage sites: Hadrian's Wall and Durham Cathedral (see Figure 5.4). The area has experienced tremendous industrial dereliction, but mostly this has been cleared away in regeneration projects, which have renewed the physical fabric and created new landscapes, cultural centres and museums based on the industrial past. The National Glass Centre (see Figure 5.5) was developed in Sunderland and is a museum, a place for artists and designers to work in, and an organization to support new, small to medium-sized businesses and business start-ups.[42] The region also has three major football teams, two with new stadiums, one of which, the Stadium of Light, home of Sunderland Athletic Football Club,[43] was constructed on the site of Wearmouth Colliery following its closure. There have also been major cultural developments in and around the city centres and the river, such as the Newcastle Quayside, linked to the Gateshead Quayside by the new Gateshead Millennium Bridge, the conversion of the Baltic Flour Mills 'from derelict hulk to art gallery' and the development of the new Sage Music Centre, designed by Norman Foster. Collectively these have resulted in a visually spectacular waterfront.[44]

*Figure 5.5* Cultural regeneration in north east England: the National Glass Centre.
*Source*: National Glass Centre.

Additionally, there is the Angel of the North, the largest sculpture in Britain, constructed on a former colliery pithead at the entrance of Tyneside in 1998 and visible to 90,000 motorists a day. The sculptor commented:

> The hilltop site is important and has the feeling of being a megalithic mound. When you think of the mining that was done underneath the site, there is a poetic resonance. Men worked beneath the surface in the dark. Now in the light, there is a celebration of this industry.
>
> (Anthony Gormley, undated)[45]

Whether the former miners appreciate this celebration, or whether this and the other cultural developments will directly or indirectly raise the material living standards of people in the area and fulfil the expectations of the promoters that 'by 2008 the north east will be the most creative region in Europe' is less clear.[46] This self-promotion has almost become mandatory among local economic development agencies and though it is obvious that not all regions can fulfil this type of claim, few can afford 'principled non-involvement in the game'.[47] Retail centres, such as the Metro Centre in Gateshead, which in 2001 was the largest indoor shopping centre in Europe, are also promoted as tourist attractions.[48]

Cultural activities have become central to many economic regeneration policies, reflecting the globalization of ideas. Indeed they have almost become a worldwide template and panacea for regeneration, partly modelled on Barcelona, where the spectacles have indeed had a spectacular impact, though not without problems, discussed in more detail in Chapter 9. Evidence of the effectiveness is mixed so their widespread adoption reflects the prevalence of neo-liberal orthodoxy on the role of the state. In contrast to Fordism when the state played a redistributive role, now the emphasis is on state assisted market-led physical regeneration, together with supply side initiatives to raise the levels of individual human capital. In the late 1980s David Harvey (1989) depicted this change as one from a managerial to an entrepreneurial state, and critically asked just how many spectacles can there be, points developed further in his discussion of the redevelopment of Baltimore (Harvey 2000).

In contrast to manufacturing, these activities create neither the scale nor the quality of employment necessary to redress regional inequalities. The cultural sector accounts for about 2 per cent of employment in the region, a figure boosted to 10 per cent if related bars and restaurants are included (Culture North East 2001) and the pay rates are barely sufficient to enable employees to be cultural consumers. So in contrast to the Fordist era, there is no link between domestic production and consumption, as many workers cannot afford the products they are producing. Moreover, so far, the external market remains limited as there are fewer visitors to the north east than any other English region (Culture North East 2001).[49] Thus while some stunning physical landscapes and monuments have been created, it is not clear how these 'spectacles' will reverse regional decline.[50] On a more mundane level, there has been a significant increase in employment in a service sector also sometimes linked to the new economy, call centres.

Call centres are believed to be one of the fastest growing sources of employment nationally and in the north east, but similarly there are limitations in terms of their capacity to reverse the relative economic disadvantage of the region. There is no universally accepted definition of what constitutes a call centre and correspondingly estimates of how many there are, where they are and how many people are employed within them vary. The consensus is that there are an equivalent of about 400,000 full time jobs in UK call centres, or 1.8 per cent of the overall workforce (IDS 2001). Some estimates suggest that this figure will double in the next five years but newer technologies, such as voice recognition software and direct use of the Internet by consumers (see Chapter 6), together with lower cost labour sites elsewhere in the world could displace this employment. British Telecom (BT), for example, is planning to move 2200 jobs to Bangalore in 2004, because wages are only 20 per cent of the UK level, even though they would be paying above the local norm. BT emphasizes that no permanent staff will be made redundant, as the losses will be borne by agency workers,[51] but this still means that employment will be lost which illustrates that this employment is also vulnerable to low wage competition.[52]

At a basic level, call centres are work environments which primarily consist of agents who communicate with clients via the telephone and simultaneously enter data into or supply information from display screen equipment or computers.[53] Employment in call centres is growing because a wide range of organizations have adopted these technologies and labour processes to deliver their services more efficiently. Banks, insurance companies, local authorities, public and private utilities all use call centres to respond to routine enquiries, to make sales and to deal with billing. Additionally, there are more knowledge based activities, such as software and computer sales firms, who have call centres to provide help for their clients, and the National Health Service has recently established NHS Direct, which provides medical advice. The nature of work therefore varies, and requires different degrees of knowledge, from being able to make routine responses to providing more complex information.[54] One unifying feature, however, is that the division of labour has been carefully established to ensure that employees are used efficiently. Typically clients respond to a series of automated questions before their calls are directed to the relevant agent. If the agents are unable to deal with the calls, they reroute them to people with the relevant expertise or authority. Even on computer help desks, agents use expert systems to diagnose the problem or else guide the caller through standard help procedures and should these fail to resolve the problem, redirect the call to an agent with a higher level of training. This division of labour, on the basis of skill, creates efficiency savings. In combination with contemporary ICTs, additional savings can be made by decentralizing work geographically to take advantage of different time zones and labour costs in different regions, enabling firms to maintain competitiveness in an increasingly globalized market. In principle, decentralization could lead to people working from home, and some people do, but the collectivized workplace helps sustain motivation, likewise necessary to retain a competitive edge, in what can be demanding and yet rather dull and repetitive work.

In the discussion below emphasis is placed on dedicated or specialist call centres, that is, ones that are located separately from a firm's other activities or run by specialist firms, which regional development agencies have targeted in order to bring the employment to their localities. These call centres are equivalent to back offices, that is branches set up in lower cost locations to do routine work, and indeed some back offices have simply been renamed call centres, so not all of the call centre employment is actually new.[55] Other firms subcontract work to specialist call centres, which literally deal only with calls. Either way, much of the work is highly bounded with little potential for career development, as there are few if any connections with the organization as a whole.

Within specialist call centres, typically, there are only four tiers, from an agent/operator (the lowest level), to team leader, supervisor and manager. The earnings gap between average employees earning £12,400 p.a. in 2002 (48 per cent of the average male and 65 per cent average female wage) and managers with £29,701 (16 per cent higher than the average UK male wage)[56] is much smaller than for many other forms of employment. Moreover, if the firms are inward investors the top management is likely to come from outside the region, so opportunities for career progression are limited within the region.[57] The work is generally considered semi-skilled, but it involves interaction with the public, and considerable ingenuity is required to deal with difficult and even abusive customers while remaining professional and sustaining the firm's reputation. Thus the work can be simultaneously routine, creative and stressful and the comparatively low pay reflects the overrepresentation of women, who are often considered to possess the necessary skills naturally. Overall between 60 per cent and 70 per cent of call centre jobs are thought to be female, although young men are widely employed especially in the specialized and higher paid areas such as computer help desks.[58]

In the high volume call centres work is highly pressurized. Trade magazines even offer advice on how to script calls to make them shorter: 'save seconds: teach them how to say goodbye'; 'cutting the length of the average call by 16% means built in growth – they can handle thousands more calls per agent' (Green 1998: 28). Work is consequently very intense as employees are constantly logged on to their computers, the time between phone calls is minimized by automatic dialling systems and individual work performances are monitored. A national TUC (2001) hotline set up for a two month period in 2001 to investigate employee concerns, received 400 calls a day. The main complaints were about the intense monitoring and scrutiny of work and of the workers' time. Thus while the workplaces are usually modern and well equipped and seen as part of the knowledge economy, they have also been referred to as the contemporary equivalent of Foucault's panoptican (Fernie and Metcalf 1997) owing to the intense surveillance and the way that the jobs have been constructed in a highly fragmented way, closer to Taylorism than to quality circles and empowered employees. This characterization has been criticized, especially by Peter Bain and Phillip Taylor (2001: 41), for failing to recognize the varied forms that call centres take. Nevertheless they concede that at the high volume end much of the work is Taylorized and current managerial practices mean that the:

> extensive and multifaceted monitoring practices, scripting, the simplification of query/response screen menus, and supervisory intervention have defined new frontiers of managerial control in service work, in which the statistical measurement of output and performance is combined in various ways with subjective evaluations of an agent's manners, behaviour, expression and style.

Although they are keen to emphasize that such practices do not preclude worker resistance and union organization.

Even supervisors, who have more autonomy and a wider range of tasks to perform, are under pressure but here the forms of control are even less visible as they are internalized by the use of human resource management techniques. These techniques appear to give employees autonomy, and a capacity for self-actualization, as they are given discretion over managing their time and tasks. In reality, however, this is constrained by broader targets set by senior managers, which results in employees internalizing the responsibility and blaming themselves for lacking organizational skills, if they fail to meet them. Julia Brannen (2002) describes the case of JJ, a call centre supervisor for a bank:

> At work JJ had to answer calls from the public as well as supervise her team's performance. Such was the pace of the day, she said she often forgot to take breaks, ignored lunchtime, and even put off going to the loo. . . . Most nights she also swatted up on work manuals and memos since she had no time to do this at work.

Interestingly, JJ who was 40 and had worked for the bank for a long time considered it to be a better employer than most. She had seen many changes that reflect the change from Fordism to post-Fordism and the new economy including:

> the disappearance of the old protective paternalism of the bank, the demise of 'jobs for life', which in the former days were secured in return for employee loyalty; the replacement of the ethic of service with the ethic of sales; [and] the introduction of individual responsibility for staying employable.
>
> (Brannen 2002: 6)

Despite these concerns about the nature of the work, development agencies tend only to look at call centres' potential for employment generation. In the north east they have developed special sites for call centres and market the region to attract new inward investors. At Doxford International Business Park in Sunderland, 6500 people are employed in a number of call centres and there are other clusters in and around the region. One business park in an enterprise zone in Easington, near Peterlee, established in response to coalfield closures, was particularly open about labour availability:

> Within the travel to work area . . . there are over 31,000 people unemployed. . . . One in every 70 people unemployed in the United Kingdom . . . lives in the travel to work area for Bracken Hill Business Park [Bracken Hill Business Park, East Durham: an ideal call centre location].
>
> (Belt and Richardson 2001: 78)

According to data from the UK Call Centre Directory, however, 24 per cent of call centres are located in the south east and just 2 per cent in the north east of England. If only the larger call centres of 250 agents or more are considered, then more have been located in the less developed regions, 37 per cent in northern England as a whole and 7 per cent in the north east (CCMA 2002).

This estimate roughly corresponds with the figure of 30,000 call centre workers in the north east, that is 6 per cent of total employment in the region (GONE 2002). One reason for their more widespread distribution is that although call centres are primarily volume employers and attracted by low cost labour, they offer flexible working hours during weekends and the evening, which, given constraints in relation to the availability and affordability of childcare, are attractive to people with caring responsibilities and also to students.[59] Thus they can tap into the much wider pool of economically inactive labour throughout the country. Some call centres also require specific skills, such as foreign languages, which are more likely to be available in big urban centres where populations are diverse and student populations more significant.

At present call centres are an important source of employment in the region but despite use of contemporary ICTs they generally reflect the continuation of specialization at the lower end of the employment hierarchy. They are vulnerable to competition from low waged regions elsewhere in the world and to further changes in technologies. Furthermore the 'life' of a call centre worker[60] is relatively short, only two years on average, partly because of the routine nature of the work and the conditions in which it is carried out, though some call centre workers rotate between call centres just to relieve boredom and the lifetime may be longer in less developed regions, where there are few alternative employment opportunities. Given the positioning of call centres, especially the large volume specialist centres, in the value chain, the capacity for both individual career progression and upgrading and wider regional growth are limited. Correspondingly although they do provide employment, whether state incentives available to firms setting up in the special zones, designed to promote development, are justified is less clear.

During the transition from Fordism to post-Fordism and now to the contemporary new economy, north east England has been transformed in terms of the kinds of goods produced, the nature of work performed and in the gender balance of employment. While 20 per cent of the population still work in manufacturing it is less likely to require traditional male manual craft skills. The closer integration of the region with the global economy has led to the expansion of inward investment and development of high tech sectors but these too are vulnerable owing to the corresponding intensification of competition. Many people have a sense of insecurity; indeed the redundancy rate, 40 jobs per 1000 per annum, is about one-third higher than the UK average (ONE North East 2002). Where jobs remain in the traditional sectors they are more likely to be flexible;[61] jobs last as long as the contract, and the concept of lifetime employment in a single sector belongs to the past.

Work in the decentralized factories is very different from the traditional industries in the sense that there is much greater direct supervision, more discipline and higher work intensity. Workers from the old traditional sectors experienced problems of adjusting to new working patterns and in general the new firms draw upon younger workers, which is possible because of the excess number of applicants. In Nissan for example workers are selected via psychometric tests and the average age is about 30. Women seem to be more adaptable and some former textile workers have become call centre agents, but the men who were employed in the traditional industries are displaced not only because their own jobs disappeared but because their skills do not match those required by the new forms of employment (Hudson 2000). In single industry communities, based on mining for example, the structural coherence has been undermined, making many people structurally irrelevant. Unemployment however is not always noticeably higher than elsewhere as many people leave the area altogether or else become economically inactive.[62] Effectively here as elsewhere there has been a shift in the balance of power between labour and capital. Work has been individualized. Decentralized wage bargaining based on individual/plant productivity has displaced collective bargaining and local wage cuts often have to be accepted as the price for keeping a job. In these circumstances there is competition between plants and places for work, rather than solidarity across sectors (Hudson 2000), even though the level of unionization here remains higher than in the south east or London (see Table 5.2). All these changes reflect the demise of Fordism and the closer integration of the region with the global economy.

These circumstances are not unique to north east England. Traditional industrial regions elsewhere in Western Europe and the United States have similarly been affected by the closures of basic heavy industry, and have experienced declines in male employment. Likewise in Latin America the opening of the economies to global competition in the context of structural adjustment policies from the 1980s has undermined many of the plants established under the import substitution strategies, which had relatively good and secure working conditions (see Chapter 4). The specific forms of change vary across regions, and in some places the state actively ameliorates the adverse effects. For example, in the North Rhine-Westphalia, an old industrial region in Germany, there was an active restructuring strategy, based on developing new technologies for countering industrial pollution together with relatively high levels of unemployment compensation, which makes the changes easier to bear. This German illustration reflects the way that despite globalization and the pressure to conform to the neo-liberal agenda not all states respond in the same way, and in this case there has been a greater willingness to retain a more corporatist tradition. In the UK and USA by contrast physical regeneration and supply side initiatives aimed at increasing people's preparedness for work, irrespective of the jobs available, have dominated regional development strategies.

Thus overall there has been a change in working conditions, a narrowing of the gender gap in paid employment and a potential reversal of traditional roles in communities characterized by industrial decline. The changes in working patterns and practices, the feminization of employment and the increasing need for more than one income to sustain a family are however much more general. The section below returns to the question of the effect of changing gender roles on gender relations.

# ECONOMIC TRANSFORMATION, GENDER
# RELATIONS AND MASCULINITY

Given this economic restructuring how and in what ways have gender relations changed and in particular how have men responded to this change in their economic and social positioning? The feminization of employment is limited in terms of the quantity and quality of paid employment but the male role as sole and even main provider has clearly been undermined, while women's contributions have become more visible or public.

In old industrial regions, such as the north east of England, economic restructuring has undermined some of the traditional conceptions of masculinity in terms of being the main worker, principal economic provider and macho man, doing physically demanding work. However, whether this will lead to a change in the gender order or to the end of patriarchy as Manuel Castells (1997) suggested is far from clear. For just as the feminization of the labour force is more complex than is portrayed by simple aggregate statistics, so conceptions of masculinity are varied and dynamic,[63] such that even when the foundations of hegemonic masculinity have been undermined, men find new interpretations, which sustain gender imbalance.[64]

Lydia Morris's (1995) study of the impact of economic restructuring in the late 1980s on gender relations in north east England found the existence of dual households with a single female earner very rare, that is instances of total role reversal were uncommon. Unemployment principally affected unskilled workers, whose wives or partners were also likely to be low skilled, and so similarly vulnerable to unemployment. Moreover, at the time of the study, the benefit system meant that female earnings would seldom increase overall household income, as they would be subtracted from 'his' social benefits. A further factor was the desire not to disturb the traditional male economic provider model especially as the unemployment was assumed, realistically for many individuals, to be a temporary, albeit recurring phenomenon. Where women did paid work, their earnings were generally spent on household rather than personal goods and though there were some changes in the domestic division of labour, these occurred mainly when both men and women were working full time. Moreover, women were more likely to have control over the household budget only when it was very limited and therefore more of an onerous task than a source of empowerment; overall there were only marginal changes in gender relations.

More recently challenges to traditional concepts of masculinity have been portrayed very poignantly in the films *The Full Monty*, and *Billy Elliot*,[65] both set in communities associated with profound restructuring and losses of male employment, which indicates how the threat to masculinity has become an issue of widespread social concern and interest.[66] In *The Full Monty*, the key character is unemployed and separated from his wife, so for him the traditional concept of masculinity is purely aspirational, yet throughout the film he clings on to this concept and uses it continually to judge his own and others' conduct.[67]

UNEVEN REGIONAL DEVELOPMENT IN THE UK

In particular he clearly suffers from being unable to play the provider role in relation to his son;[68] yet he is highly critical of his friend for taking a security job at 'women's wages' and regards the idea of himself working in retail at the minimum wage laughable. Doreen Massey (1984) also highlighted the impact of the different nature of work arising from restructuring in the 1980s following a round of closures in the mines and steel industry in Wales:

> Further when new jobs are made available to men we hear, as though it were patently funny, that Welsh ex-miners cannot be expected to turn their attention to making marsh mallows, or underwear. What is at stake is the maintenance not just of a social structure in which men are the breadwinners, but also of a long-held self-conception of a role within the working class – the uniqueness, the status and the masculinity, of working down the mine.
>
> (Massey 1984: 211)

These images are vivid, but more recent research, while finding some resonance with these concerns, also demonstrates men's resilience and ability to redefine masculinity in ways which reproduce unequal gendered power relations despite their relative economic demise or change of work. In an unnamed region, an in depth study of 10 men who had moved into stereo-typically female jobs, including miners who had become nurses, found that while they lied about their work when socializing with new people, nevertheless maintained their masculinity at work in several ways. They ascended the hierarchy more rapidly than their female equivalents, and did rather different tasks within the same job. In nursing they focused more on paper work and organizational issues rather than direct care, and entered what were perceived to be the more macho branches, for example psychiatric nursing, which sometimes involves restraining violent patients and thereby retains traditional masculine traits, in contrast to occupational therapy which was referred to dismissively as 'basket making' (Cross and Bagilhole 2002). Similarly in an area of industrial dereliction in the West Midlands, Sara Willott and Christine Griffin (1996) illustrate how a group of working class men renegotiated or re-defined masculinity as they adjusted to long-term unemployment. Loss of work reduced their incomes and freedom to socialize so they were confined much more to the domestic sphere and those with working partners felt their status was undermined, as Gavin one of the respondents comments:

> If she gives the word when she gets home 'I'm going out at the weekend' and you goes 'who with like?' and she turns round and says 'you can't stop me because its my money, I'm earning it, its my money' [repeats]. When you're sitting at home watching the telly you're thinking 'look at the time, she's out like, where is she and that like?' It all boils round in your head like and the next thing

you get on to is 'she's seeing someone else now, it's two o'clock in the morning like', you know what I mean. . . . She buys you a four pack.

(Willot and Griffin 1996: 85)

Parallel to the cooperative conflict model, Gavin clearly believes that earning money gives an entitlement in terms of how it is spent, and therefore feels unable to challenge his wife's socializing, but at the same time it is clearly a source of deep concern. He also reveals his dependence when he points out that his wife 'treats' him by buying him a 'four pack', i.e. four cans of beer.

The study found a wide gulf between the continuing dominance of the male provider discourse and the reality of the decreasing proportion of men able to play this role. At the same time they found that the men resisted exclusion from the dominant discourse by developing new powerful roles for themselves. They either treated what they had previously dismissed as 'women's work' as 'real work' by investing it with technical concerns such as the relative merits of different cleaning machines and by claiming to be more professional house workers and carers than their partners, or else tried to maintain the provider role through various forms of informal work. Thus as individuals they may have lost the material foundations for a traditional hegemonic status and, similar to Gavin cited above, were keenly conscious of their individual powerlessness, but they still endorsed and tried to approximate prevailing conceptions of masculinity (Willott and Griffin 1996).

In a rather different context in Vermont, in the United States, where women had recently entered paid work on a much larger scale, employed male partners of women with full time jobs responded to this challenge to their provider role by taking on additional informal entrepreneurial work. This work frequently involved costly tools and equipment and although cash was given for the jobs done they were not always economically viable. Nevertheless it gave the men seemingly inflexible obligations and a corresponding freedom to absent themselves from housework and childcare, leaving the responsibility and limitations arising from running the household to women. By so doing they maintained a 'traditional division of labour in untraditional circumstances' (Nelson and Smith 1999: 147). These examples are similar to the way that men defined a role for themselves in the micro-credit businesses of their wives or partners in Colombia (Pineda 2001) discussed in Chapter 4.[69]

In these old industrial regions where the role of economic provider has been undermined and the forms of work have undergone profound change men seem to retain their superior position to women by redefining their masculinity, that is when everything appears to change new ways of being masculine are created to sustain old power differences, albeit in new forms. Thus rather than the transformation of work leading to an end of patriarchalism, patriarchalism can be reconstructed taking into account the new forms of work or non-work. These more detailed analyses highlight the risks of making loose associations between aggregate statistical change and social behaviour.

The question then becomes why, despite the new circumstances, are traditional gender ideologies sustained and gender divisions and inequalities reproduced? One possible explanation is that capitalist societies sustain profitability partly by minimizing the costs of social reproduction. Historically this has been achieved through the use of women's unpaid domestic labour following on from their role in bearing and caring for children, which reflects a certain path dependency reinforced through social processes, but there is no logical reason why these gender divisions should be sustained unless perhaps it is simply in the interests of men so to do. The main challenges to existing gender inequalities will perhaps come from the increasing individualization of work, which may threaten the rationale for coupledom and relatedly undermine social reproduction. Referring again to north east England, a study of young people's identities in Newcastle upon Tyne (Hollands 2001) found both continuity and change, but young women in particular had found a degree of freedom from delayed marriage/household formation as Meg (aged 31) comments: 'I have a reasonable job, a house and mortgage and a car, so I don't really need a bloke around. If I meet someone its on my own terms' (Hollands 2001: 149).

Delayed family formation may initially have been forced by the financial impossibility of setting up independent households on the more individualized wages that they and their prospective partners receive, but the more individualized lifestyles may also become a choice and turn into no marriage/partnership and family formation. This will become a serious challenge to capitalism only if fertility levels fall below those at which the population can be sustained and supplies of immigrants and migrant labour dwindle. Perhaps at that point the work of social reproduction will be more highly valued and the workers more highly paid, which could then form a material foundation for challenging gender inequality. This may seem a dismal conclusion, and the empowering effects of increased female participation in the public world of work discussed here and in Chapter 5 should not be overlooked. Nevertheless there have been significant increases in the proportion of women in paid employment and decades of equalities policies, which have become ever more sophisticated, but so far little change in the three key barriers to equality, the unequal division of domestic labour, gender segregation in employment and the lower valuation of work disproportionately carried out by women. The new economy is associated with deteriorating terms and conditions of employment for many employees. This arises as a consequence of privatization and subcontracting in the UK and more generally throughout the world with the expansion of the informal sector, in which women are overrepresented. Thus as the feminization of employment increases, the nature and meaning of employment has changed in a way that works against greater gender equality.

## CONCLUSION

Capitalist societies are characterized by uneven development over time and space. This chapter has focused on the broader processes of economic and social change which shape regional well being. The second half of the twentieth century experienced an economic boom in the third quartile and relatively slower levels of growth associated with the development of post-Fordism and the new economy in the fourth quartile and beyond.

Emphasis has been placed on north east England, which has been character-ized by relative deprivation in comparison to the rest of the UK. This deprivation arises from the industrial structure, first because of the focus on heavy industries, then as a branch plant economy. With the increasing interconnectedness of the global economy, the fortunes of the region are increasingly shaped by external factors, and specifically changes which have given rise to further losses of employment even in the more recently established high tech industries. State policies designed to promote economic regeneration through physical regeneration have created dramatic landscapes but so far not resulted in sufficient trickle down effects to narrow the economic and welfare gaps with other regions.

Economic restructuring is also accompanied by social change, and this region has been characterized by a decline in traditional male manufacturing jobs and a feminization of employment, in particular with the growth of a branch plant economy and subsequently with the relative expansion of the service sector, especially in care work, the cultural sector and more recently in call centres. While it has been argued elsewhere that the feminization of employment is associated with the end of patriarchy, this chapter has demonstrated that a more detailed examination of the aggregate statistics, which focus on the extent of employment and the levels of pay, suggests that despite tendencies towards parity on the crude employment rate, deep seated inequalities remain. These tendencies are not unique to this region but were also found in the discussion of the feminization of employment on a global scale in Chapter 4. Almost universally, employment continues to be highly segregated both vertically and horizontally and the levels of pay given to jobs where women are over-represented remain comparatively low. Furthermore even when the material basis to traditional conceptions of masculinity has been undermined by employment changes which threaten the male role as sole or even main provider men seem to be capable of inventing new forms of masculinity to preserve their relatively superior social status. Thus the idea that patriarchy is about to end is currently premature.

## Further reading

R. Hudson (2000) *Production, Places and Environment: Changing Perspectives in Economic Geography*, Harlow: Prentice Hall.

J. Tomaney and N. Ward (eds) (2001) *A Region in Transition: North East England at the Millennium*, Aldershot: Ashgate.

## Films

*Billy Elliot* (2000) dir. Stephen Daldry.
*Modern Times* (1936) dir. Charlie Chaplin.
*Stepford Wives* (1975) dir. Bryan Forbes.
*The Full Monty* (1997) dir. Peter Cattaneo.

## Websites

For details of contemporary changes in the north east region see ONE North East (2002). Regional profile: http://www.onenortheast.com.

For statistics relating to the UK visit the official website http://www.statistics.gov.uk/.

For images of cultural regeneration and live web cams see http://www.gateshead.gov.uk/bridge/bridged.htm (last accessed July 2003).

## Notes

1  See Marx (1976) and Keynes (1947).
2  There is a rather similar explanation of the economic restructuring of this period developed by Michael Piore and Charles Sabel (1984) who refer to the transition from mass production to flexible specialization. Their analysis is much less rooted within Marxian political economy and relates the crisis to market exhaustion for mass produced goods and the oil crisis. Moreover it is much more confined to discussing patterns of industrial organization rather than dealing holistically with economic and social change.
3  See for example Aglietta (1979), Dunford and Perrons (1983) and Lipietz (1992).
4  These principles were based on the work of F.W. Taylor who first published his book in 1911 – see Taylor (1967). For an interpretation and secondary accounts see Braverman (1974) and Dunford and Perrons (1983).
5  See Sennett (1998) in relation to the United States.
6  See Daniels and Rycroft (1993) for a more detailed discussion.
7  *Stepford Wives* was directed by Bryan Forbes and released in 1975. See also Friedan (1963) and Silverstone (1996).
8  The narrowest income differentials in the UK were reached in the late 1970s (Atkinson 2002).
9  See Westwood (1984) and Cavendish (1982).
10  See Metcalf (2003).
11  See Baumol (1967), O'Connor (1973) and Chapter 1.

12 See Aglietta (1979) and Palloix (1976).

13 Measured by both household disposable income and earnings, income inequality increased in the UK from 1977 until 1999. On the first of these measures the increase was particularly marked in the 1980s (see Atkinson 2002).

14 See Palmer, Mohibur and Kenway (2002). The proportion of children in low income households has fallen by half a million since 1995/6 but still remains higher than the 1990 level. Moreover some of the decline is due to the fact that while the numbers in workless households (now 15.5 per cent of all children on the basis of the wider national definition – those below 60 per cent of median income) have decreased, those in low income households have increased (30 per cent of all children), reflecting the decline in real earnings of low paid workers.

15 Joint report by the End Child Poverty campaign and the Work Foundation (2003).

16 Data calculated from ONS 2003.

17 The employment rate is the number of people in employment expressed as a percentage of the relevant population; e.g. the working age employment rate is the number of people in employment aged between 16 and 59 for women and 16 and 64 for men as a percentage of the total population in that age group.

18 See Duffield (2002) for more statistics.

19 The economic activity rate is the number of people aged 16 and over who are either in employment or ILO unemployed expressed as a percentage of the relevant population. The ILO unemployment rate refers to people who are without a job and who were available to start work in the two weeks prior to the LFS (labour force survey) and who had either looked for work in the four weeks prior to the survey or who were waiting to start a job they had already obtained expressed as a percentage of the relevant economically active population.

20 The economic inactivity rate is the number of people who are neither in employment nor unemployed on the ILO measure.

21 The comparable figures for 10 years earlier, in 1992, are 38.2 per cent for women and 26 per cent for men, indicating that women's economic inactivity at this age is continuing to fall while the male rate is increasing. Figures are from http://www.statistics.gov.uk/ – the official website for UK statistics.

22 See Barham (2002).

23 See Beatty and Fothergill (2002).

24 For women aged 25–34 72 per cent were inactive because of family reasons compared to 11 per cent of men (Barham 2002).

25 See Atkinson (2002).

26 See Toynbee (2003).

27 See Rubery et al. (2003); Toynbee (2003).

28 See Nolan and Slater (2002).

29 Regional policy was first introduced in 1928 with the Industrial Transference Board, which was a supply side measure designed to encourage people to leave areas of surplus labour. It was strengthened as part of the post-war raft of welfare legislation and again in the 1960s.

30 See Austrin and Beynon (1980), Beynon, Hudson and Sadler (1993) and Massey (1984).

31 See Austrin and Beynon (1980).

32 See Turok (1993), Amin and Tomaney (1995) and Porter (1998a).

33 This led to Doreen Massey's classic article 'In what sense a regional problem?' (see Massey 1978).

34 See Beynon, Hudson and Saddler (1993) and Morris (1995).

35 The lead time is the time taken between placing an order and its delivery.

36 For further details see http://news.bbc.co.uk/hi/english/business/newsid_1135000/1135700.stm.

37 There is a clear air of uncertainty. While I was taking a photograph in 1998, shortly after the closure of Fujitsu had been announced, a passing worker smiled and said to me ironically 'You're not from real estate are you?'.

38 The Fujitsu closure led to 600 immediate job losses and Siemens, closure to 1100.

39 Personal interview with Fujitsu manager and see BBC Online (1998).

40 See Pike (1999) and Pike and Tomaney (1999).

41 See Phelps et al. (2003).

42 For more details see http://www.3k1.co.uk/ngc/general/generalmain.htm (last accessed January 2003).

43 The stadium was completed in 1997.

44 The bridge is spectacular. It 'resembles the opening and closing of a giant eyelid' and uses 'a world-first tilting mechanism to open, turning on pivots on both sides of the river to form a spectacular gateway arch, and to allow ships to pass through (Gateshead Council 2003, Millennium Bridge Factsheet); http://www.gateshead. gov.uk/bridge/facts.htm. The music centre was financed by Sage – the producer of accountancy software.

45 See http://www.gateshead.gov.uk/angel/backgrnd.htm (last accessed July 2003).

46 Bill MacNaught, head of cultural development in Gateshead, speech to RICS 2003 (cited by Beckett 2003).

47 Peck and Tickell (2002: 389).

48 See http://www.gateshead-quays.com/main.html to view the new cultural buildings, including a live web cam (last accessed July 2003).

49 The combined Newcastle–Gateshead bid to become the European City of Culture in 2008 failed. Had it succeeded, it may have increased the number of external visitors.

50 See Harvey (2000) for a discussion of the impact of spectacular projects in Baltimore.

51 Agency workers are another aspect of the new economy discussed further in Chapter 7.

52 See Maguire (2003).

53 This definition has been adapted from HELA (2001).

54 See Glucksmann (2002).

55 For some historical detail on the development of call centres in the United States which predated those in Europe see Belt and Richardson (2001).

56 The average UK call centre manager's salary is 36 per cent higher than average female wages in the UK. Agents earn only 48 per cent of the average male wage or 65 per cent of the average female wage. Own calculations based on IDS (2003) and NES (2002) data.

57 See Belt and Richardson (2001).

58 See Belt, Richardson and Webster (2002).

59 As with retail work, partnered parents often work during unsocial hours, that is one does paid work while the other does unpaid work in a serial fashion throughout the day or week.

60 Other reasons are because call centre work is taken up by students and others not seeking permanent employment.

61 See Beynon, Saddler and Hudson (1993) and Morris (1995).

62 See Beatty and Fothergill (2002).

63 See Wetherall and Edley (1999) for a discussion of different conceptions of masculinity.

64 See Willott and Griffin (1996) and Nelson and Smith (1999).

65 The Full Monty (dir. Peter Cattaneo, 1997) was set in Sheffield; Billy Elliot (dir. Stephen Daldry, 2000) is based in a mining community during the 1984 coal miners' strike.

66 See McDowell (2002) for a discussion of masculinities in the context of economic change in Britain.
67 See also Connell (1995).
68 For example the father suggests they might watch some football and while his son envisages a trip to a major team, the father was thinking of watching local players in the park, as he could not afford anything else.
69 See Pineda (2001).

# Part III

# THE NEW ECONOMY, GLOBALIZATION AND GEOGRAPHY

Some people live in a new economy, many more people are affected by it and others are excluded altogether. But what is the new economy? Is it just another buzzword that came and went with the dot.com boom or does it represent a more fundamental change in the economic and social structure of society? Part III reviews some contrasting understandings and considers whether there are new opportunities, new risks or a mixture of both and how they are distributed between people within and between places, before developing a more composite understanding of the new economy as a new wave of development and assessing its sustainability.

The new economy reflects the outcomes of contemporary restructuring processes but there are narrow, wide-ranging, optimistic, pessimistic and sceptical understandings as well as various combinations of these. For some writers and politicians the new economy is characterized by ICT, knowledge and the 'e' economy. They focus on the economic and social impact of ICT, variously arguing that the new information or knowledge based society brings economic growth, through new ways of organizing business and working patterns, greater democracy through increased access to information and a potential challenge to the capitalist system by facilitating profitless exchanges of dematerialized goods between consumers. Other, more pessimistic, writers place emphasis on the growth of global and social divisions. They point to the uneven development of ICT within and between countries and to the development of more polarized and precarious forms of work arising from the intensification of competition in the new global context. These different dimensions are however organically connected as the ICT revolution, which has brought immense changes in the organization of the economy and opened up new opportunities, also sustains and reinforces existing global divisions because it consists of both highly paid 'self-programmable knowledge workers' and low cost 'generic' labour (Castells 2001). These divisions have been theorized by writers such as Danny Quah

(1996; 2003 – see Chapter 1), and the effects on divisions in the labour market have been illustrated by a number of writers.[1]

Other writers have taken a broader perspective and suggested that the ICT revolution marks the beginning of the fifth Kondratiev wave, following on from past waves based on the technologies of cotton, steam power, steel, engineering and electrification, the internal combustion engine and oil respectively with widespread implications for the organization of economic and social life. There are broad parallels in the timing of the ICT revolution and the eras of post-Fordism identified by the French Regulation School and the second modernity identified by Beck (2000a), although these latter two approaches pay greater attention to the social context and social determinations and implications of change. Cyber sceptics, by contrast to all of the above, doubt the significance of the information technology revolution altogether.

In Part III I try to integrate the technical and social ideas in a synthetic way to provide a more composite framework for understanding contemporary changes and their likely impact on people in different social and geographical positions and to suggest that both globalization and the new economy are profound changes rather than blips, that do indeed lie at the heart of a new wave of development, but rather like the early years of Fordism this wave is not currently sustainable because it is not sufficiently inclusive. I hope to illustrate that until the boundaries of consumption are widened, increasing inequality within countries and between countries will remain, creating problems for social and political sustainability despite the revolutionary technical advances.

To be entirely inclusive the levers of capitalism will undoubtedly require further change. Nevertheless advanced countries have already moved a long way from the competitive capitalism of the nineteenth century leading to unprecedented opulence, in which significant numbers of people share but from which far more on a global scale are excluded. For continued political and social sustainability however the boundaries of inclusion need to be widened. Ideally, leading financiers, industrialists and politicians will recognize that they have much to gain from widening inclusion, similar to the way that post Second World War politicians introduced a range of welfare measures to enable Fordism as a regime of accumulation to take off in the richer countries of the world through a wider sharing of the inherent productivity gains. Even now the extent and rate of growth of inequality are not uniform throughout the world indicating that there are different ways of managing the new economy within capitalism. In richer countries the levels of inequality are lower in Sweden and Germany than in the US or the UK; and in poorer countries lower in Vietnam than in Guinea (see Chapter 2). Thus the new economy can be managed in more inclusive ways. This time however the redistribution would have to be international as well as internal, but again there are precedents, such as Marshall aid to assist post-war reconstruction and to counter the alternative political pathway of communism. This way perhaps capitalism may be changed peacefully and without undermining its dynamic and technologically progressive nature. Currently this optimistic scenario drawing on the positive aspects of capitalism seems rather remote. With the war against Iraq, it seems as if there has been a retreat to a more basic and primitive form of capitalism in a bid to capture and control key resources, irrespective of the human cost. These issues are addressed further in Part IV. Meanwhile in Part III, the uneven development of the technology underpinning the new economy is discussed in Chapter 6 while Chapter 7 focuses more on the processes underlying the social and spatial divisions, especially by considering the development of global cities, global city regions or superstar regions and examining how different groups of people there experience life in the divided new economy.

## Note

1 See for example Beck (2000a), Reich (2001a), Nolan and Slater (2002), Mishel *et al.* (2003).

# 6

# THE NEW ECONOMY AND THE DIGITAL DIVIDE
## Global and social divisions

Optimistic accounts of the new economy focus on technological progress, the increasing use of computing and information technologies, the expansion of knowledge goods, increasing opportunities, productivity and well being. Manuel Castells (2001) argues that we have entered a new technological paradigm centred on microelectronics based information/communication technologies (ICTs) and genetic engineering. The ICT revolution facilitates increasing global integration but is also geographically uneven. In particular Internet access is differentiated by location, social class, gender, ethnicity, age and education, collectively referred to as the digital divide. Existing divisions are often built upon and reinforced by the new communications technologies but even so they still bring new opportunities to disadvantaged groups, which should not be dismissed. In particular it has been suggested that they allow less developed regions to leapfrog stages of development.

Advanced developments in ICT technologies have undoubtedly facilitated increased connectivity between places. Indeed computer power has been doubling every 18–24 months due to rapid developments in microprocessor technology, thereby realizing the prediction made in 1965 by Gordon Moore, the founder of Intel (Moore's Law). At the same time real costs have declined and the physical space required for the same amount of computer power has fallen radically. Similarly, communications power has increased dramatically due to developments in fibre optics, which allow the expansion of bandwidth, thereby simultaneously increasing quality and capability while lowering costs. For example, e-mailing a 40 page document from Chile to Kenya costs less than 10 cents, faxing it about $10 and sending it by courier $50. Likewise, a three minute phone call from New York to London in 1930 would have cost the equivalent of $300 but today costs less than 20 cents. Correspondingly the volume of communications has risen. In 2001, 'more information was sent over a single cable in a second than over the entire Internet in a month in 1997' (UNDP 2001: 30). The increasing ease with which information can be sent between institutions, governments, firms, consumers and citizens and between places has immense implications for the organization of economic, social and political life. In particular it has provided a new means of transmitting and receiving information, led to new ways of organizing production and consumption, allowed new ways of working, and is a vital part of the increased economic, political and social integration between people and places on a global scale.

This chapter emphasizes the role of ICT in shaping uneven development within the new economy, while the effects on people's lives are discussed more fully in Chapter 7. It begins by discussing different ways in which the Internet shapes the organization of the economy through e-commerce and government through e-democracy. The emphasis of the chapter is however on the geographical or digital divide; in particular on the uneven spatial development of the Internet and continuing social divisions. The role of the Internet in shaping economic and social development and challenging some of these divisions is also discussed.

## THE INTERNET, THE ECONOMY AND POLITICS

The different forms of Internet transactions are illustrated in Table 6.1. These include e-commerce, such as transactions between:

- firms or businesses (B2B),
- businesses and consumers (B2C),
- government and businesses (G2B) and
- consumers (C2C);

and e-democracy, such as information flows between:

- government departments (G2G),
- government and consumers or citizens (G2C),
- consumers or citizens and government – providing feedback on government services or voting (C2G) and
- citizens and other citizens (C2C) – which includes any kind of dialogue between people, or between people and less formal organizations such as NGOs, from personal e-mail to music, political ideas or organizing protests.

### E-commerce

The scale of e-commerce is difficult to assess, owing to varying definitions, but is believed to be growing, although many of the optimistic forecasts about the Internet becoming a primary means of exchange, which underlay the dot.com boom at the turn of the millennium, have not yet been realized.[1] Companies specifically set up to run virtually on the web, for example boo.com to supply fashion items, lastminute.com to supply tickets or kitty.com[2] for pet supplies, that is 'e-tailors' that have no real, physical foundations, were

*Table 6.1* Internet transactions

|  | *Government* | *Business* | *Consumer/citizen* |
|---|---|---|---|
| Government | G2G<br>coordination | G2B<br>information/services | G2C<br>information/services |
| Business | B2G<br>procurement | B2B<br>e-commerce | B2C<br>e-commerce |
| Consumer/citizen | C2G<br>tax forms, voting,<br>feedback | C2B<br>web visits/browsing | C2C<br>auction markets,<br>digital exchanges,<br>e.g. music,<br>information,<br>protests |

*Source*: Modified from OECD (2000).

particularly vulnerable to stock market fluctuations, with dramatic increases in share values based on anticipated gains being followed by equally dramatic collapses.

> On 14 April 2000, the Dow Jones Industrial Average saw its biggest ever one day decline. After a period of spectacular growth, the global technology market went into free-fall. By the autumn, American Internet start-ups were going bust at around one per day, with equally high profile failures in the UK.
>
> (Coyle and Quah 2002: 3)

Some of these firms are now beginning to make progress again, such as lastminute.com and amazon.com, which recorded its first profit in the last quarter of 2001,[3] while others have disappeared altogether. However these companies were very much the tip of an iceberg and many more companies with a real physical existence have added web based activities such as web marketing, advertising, sales or booking systems to their existing operations. These are currently known as 'clicks and mortar' firms to express their simultaneous physical and virtual presence, but as Internet use becomes more generalized and just another medium of communication, they are less likely to have any kind of special name. For example, teleshopping from the large supermarkets is growing rapidly in the UK, but continues alongside traditional shopping practices.

The vast majority of e-commerce takes place within and between OECD countries where a relatively high proportion of businesses use the Internet but, so far, only for a small proportion of their transactions. Levels of use are highest in the northern European countries, where over 93 per cent of firms have Internet access and between 65 per cent and 80 per cent their own websites, and lower in southern Europe, where although around 75 per cent of firms in Italy and 70 per cent in Spain have Internet access, only between 7 and 8 per cent of firms have their own websites.[4] In Japan 91 per cent of firms had Internet access as did 86 per cent in Australia. Within each country access varies positively with firm size, with a very high proportion of large firms having high speed access, and also by sector, being particularly high in finance, insurance, business services and wholesale. In the case of business services an increasing proportion of transactions take place through the Internet; in Denmark and Finland 50 per cent of orders were received this way.[5]

The vast proportion of electronic transactions are between businesses (B2B), which accounted for between 75 and 80 per cent of all electronic sales in OECD countries (OECD 2000). They build upon existing electronic systems of communication between firms, such as electronic data interchange, which have been in place for some time, especially in the vehicle and clothing industries. Buyers can in principle save costs and lower inventories by making direct contact with suppliers thereby bypassing intermediaries and vice versa. Similarly small-scale producers can in principle bypass intermediaries by finding information about world markets and prices and even contact consumers directly (B2C), thereby retaining a larger share of the value of their products (see Chapter 3). Open marketplace transactions, in which firms use the Internet to buy and sell goods and services on line with interactive payment facilities are very rare and lie at one end of a spectrum of possible uses of new ICT technologies. This form of B2B is generally only used for very standard supplies such as pens, chairs or packaging. Operating in this way for components and inputs is very risky as issues such as quality and reliability of delivery, which depend on trust established through personal contacts and long-standing arrangements are more significant than cost. Thus the more general pattern is to build the new technologies into existing patterns of trade and by so doing increase efficiency. However, some firms do set up private networks, intra- or extranets with existing contacts and even set up open auctions between their suppliers, but these are restricted to members only, that is after some level of competence and reliability has been established.[6]

Sometimes new electronic intermediaries or e-hubs are established to facilitate exchanges on a sectoral basis, which can either be entirely commercial, as for example Cisco Systems Inc., or have a social purpose. Café Direct plays a similar role to an e-hub by coordinating small producers to supply specialist markets (see Chapter 9). Indeed as Alfred Marshall noted in relation to a different era, 'newspapers, and trade and technical publications of all kinds are perpetually scouting for [them] [the small business person] and bring much of the knowledge [they want] – knowledge which a little while ago would have been beyond the reach of anyone who could not afford to have well-paid agents in many distant places' (Marshall 1961: 285). Clearly the Internet takes this process further as it provides a vast source of information about prices, markets, processes and so on, at least for those with access and with the capacity to interpret the veracity of information found on the web.[7]

Producers can also take advantage of time differences, especially in relation to dematerialized products or 'bitstrings' (Quah 2003, and see Chapter 1) to speed up production.[8] For example, high quality connections allow film makers in Hollywood to send their film for postproduction special effects to companies in London, after a day's shooting, and find their work ready and waiting the following morning (DTI 1998). Medical transcriptions are also carried out in this way to take advantage of labour cost as well as time differences and the Internet also expands the range of potential locations for call centres, data entry operations and business processing outsourcing more generally.

Significant problems remain however for small scale producers in less developed countries where not only is the basic ICT infrastructure lacking but also physical requirements such as transport, which is still necessary to collect and deliver goods. There is also a deficiency in the institutional mechanisms necessary for managing the payments (UNCTAD 2001), although as indicated earlier on line payment systems in B2B commerce are not yet widespread. Thus many problems associated with conventional forms of transactions remain. Furthermore, it is still necessary to ensure that the product meets the buyer's technical and social specifications, such as the conditions under which products are made, which is one reason why supermarkets have moved towards explicitly coordinated supply chains rather than open markets (Humphrey 2002). Thus while e-commerce in principle opens up markets, and there are some illustrations of significant gains, only rarely are entirely new contracts made. Thus use of the Internet does not suspend all of the problems associated with conventional trade, although undoubtedly it is significant in terms of establishing contacts between buyers and suppliers, which can then be followed up in more conventional ways.

Advances in ICT also facilitate decentralized or networked production and the networked enterprise, that is an organizational form which comes into being, dissolves and reforms around specific projects within and between physical enterprises. Manuel Castells (1996), one of the first to identify the importance of the networked firm, stresses the significance of the enhanced communications provided by the Internet in their development. The most well known firm that operates in this way is Cisco Systems Inc., which has established a whole range of different kinds of relationships with other firms of varying degrees of permanency and sells most of its output over the Internet. Castells (2001) describes the structure of this organization in the following way:

> One of the most valuable manufacturing companies in the world manufactures very little by itself, having outsourced over 90% of its production to a network of certified suppliers. But Cisco controls its supply chain closely, integrating key suppliers into its production systems, automating routine data transfer through EDI's, automating the gathering of product data information from suppliers, and decentralising testing procedures at the production point, under

standards controlled tightly by Cisco's engineers. So Cisco is indeed a manufacturer, but based on a virtual global factory, of which it has responsibility in terms of R and D, prototype engineering, quality control, and brand name. Cisco has also integrated its inventory system, with a dynamic information system that has prevented major supply problems in several instances. Furthermore, the Cisco Employee Connection is an intranet that provides instant communication to thousands of employees, across the building or across the globe.

(Castells 2001: 70)

Cisco supplements these virtual connections with employees with more material communications such as providing breakfasts on their birthdays (Freeman and Louçã 2001). However this structure did not prevent the company from suffering during the dot.com collapses and neither did the enlightened human resource management practices save the employees from being made redundant. As Castells (2001: 69) points out, in April 2001 the company laid off 8500 of its 44,000 workers 'although most of them were temporary workers and the others were part of the company's usual 5% attrition rate'. Castells implies that this loss is somehow inconsequential, which rather overlooks the way that temporary workers form a structural component of the new networked enterprise, allowing the firm flexibility but with adverse implications for the employees themselves. The precarious nature of contemporary work is referred to again in Chapters 7 and 8.

Firms can also take advantage of different time and wage cost zones as illustrated above and in Chapter 4. The fragmentation of production always presupposes that it can be recombined at some point and with the Internet it is possible to track the progress of components through their various stages and thereby allow firms to make adjustments at an early stage should any problems occur. Thus, ICT effectively reinforces existing international or global divisions of labour by allowing it to be monitored far more effectively. In the case of fruit for supermarkets for example each box can be electronically tagged and traced throughout its journey from field to shelf, which makes it increasingly necessary for producers to be e-competent and have Internet access if they are to become part of global supply chains.

The global division of labour is not restricted to large firms, however, especially where knowledge goods or 'bitstrings' are concerned. One sole trader (Molly) ran her firm from her front room in Brighton and Hove (UK)[9] and subcontracted work to a programmer in India entirely virtually, through e-mail, the contract having been agreed with the third of a number of offers received. Another firm in the same town grew from four people working in two bedrooms to a global organization employing 50 people *in situ*, and a further 20 in New Delhi as well as having a network of freelancers in a number of other countries, in just under four years (see Box 6.1 and Figure 6.1).

*Box 6.1* Global connections: Brighton and India

Babel Media Ltd is a small firm in Brighton and Hove that specializes in games localization and quality assurance. Computer games is a major industry (£20 billion worldwide) and although most games are now written in English, half of all sales are in foreign languages. Thus Babel translates games into a range of languages including French, Italian, German, Spanish and Chinese and provides a more general games testing service. It has expanded rapidly since starting up in 1999 and survived the dot.com slump. It currently employs around 50 full time staff in Brighton and Hove with an average of 80 on the payroll, and a further 20 people in its New Delhi branch.

Babel reflects many aspects of the new economy in terms of the:

- Product – it is in new media, a new knowledge sector.
- Structure – it is an outsourcing services company, that is, it is a unit in the network or Hollywood model of the firm. Babel does specialist work for major multinational games publishers, for example Atari and Vodafone, who essentially make their products by coordinating output from a whole range of suppliers.
- Employment relations – it maintains a minimum headcount, that is it has a core workforce that is supplemented by drawing on freelancers on a regular basis according to specific skill requirements.
- Ownership – it is an independent company, but has received finance from venture capitalists.
- Global orientation.

Babel is global in three senses of the word:

1. It employs a multinational staff *in situ* and has a network of freelancers across the globe: 'So most of the boys and girls that you see working up here are foreign nationals, Italian, French, German, Spanish, Scandinavian. But we have also set up a specialist games translation network across Europe, Scandinavia, Asia and the Americas. This is crucial because even foreign nationals in the UK become Anglicized quite quickly. So the logic to that is you can't have people sitting here, you can't have bums on seats – they won't get the latest cultural references from the newest TV shows – we've got a satellite dish on the roof now, we can pick up any channel in Europe – but nevertheless, my lot are Anglicized.'
2. It has its own outsourcing branch in India in association with ITIL International and is soon to develop another branch in Canada. ITIL effectively provides the office space, infrastructure and a QA team, trained by Babel personnel. It mirrors the operations in the Brighton and Hove unit, and provides efficiencies in terms of lower labour costs and contributes to faster overall turnaround times as a result of the six hour time difference.
3. ITIL is itself part of a wider company – RAI Group – which has interests in IT, telecommunication, manufacturing and real estate. In addition it funds

continued . . .

> the Rai Foundation, a 13 campus university which can potentially provide Babel with an unrivalled recruitment pipeline.
>
> All of this has come from four people in a two bedroom flat in Brighton and Hove.

Moreover, on the Internet firms have only a virtual presence and therefore people are free to construct themselves as they wish, although this creates risks on both sides as another firm in this same locality pointed out:

> We work from a home so the clients have no idea of the size of the company. The beauty of working virtually is that anything is possible. . . . But there is also a danger in that you do not know who your clients are – it's quite frustrating. People send you an email and they could be anybody. So we do like at least one meeting . . . but we like to make them rather special. We sometimes meet them in a black Mercedes at midnight by the West Pier and my partner films the whole event, to make the day in Brighton special.
> (Internet entrepreneur, interview by author)

In this case the new media firm checked out its client while trying to sustain a somewhat false impression of its own significance. Cyber disguise is also a global phenomenon; in Thailand, for example, a woman entrepreneur whom a potential client had assumed to be male, allowed the delusion to continue until the deal was agreed (Kommolvadhin 2002).

Electronic markets have also been created for labour, in the form of virtual recruitment agencies, and although as with physical goods face to face contact remains important especially for final choices, the Internet plays a growing role at least in terms of short listing. For non-standardized physical products however, e-commerce is more complex and virtual exchanges of information usually have to be supplemented by more physical encounters (see Humphrey 2002 for a detailed discussion of the different forms and degrees of intensity of B2B transactions).

Governments have also drawn upon e-commerce to make savings through increased competition. In Chile there is an e-system, which registers companies seeking to do business with the government. When the government wants to purchase goods or services the specifications are automatically sent to the companies who then have an equal opportunity to respond thereby minimizing corruption and saving government funds. Savings of $200 million a year have been predicted (WDI 2001: 267). But while savings can be made for the government or taxpayer the associated heightening of competition between firms could also lead to a deterioration in working conditions for the workers involved in producing these goods and services.

*Figure 6.1* Babel. (a) Brighton; (b) India.
*Source*: Babel Media Limited.

A growing proportion of consumers have access to the Internet, leading to the development of B2C. Access is highest in Sweden (70 per cent), Finland (65 per cent) and Denmark (55 per cent), but smaller proportions of people actually place orders over the Internet (30 per cent for Sweden and lower for the other countries). In Australia, the United States and the UK about 50 per cent of people have access to the Internet but only 10 per cent or fewer order goods or services, though these figures are changing very rapidly. Thus, so far Internet sales in OECD countries remain low, nowhere accounting for much more than 2 per cent of all retail sales. However, given the scale of the United States market this represents $10 billion (OECD 2002a), nearly double the 2000 figure and, given the considerable gains for both producers and consumers, this form of trade is likely to increase. Survey data for the UK finds that as Internet use in general expands, then people are increasingly likely to use it for making purchases, especially tickets for travel and events, music, CDs and books (National Statistics 2003a).[10] Elsewhere in the OECD the most frequent use is downloading digitized goods such as games, music and free software. In the Nordic countries the Internet is also used to access on line services such as banks. In Finland 64 per cent of the population used the Internet to carry out at least one banking transaction in 2001, as did 50 per cent of Danes and 30 per cent of the population in Portugal, Sweden and the UK, and it is also used to access public authorities (OECD 2002a). Overall if the current rate of expansion continues then e-commerce will account for 5 per cent of business transactions by 2005 (OECD 2000).

In terms of B2C, producers can make savings on showrooms but their overall delivery costs could rise as distribution would switch from high to low density routes, i.e. instead of journeys from warehouses to shopping centres more diverse journeys from factories to residential areas would be made (OECD 2000). Moreover this illustrates how virtual communications and the expansion of the knowledge economy do not dispense with hard physical labour; supermarket teleshopping for example still has to be delivered to the doorstep.[11] This problem would not arise in the case of weightless, aspatial or digitized products or bitstrings which can be distributed very cheaply and incur minimal storage and inventory costs. Producers can also target their marketing more effectively and build up profiles of their clients, allowing a form of mass customization. Amazon.com for example recommends new purchases to clients by comparing their purchase records with other clients. Babycenter.com tailors its information to parents according to the stage of pregnancy or the child's age (Borenstein and Saloner 2001). Thus by using their electronic databases, largely constructed by the consumers as they enter their details, firms can provide individualized and what appear to be personalized services far more efficiently than by traditional face to face contact.

Cases such as babycenter.com above have shades of 'Big Brother', with firms anticipating consumer wants, but people with Internet access can save on search and travel costs, by comparing prices and purchasing directly from home. Potentially they have immediate access to suppliers worldwide, which could stimulate competition, increase efficiency at the macro scale through enhanced economies of scale, as well as enabling wider choice and more customized products. Robert Reich (2001a) gives an interesting personal example:

> I'm all of four feet ten inches tall, with a waistline significantly larger than that of a ten year old boy, which means that if I'm to look even vaguely respectable, anything I wear has to be custom-tailored. It's a royal pain, and often I don't bother. But recently I discovered the Web site of a clothing manufacturer on which I can enter all of my size specifications and select the shirts and trousers (along with fabrics and styles) I want. Within days the garments arrive at my front door. When I first ordered, I expected the tailor who received my improbable measurements to assume they were mistaken, and change them (this had happened before). But the shirt and trousers fit perfectly. And then it hit me: I wasn't dealing with a tailor, I was transacting with a computer that had no independent judgement.
>
> (Reich 2001a: 11–12)

Thus the Internet expands potential choice for consumers and where transport costs are low or not directly related to distance, as in the case of many postal services, allows people in remote regions to obtain more exclusive products. Clearly, this could also have detrimental effects on local and regional suppliers, as reducing the distance constraints on market areas means that more popular or competitive firms can expand into them and capture a larger market share, which is another illustration of the superstar effect (see Chapters 1 and 7) and in the longer term could also reduce choice. In a promotional document designed to highlight business gains from the Internet, the UK's Department of Trade and Industry (DTI) provides an example of how a small father and son firm was transformed into a 24/7 global business as a result of going on line. The company, which makes equipment for creating ice cream parlours, can overcome the inherent seasonal limitations by supplying countries as far apart as New Zealand, India and Tanzania (DTI 1998: para 2.72). Clearly the impact of this expansion on producers in these other countries is of no concern to the DTI.

Thus the worldwide nature of the web allows greater connectivity between more and less developed regions. While this often takes the form of a one way flow of information from the more to the less privileged, in some cases the reverse can also happen and firms can find new markets through the Internet. For example in Gujarat, one of the poorest regions of India that was further devastated by the earthquake in 2000, the Self-Employed Women's Association[12] (SEWA), which had already organized the collection and marketing of gum to reduce the number of intermediaries and ensure that the women received a higher share of the value, received an order through the Internet which tripled the volume of its sales. Clearly this involved additional work but the fact that the direct producers were receiving a larger share of the value meant that the additional sales resulted in a significant increase in income (SEWA 2002).[13] SEWA has also been establishing Technology Information Centres in this district.

The precise effects of e-commerce on economic and social development are contingent, varied and rapidly changing and therefore need to be explored empirically. The same processes or technologies that allow small firms access to world markets, which may enhance local development, simultaneously give peripheral consumers access to firms in the centre thereby facilitating the development of 'superstar' firms, and in turn potentially promote superstar regions, as discussed further in Chapter 7. The Internet is a technology which has been developed within capitalist social relations of production and while it facilitates and indeed induces change in the ways in which commerce is organized, and has tremendous potential to lower the costs of accessing information, some of which are discussed below, it does not suspend these social relations, which as argued earlier promote uneven development.

## E-democracy

The falling costs of communication potentially give people in geographically remote regions access to the most contemporary sources of scientific, political and academic knowledge, and thus the possibility of raising knowledge about a range of issues such as health, including how to tackle HIV/Aids, employment, human rights and so on. Such knowledge is a crucial foundation for expanding human capabilities, democracy and human well being. Amartya Sen (2000) has argued that there has never been a famine in a democratic regime and while the Internet cannot create democracy where authoritarian regimes prevail, it is nevertheless increasingly difficult for them to withhold information. The increased communications offered by ICT have also enabled protest groups to coordinate more effectively, for example the anti-globalization movement or the anti-sweatshop and the fair trade movements which can trace the activities of companies more effectively and provide consumers with more detailed information about the products they are buying. Likewise local protest movements can either directly or by establishing links with similar movements in other localities bring their cause to the global arena, as in the case of the Zapatista movement in Mexico, discussed further in Chapter 9. Similarly, the women's anti-liquor movement in Andhra Pradesh in India was able to project its case to women's civil rights and other NGOs throughout the world.[14] Thus ICT or cyberspace potentially enables people in different local environments to gain support for their struggles and connect with locally based struggles in other villages, neighbourhoods and cities, such that what may in the past have been a rather marginal activity in which women's domestic isolation is reproduced becomes instead an 'anchor for their participation' in struggles with a 'global span' (Sassen 2002a: 379). Thus while contemporary ICTs do not end power inequalities, less powerful groups are able to project their cause to much wider audiences than would have been possible through older technologies.[15]

At the same time it is important not to be deceived by the sophistication of websites and the potential of global communications. Whether a website leads to increased awareness and activism depends also on the receptiveness to the information provided. Just as the dominance of superstar firms depends on people's endorsement of their brands, so too do local struggles. Referring to my own research on Brighton and Hove, there has been a campaign to increase awareness about the health and safety of agency workers, following the death of Simon Jones, a University of Sussex student, who was killed on his first day at work for a temporary employment agency in 1998. Both the agency and the employing firm denied responsibility. Eventually, after a long legal campaign managers at the employing firm, Euromin, were cleared of manslaughter but fined £50,000 for two breaches of the health and safety law. In May 2002 despite the work of the Simon Jones Memorial Campaign and other groups the government postponed consideration of a new crime of corporate killing. The campaign has a sophisticated web presence yet local demonstrations in the town have been relatively small with at most 100 protestors.[16]

Some local and national governments are investigating the possibilities of making greater use of the Internet to increase citizen participation in political affairs, including voting, but these developments are in their early stages and so far results have been mixed (see for example Finquelievich (2001) on the development of e-democracy in Buenos Aires and Montevideo). One concern is that given existing gender divisions in Internet use e-democracy will become he-democracy, though again as the range of activities available through the Internet widens, gender differences are narrowing.

One interesting application of C2C and C2G was developed in the UK in relation to domestic violence as part of a wider programme to examine the effectiveness of on line communications. The government wanted to see whether people would be willing to become involved in e-discussions, whether there would be genuine exchanges of information, whether such discussions would be representative of the population, and indeed whether it would be possible to develop a system that would be secure and usable by people with little experience of computer or Internet use, and finally to see if the discussions would be used by politicians to effect policy outcomes. The Womenspeak project recognized that women are underrepresented as Internet users in the UK, so efforts were made to provide information about access to and training for using the Internet, together with the necessary security to ensure that their abusers would not be able to trace their messages. Over 1000 messages were posted during the month long operation of the scheme which not only provided MPs with information about the direct experiences of how people survived domestic violence but also how they thought judges and the police might be better informed about the nature of domestic violence. Subsequent to the month long discussion some of the contributors set up their own e-mail list so that they could continue to exchange views and provide support for one another after the project ended (Bossy and Coleman 2000). Many people contributing to the project were from women's refuges who emphasized how valuable these had been, but also how the Internet could provide a way of maintaining support after they returned to the wider community. Overall, electronic forms of communication were considered of great value for exchanging ideas but as additional resources, not as substitutes for personal contacts or for the physical space of the refuge.

E-democracy is enhanced by the decentred architecture of the Internet; this derives from its origins, which lay in the perverse combination of the Pentagon's wish for a communications system capable of withstanding any kind of disruption and the early code writers' libertarian desire to make their knowledge freely available in the public domain. Thus in principle people anywhere in the world with a basic PC, mobile phone or other handheld device can communicate with each other. In India, academics from the Indian Institute of Science and engineers from Encore Software in Bangalore designed a handheld Internet appliance (the Simputer based on the Linux operating system rather than Microsoft) for less than $200, which will provide e-mail and Internet access in local languages and touch screen applications for illiterate users. A non-profit trust has been set up to license the technology to manufacturers – in order to provide the product cheaply (UNDP 2001). However, $200 is a large sum of money when annual incomes can be as low as $1 a day and literacy skills are necessary for all but basic applications. Other ways of extending use are through more collectivized facilities like cybercafes or Technology Information Centres such as those being developed by SEWA in the remote villages of Gujarat, in which costs can be shared and skills transferred. People without direct access to the Internet can also use mobile phones to make contact via people who do have access.

Increased access together with the dramatic fall in real costs of transmitting information throughout the world create the necessary conditions for increased democracy, through facilitating a two way exchange of information and ideas. Clearly, at the same time increased integration also provides the potential for domination by media magnates and mass culture. Which route comes to dominate depends in part upon whether attention is given to developing ways of enabling people to project their own information on the Internet and develop skills for interpreting the information they find there, rather than simply increasing access to the Internet on which there is a tendency for both commercial organizations and official institutions to broadcast their own messages (see Mansell 2002).

Access to the Internet is generally provided through commercial Internet service providers (ISPs), many of which are multinational such as AOL or Freeserve and the market is becoming more concentrated. At present there is still some competition but in the desire to provide a high quality service to their users the ISPs seek control over other networks to allow an even quality throughout the web, generating an uneven quality of service between those included within and those excluded from the higher quality services, generally developed between the major cities.[17] Moreover, links are being established between the larger ISPs and existing media magnates such as TimeWarner,

Murdoch and Berlusconi.[18] Concentration among ISPs is important because these provide access to the Internet and connections between ISPs and traditional media monopolies potentially results in a small number of organizations having control over access to, as well as material available on, the Internet (Freeman and Louçã 2001). Likewise in software Microsoft has become very dominant, partly because it has developed very easy to use software, but also becuase it has tried to lock people into its products by linking their applications. Microsoft tried to link its widely used Windows operating system with its own Internet browser, Internet Explorer, which makes it very difficult for competitors to survive. These software systems are generally more expensive than the Linux system referred to above, and certainly prohibitive for many people in poorer countries, but the standardized user friendly interface also widens access by enabling people with comparatively basic computer literacy skills to use a wide range of applications.

In addition to facilitating the flow of information between people, the economic properties of knowledge goods, weightlessness and non-rivalness (which as indicated in Chapter 1 means that they can be consumed more than once or simultaneously by different users), together with the user friendly file exchange systems distributed freely over the Internet, have enabled people to exchange or more accurately give goods directly to each other.[19] The downloading of music files has been particularly popular especially among young people and piracy of DVD movies has been increasing. Large media firms have tried to prohibit these developments through legal action but so far, as one system has been challenged another has emerged. However, while growing, especially as more people have broadband rather than dial up access, the scale of these developments remains relatively small and the increased exposure through the Internet may also lead to increases in physical sales, thus so far these systems are probably 'merely irritants' (Leyshon 2003) rather than major threats to corporate capitalism.

The resources necessary to integrate with the new global economy via the Internet are comparatively modest and costs of equipment and access are continuously declining, allowing an increasing dispersion of ICT at the same time as facilitating an increased level of connectivity on a global scale. Even so the Internet remains inaccessible to the vast majority of people in poorer countries and to poorer minorities in richer countries, leading to a highly uneven pattern of use.[20] These global and social divisions are discussed in the next two sections.

## GLOBAL DIVIDE: GEOGRAPHY AND THE INTERNET

Underlying technological developments in ICT, which have stimulated new ways of organizing economic, social and cultural lives, are tremendous advances in human ingenuity and skill, rather than physical resources or financial capital.

Consequently it is potentially more inclusive than past trajectories of development and may enable poorer countries to leapfrog stages of development. Kim Dae-jung, President of South Korea, encapsulates this optimistic scenario in the following comment:

> During the 20th century such tangible elements as capital, labour and natural resources were the driving force behind economic development. But in the new century such intangible elements such as information and creativity will give nations a competitive edge. Consequently, if we succeed in developing the potential of our citizens by fostering a creative spirit of adventure, individuals and nations will become rich, even if they are without much capital, labour or national resources.
>
> (UNDP 2001: 24)

Thus, creating value depends less on physical resources and more on human capabilities, although the physical and social infrastructure required to bring these capabilities to fruition should not be underestimated as the Internet is a general purpose technology that requires a range of complementary investments. Indeed, despite this optimism and the dramatic expansion of the Internet, so far worldwide access is very uneven.

There has been an exponential increase in Internet use from fewer than 20 million users in 1995 to over 605 million in late 2003, that is 1 in 10 of the world's population has accessed the Internet. By 2005 it is predicted that there will be 1 billion users or approximately 1 in 6 of the world's population. Figure 6.2 illustrates this exponential growth and shows that although the majority of users are in the USA, their proportion of the total has been declining as the total has increased.[21]

This expansion has occurred as real costs have fallen and the Internet has become increasingly accessible to the general user with the development of user friendly browsers and plug in and play hard and software. As more people use the Internet, there has been a corresponding increase in the range of applications, which has expanded utilization in a cumulative way. Thus e-mail has displaced the telephone for many routine business communications, e-commerce is expanding, web addresses are routinely given in TV and radio programmes. Correspondingly, users vary from silver surfers planning holidays, mobile mechanics receiving work schedules and entering invoices, children and teenagers chatting, downloading music and playing games, students downloading data, small-scale producers looking for markets, through to the stereotypical computer boffins exchanging computer code.

Access is however highly uneven. Overall, in 2003 62 per cent of Internet users were located in Europe, Canada and the USA, with only 16 per cent of the world's population, yet only 1 per cent lived in Africa with 13 per cent of the world's population. Figure 6.3 shows the proportion of households with Internet access in 2001 between different regions of the world and Figure 6.4

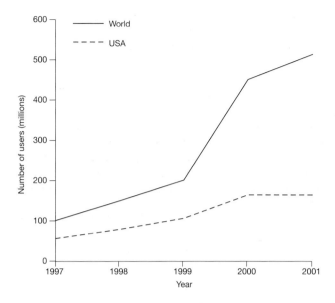

*Figure 6.2* Growing worldwide use of the Internet.

*Source*: Data from NUA (2002).

is a more detailed breakdown by country, both indicating the uneven pattern. Thus far from the 'borderless world' (Ohmae 1990) and the 'end of geography' given the economic advantages arising from ICT, the currently uneven access suggests that existing inequalities will be intensified within as well as between countries, socially and spatially, and while stages of development may be leapfrogged, uneven development will remain.

Despite these stark contrasts there has been a rapid increase in Internet access in some of the poorer countries, especially in Brazil, China and Malaysia, as cyber centres or corridors are being developed. These centres bridge the digital divide between richer and poorer countries but simultaneously create even starker spatial and social divisions and contrasts internally as people and places are selectively connected or bypassed.[22] The development of a new multimedia super corridor in Malaysia indicates how rapidly poorer countries can move directly to the technological frontier. This multimedia super corridor (MSC) project[23] stretches from Kuala Lumpur city centre for a distance of about 50 km to a new administrative capital at Putrajaya in the south, at the heart of which is Cyberjaya, a new city which not only forms a site for IT companies but also uses information technologies to manage the functioning of the city – in the form of an automated traffic management system and house surveillance.[24] The need for house surveillance suggests however that while the global divide may be narrowing between superstar regions across the globe, social divisions remain. Penang, in the north of Malaysia where there is a similar IT focus with companies such as Intel, Cisco and Seagate, is known as Silicon Island and is

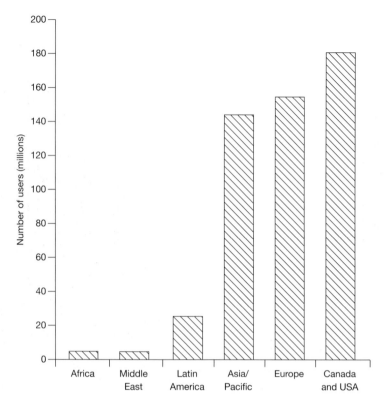

*Figure 6.3* Internet use: world regions.
*Source*: Data from NUA (2002).

the second richest region of the country. These developments have been based on the past involvement of Malaysia as a manufacturer of electronics (see Chapter 4) and the state has played and continues to play a significant role in these developments. Over $7 billion of state expenditure has been invested in the MSC and the state has also consistently supported the 'I-land' of Penang as reflected in the comments from the Chief Minister:

> The state has to initiate further accessibility, improve connectivity and deploy broadband infrastructure as well as to position itself as an 'e-manufacturing hub'. . . . Penang hopes to achieve a new status as the I-land (intelligent land) with a vibrant and viable K-society and K-economy through the promotion of Information and Communication Technology.
>
> (Yan Sri Dr Koh Tsu Koon, Chief Minister of Penang in PDN 2001)

While Intel has been upgrading and including design functions and Cisco has been involved in setting up Cisco Networking Academies to train knowledge

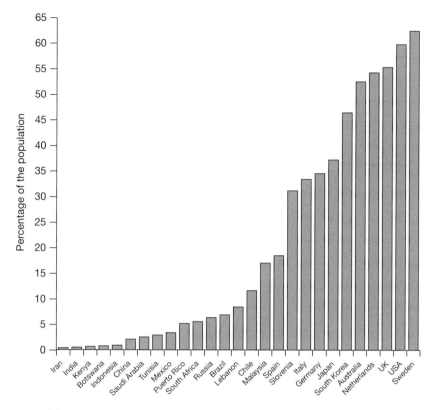

*Figure 6.4* Internet access: percentage of households in selected countries.
*Source*: Data from NUA (2002).

workers especially in networking technologies, Seagate, a disk drive producer, has announced redundancies which reflects the continuing decline in employment associated with the increased capacity and falling price of memory chips, discussed earlier in Chapter 5. This example illustrates both the gains from ICT but also the vulnerability of firms and state development strategies that rest upon the rapidly changing high tech sector. Clearly education and training facilities are necessary so that the population can move forward with the development of new technologies, although as illustrated in Chapter 4, employees' responses to working for these companies are varied.

Even in Africa, the least connected continent, every capital city now has the Internet and cybercafes are widely dispersed, appearing in smaller cities, towns and even villages. These collective forms of provision greatly extend accessibility; nevertheless, there is an extremely long way to go. Even allowing for each Internet connection to be used by three to five people on average, this would still mean only one connection per 200 people compared to one in three in North America or Europe. Within Africa there is a wide disparity with the vast majority of users being at either end of the continent, especially in South

Africa (0.75 million out of a total of 1.2 million), with lower levels of access in between (Jensen 2001).

One of the main constraints on Internet use, which corresponds with the uneven development between richer and poorer countries, is the lack of basic communications infrastructure and even where connections are established they can be extremely costly. Mobile phones can to some extent bypass older forms of communication and connect with the new. Again in Africa, where landlines are sparser than elsewhere, the number of mobile phones has rapidly overtaken the number of fixed lines (Coyle and Quah 2002) but can still be costly. In Uganda the remote Kisiizi hospital has a satellite connection but the cost of a one minute phone call to Kampala is $2.50, equivalent to five days' work (WDI 2001: 265). Similarly in Thailand, Internet charges, especially for high speed connections, are considerably higher than for other countries in South East Asia, almost twice as high as South Korea, three times as high as Malaysia and four times as high as Japan. These charges may explain the comparatively lower numbers of Internet hosts in Thailand, than would be expected from the comparative level of GDP. The higher charges are attributed to the existence of a state monopoly over access to international lines and the way that the state appropriates a one-third equity stake in private ISPs (Tangkitvanich 2001). Even so use of the Internet is becoming widespread and reaching remoter areas as indicated in Figure 6.5. However, even in countries where deregulation has taken place there is growing evidence of concentration among private companies; thus 'the old publicly owned state monopolies have gone, only to be replaced by new giant global multinational corporations' (Freeman and Louçã 2001: 330). Given the centrality of information to contemporary society, if it is the large corporations rather than nation states that control the basic infrastructure, these developments represent a serious challenge to democracy and another dimension of the weakening powers of nation states discussed further in Chapter 8.

## Internet and development

The global divide in Internet access is linked to the levels of GDP. Figure 6.6 plots the relationship between per capita GDP and percentage of households on line for all countries where at least 0.5 per cent of households have Internet access. Regression analysis shows that the relationship is positive with 78 per cent of the variation in Internet access attributable to GDP. This relationship however is non-linear, suggesting that a certain level of GDP is necessary before Internet access exists, but after this level, Internet access increases at a faster rate than increases in GDP, before tailing off as some kind of ceiling is reached. However the model does not explain all of the variation in the data as indicated by the difference between the observed and predicted values on the regression line. More specifically, there are countries with higher and lower levels of Internet access than would be predicted by their level of development as

*Figure 6.5* Internet cafés everywhere.

*Source*: Photograph by Alexa Koller.

measured by GDP per capita. Sweden, for example, has a much higher than expected level of Internet access (63 per cent of households) for its GDP per capita figure ($27,000) as do Estonia, Slovenia and South Korea. These countries are also ranked higher on the HDI than on GDP per capita, which suggests that

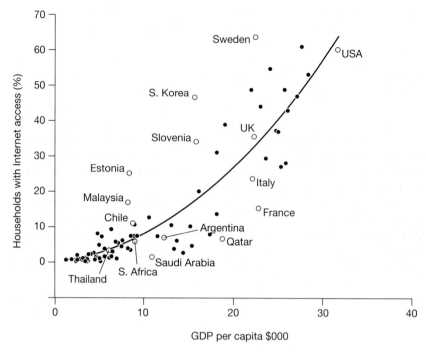

*Figure 6.6* Internet access and GDP per capita.

*Source*: Data from NUA (2002) and UNDP (2001).

*Note*: The regression equation is IA = 0.494 + 0.21 GDP + 0.06 GDP$^2$, R$^2$ = 0.78 (adjusted R$^2$ = 0.77) and the regression coefficients were significant at the 1% level. The regression line illustrates that for each $1000 increase in GDP per capita a further 2% of households are likely to have Internet access. As the level of GDP per capita rises further there is a more than proportionate increase in Internet access, thus the relationship is quadratic rather than linear. The deviations from the regression line indicate exceptions to this general trend and need to be explained by other factors.

their resources may have been directed more towards expanding human capabilities such as education or that income is more evenly distributed. Alternatively the deviations might be explained by more specific factors such as lower Internet connection charges, greater familiarity with English (currently the dominant language of the web) or specific state policies.

Other countries have lower levels of Internet access than would be predicted from their GDP per capita, for example Japan (37 per cent access with a GDP per capita of $25,000, France (33 per cent access with a GDP per capita of $22,000) and Qatar (only 6.2 per cent Internet access with a GDP per capita of $18,000). Possible explanations include unfamiliarity with or antipathy towards American/English, restrictions by authoritarian governments, high levels of income inequality, or the high cost of telephone services. In Japan the strong tradition of centralized power and low use of credit cards is thought to be important (Rimmer and Morris-Suzuki 1999; Zook 2001). Even so economic growth, measured by GDP per capita, was also found to be by far the most significant factor in accounting for differential Internet use between countries in a more detailed analysis (Norris 2001).[25]

While it is clear that levels of economic growth can explain a large amount of the differential access this does not mean that providing Internet access would necessarily lead to increased development. Internet access could just as well be a reflection of higher levels of development and not a cause of that development, even though certain productivity advantages can be broadly associated with the development of ICT (Coyle and Quah 2002). This then raises the question about whether poorer countries should use their scarce resources to boost ICT infrastructure or whether other programmes in relation to basic education, health and welfare might be more appropriate.

To investigate whether there are specific barriers to use of the Internet or whether the low use reflects low levels of human development Pippa Norris (2001) explores the relationship between access to old (TVs, newspapers, etc.) and new media in different countries and finds a strong correlation. That is, countries that have a low on line population also have a low level of access to newspapers, TV and land based phones and vice versa which suggests that low Internet use is as likely to be a consequence of low income and education as a lack of technology. Norris (2001) concludes therefore that these issues would need to be addressed directly before much use could be made of the Internet, that is while more computers and greater connectivity would probably bring positive benefits a deeper social restructuring is necessary to overcome the global digital divide. Interestingly Bill Gates shares this view and he rather undermined the rationale of a conference of high tech companies, venture capitalists and governments on 'Creating Digital Dividends' by arguing strongly that basic literacy and health care rather than IT represented the best form of assistance to the world's poorest people (Richman 2000).[26] Similarly, the emphasis on the development of costly ICT infrastructure, in particular of secure servers, has been criticized for failing to recognize that while a basic level of ICT access enables firms to participate in global supply chains, most transactions are established and paid for in more traditional ways (Humphrey et al. 2003). Norris also suggests that the solution to the digital divide does not rest with technology alone:

> The explanation for the digital divide is often assumed to lie in certain characteristics of this technology, such as the need for computing skills and affordable connections. The policy solutions designed to ameliorate the digital divide commonly focus on just these sorts of fixes, such as wiring classrooms, training teachers, and providing community access in poorer neighbourhoods. Certainly this can do no harm. But will these initiatives work in terms of diversifying the online population? The results of this analysis suggest that, unfortunately, it seems unlikely. The policy fixes are too specific, the problem of social inequalities too endemic.
>
> (Norris 2001: 91)

Clearly, basic well being is vital, but development strategies need not be counter-posed in this way. Some 'stages of growth' could be skipped, if the potentialities of the Internet were utilized in literacy, education and health programmes, especially if cheaper access technologies became more widely available. Further, Internet access can widen markets and incomes as the case of SEWA acting with the gum collectors in Gujarat demonstrated. Furthermore, the question of the form of access also needs to be considered. Robin Mansell points out how new media literacies are also necessary in order to comprehend the veracity of information available on the web given its varied provenance, and that it is important to consider whether the technical configurations of new software and services are consistent with expanding citizens' capabilities or empowerment rather than simply enabling them to better receive the messages of the already powerful. More specifically, she argues that while there are some organizations that provide toolkits to assist people with a minimum level of technical expertise to develop their own websites, these tend to be underfunded in contrast to those that simply provide one way access and 'push information at viewers' (Mansell 2002: 412). Thus it is important to go beyond the question of access alone to consider the nature and forms of access as well as training that are being developed. In addition the localization of software into a wide range of languages is necessary so that use of the Internet can extend beyond the elite.[27]

One crucial aspect of the technology is to speed up the flow of information, especially where other transportation and communication media are lacking. The use of satellite communications can transform rural as well as urban economies; messages that might have taken a whole day to convey on foot can now be delivered almost instantaneously. In this respect there have been some interesting applications in India, where efforts have been made to adapt the technologies to make them usable by people in remote rural areas with little formal education, in some cases by using local languages and multimedia interfaces. SEWA has utilized this technology to enable community groups and organizations to share knowledge between centres and allow local people to communicate with district and state levels (Patel 2003). Elsewhere, in Pondicherry, South India, a series of knowledge centres have been established, in which local people gather information and put it on the intranet set up between them, rather than simply receiving global information. In Veerampattian, a fishing village, the system is used to provide information about weather and sea conditions as well as fish marketing. Additionally software has been developed to manage the micro-credit scheme. Another centre at Kizhur provides agricultural information and water management but as a consequence of using the Internet people have acquired the necessary skills and information to package and market their incense sticks to distant locations themselves and thus upgrade from being simple subcontractors. Thus once knowledge centres have been established the range of applications seems to proliferate. Women have been central in many of these centres and the effect has been to increase their status within the community (Rajasekarapandy 2003). Voluntary labour is central to many of these schemes and so as Mansell (2002) argues, they are underfunded compared to larger-scale commercial projects. To be widely effective however such programmes would still need to be combined with deeper social and economic restructuring on a global scale.

## Social and spatial divisions

The global digital divide between nations coexists with complex divisions within countries by location, social class, gender and ethnicity. The production and supply of Internet content is concentrated in a limited number of urban areas or 'on line spaces' with many 'off line places' in between (Mansell 2001). That is, a new urban dualism is emerging from the opposition between: 'the space of flows that links places at a distance on the basis of their market value, their social selection, and their infrastructural superiority' and 'the space of places that isolates people in their neighbourhoods as a result of their low incomes and lack of connections' (Castells 2001: 241).

Somewhat paradoxically it is the power of ICT, which in principle increases connectivity between places, that also strengthens existing urban hierarchies, in a spatial parallel to the superstar effect that widens income division between producers of knowledge goods – such as the opera singer discussed by Quah (1996) (see Chapter 1) – as increased connectivity also allows the most well known regions to capture a larger share of the market. As Castells argues:

> it is precisely because of the existence of telecommunications networks and computer networks, that these milieux of innovation . . . can exist in a few nodes . . . reaching out to the whole world from a few blocks in Manhattan, in San Francisco, in the City of London, in Paris' Quartier de L'Opera, in Tokyo's Shibuya, or in São Paulo's Nova Faria Lima. While concentrating much of the production and consumption capacity of a vast hinterland, these territorial complexes of knowledge generation and information processing, link up with each other, ushering in a new global geography, made up of nodes and networks.
>
> (Castells 2001: 228–9)

Matthew Zook (2001) has demonstrated the urban concentration of Internet content producers by mapping domain names, that is companies, organizations, network providers, government or education institutions that have registered their web presence (see Table 6.2). Most of the top 25 cities were in the United States, with the exception of London (fourth) and Toronto (twenty-first). He also found that when measuring the ratio of domain names to population, some of the smaller cities had a distinct web presence especially in Silicon Valley with San Francisco having the highest number of domain names per 1000 people (43) followed by San Jose (32). Zurich (26.8), Oslo (16.8) and New York (16.8) had higher levels than London (8.6). In Germany the distribution of domain names between cities was also more even, reflecting the more decentralized German urban and regional structure. Figures outside Europe and North America were much lower, for example Tokyo (1.3), Seoul (1.6) and Hong Kong (3.6). In the US as a whole Zook (2001) finds that 86 per cent of Internet delivery capacity was concentrated in the affluent suburbs and business centres of the 20 largest cities, with the Internet content production industry, measured by location quotients,[28] geographically concentrated in San Francisco, New York and Los Angeles. Thus far from the end of geography uneven development between and within countries remains and may be reinforced by the uneven development of ICT.[29] With Internet use becoming more widespread, however, an increasing number of firms and organizations will have web addresses, so mapping the distribution of the Internet in this way will lose its value. In 2001, for example, 50 per cent of UK businesses had their own website (National Statistics 2003a) and as discussed earlier higher figures exist in northern Europe. Nevertheless, so far it would seem that the Internet tends to reinforce the dominance of already established centres provided that they keep up to date with contemporary technologies and similarly emerging centres have to be technologically modern. The regional landscapes associated with the new economy are discussed further in Chapter 7.

In addition to these geographical divisions between countries there are also important social divisions within countries, with wealthier, more highly educated males belonging to dominant ethnic groups having higher levels of Internet access than people with opposite characteristics. In Panama for example

*Table 6.2* Urban concentration of Internet domain names

| City rank (by no. of domains) | World's population (%) | World domain names (%) |
|---|---|---|
| Top 5 | 1.1 | 17.5 |
| Top 10 | 1.6 | 23.9 |
| Top 50 | 4.7 | 43.0 |
| Top 100 | 6.7 | 51.4 |
| Top 500 | 12.9 | 63.7 |

*Source*: Zook (2001).

*Note*: Zook bases his figures on a combination of data sets for the registration of CONE and CC domain names (CONE representing .com, .org, .net and .edu, i.e. businesses, non-profit organizations, computer networks or education institutions respectively and CC country codes) collated in January 1999. Zook (2001) argues that there is a geographical correlation between the registration address of a domain name and the physical presence of the owner organization and that the issue of scale is in part overcome by the fact that large organizations generally register many variations of their domain name partly to reflect the variety of activities they offer and also to protect their Internet brand. However, as Internet use becomes more generalized this problem may become more serious. His data relate to 1999 and given the way in which use of the web is becoming more generalized and companies develop a web presence as a matter of routine this measure may simply begin to reflect the distribution of organizations in general.

the richest quintile are over 40 times as likely to have access to a telephone as the poorest quintile. The most extensive analysis of social divisions however has been carried out for the US by the Department of Commerce (US 2002) but parallel divisions were found for the UK and more generally throughout OECD countries (see Figures 6.7 and 6.8).[30] Figure 6.7 shows that the difference in Internet access for the highest and lowest income quartiles in a selection of OECD countries is highest for the United Kingdom and the United States and is lower for Denmark where the overall average rate of access is higher. Figure 6.8 illustrates differential access to the Internet in the US in terms of the characteristics of the householder and indicates that although Internet access is increasing quite rapidly for all groups there are wide differences by income, education and ethnicity.

The measures in Figure 6.8 give no indication of the extent or purpose of Internet use. They are also based on the characteristics of the 'householder',

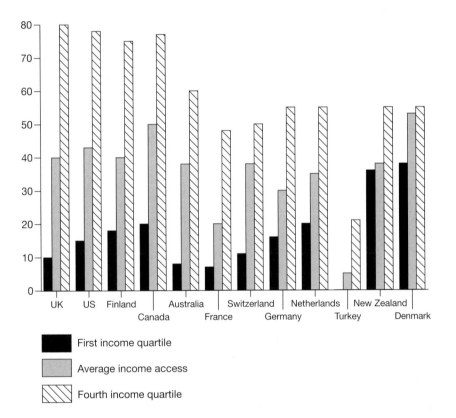

*Figure 6.7* Internet access and income group: OECD countries.

*Source*: Data taken from OECD (2002a).

*Note*: The data for the UK are the first and last deciles, and for Germany and New Zealand the first and last income brackets.

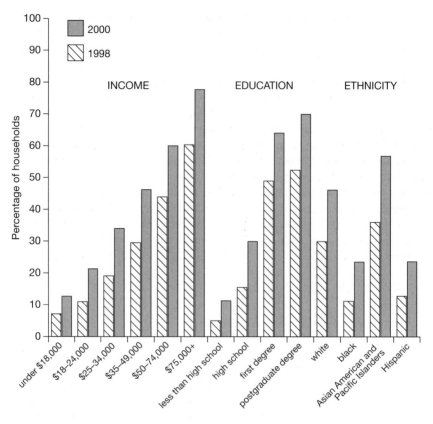

*Figure 6.8* Internet access in the US by income, education and ethnicity.
*Source*: Data from US (2002).

which may be one reason why gender differences are not very marked; alternatively the gap between women and men may be narrowing as the range of applications increases. Income is linked to education and ethnicity, and is the most decisive factor. Households with incomes above $75,000 (approx 83,000 Euros or £52,000) were 20 times more likely to have Internet access than those on the lowest incomes (see also Norris 2001:10). Similarly in the UK use of the Internet is greater among higher income, more educated and younger groups, and among men compared to women (National Statistics 2003b). More specifically in 2002, 45 per cent of all households had access to the Internet at home (five times the figure four years earlier) and 62 per cent of all adults had accessed the Internet at some time, with 95 per cent of people aged between 16 and 24 compared to only 15 per cent of those over 65 years having done so. The figure for men (66 per cent) was higher than for women (58 per cent) and men were also more likely to access the Internet from home and more frequently than women. The most striking factor is the speed of growth over the last four years. For example in Portugal, Internet access increased by 125 per cent between

2000 and 2001 and for the UK the figure was 110 per cent (OECD 2002a). Moreover as use increases, so does the range of purposes for which it is used, and the method of access becomes more sophisticated. While most households currently use dial up access, high speed access such as broadband which dramatically increases access quality is growing. These relations between the scale of use and access mode reinforce one another in a cumulative way.

On a wider scale gender differences remain. The new ICT technologies have often been heralded as a way of increasing opportunities for women because they do not require physical strength, can be operated from home and enable people to work flexible hours. In the UK, the Women's Unit argued that ICT represents 'one of the biggest opportunities for women in the 21st century to earn more, have more flexible working practices and adapt their current business or try a business start-up' and further that 'self-employment and enterprise offer women a real alternative means of earning good income and achieving greater flexibility in their working lives' (Women's Unit 2000). These sentiments were endorsed by Molly, the sole trader referred to earlier:

> 'IT's WONDERFUL!' [her emphasis] 'As I own and run my own business in the home, my work/life balance could not really be improved. I have the flexibility I need which is why I set up the business in the first place.'
> (Molly, aged between 35–44, with two dependent children)

The studies referred to above from India suggested that the engagement with new technologies increased women's status. Technologies develop within particular sets of social relations and while they do provide new opportunities at a variety of levels, they also provide a way of combining paid work with caring responsibilities generating more marginal changes in gender relations. From the new media study in Brighton and Hove there were an equal number of comments about the oppressive nature of working at home, indicating how it is possible to be communicating in cyberspace while simultaneously remaining trapped in patriarchal domestic relations. These developments then do provide a way of entering into waged employment and becoming a political actor on the global stage but are by no means a panacea for unequal gender relations.

The UK government is also concerned about the digital divide and has developed a strategy to expand the use of computers and access to the Internet in schools (SEU 2000). Of the 38 per cent of people who had never used the Internet in 2002, 42 per cent said they were not interested but 36 per cent said that either they could not afford to use or had no access to the Internet (National Statistics 2003b). Similarly to the issue of the role of the Internet in economic development, access to the Internet will not be a panacea for deep-seated economic and social deprivation or for the existence of social divisions in a capitalist economy. Whether it will assist in overcoming such divisions will depend on how it is used, that is, it is very much an enabling technology and so the outcomes will depend on factors relating to the regulatory environment, issues such as cost, mode of access, and on training in its use. Given however the efficiency and capabilities of contemporary technologies, familiarity with this key technology is likely to be a necessary although by no means a sufficient condition for economic and social well being.

## CONCLUSION

There are optimistic and pessimistic interpretations of the nature and impact of the new economy and cyber sceptics who see no real change. In this book I have used the new economy as a descriptive term to identify the specificity of the contemporary era, in contrast to other definitions such as post-Fordism or late modernity. The contemporary global economy is profoundly shaped by new ICTs that have allowed increasing integration between countries, but as these new technologies have developed within capitalist social relations of production and unequal gender relations, they have built upon and reinforced existing spatial and social divisions. Thus the new economy has enormous potential to increase human welfare, but the realization will depend upon broader social and political change.

This chapter has focused on the ICT architecture underlying the new economy and its potential uses in e-commerce and e-democracy. Contemporary ICTs allow unprecedented speed in the transmission of information and generate tremendous productivity increases in some spheres. As a general purpose technology, which diffuses slowly and cumulatively, productivity increases are difficult to measure,[31] leading to the paradox of computers being everywhere except in the productivity statistics (Solow 1987). Moreover, as in the case of teleshopping from supermarkets ICT is associated with increases in labour intensive physical work as well as knowledge based employment so the impact on productivity is contradictory. This duality also accounts in part for the widening social divisions, as these different activities have opposite economic properties and spatial patterns, discussed further in Chapter 7. In this chapter the social divisions have been described in terms of the unequal access to the new technologies, by income, gender and ethnicity.

The unequal spatial distribution of Internet access has also been described. Internet use is far more advanced within the richer countries and by richer people within the richer countries, but infrastructural developments and Internet access are necessary rather than sufficient conditions for bridging the digital divide in terms of enhancing development or tackling social exclusion. It is possible to acquire knowledge of markets and market prices much more quickly, potentially enabling people anywhere in the world to engage with world markets. However, this access works in both directions, so as well as giving access to world markets for people in remote regions it also exposes their markets to global products, thus outcomes are contingent and depend on factors other than technology. Attention also needs to be given to expanding capabilities for using and interpreting web information and ensuring that it becomes a two way channel of communication, in which the disadvantaged can project their ideas and products rather than a one way medium through which the already powerful can broadcast their ideas and products more effectively. While still requiring considerable investment the current architecture of the Internet allows open access, and illustrations have been given of people in remote rural regions using the Internet as a marketing device and as a means of organizing politically. However there is a danger that the larger Internet service providers, software producers and media companies will cease control and restrict entry. Furthermore it is possible to be a citizen in cyberspace while still being oppressed within the home. New technologies may assist social change but it is by no means guaranteed. Thus while the technology has progressive potential and in principle could transcend the social relations of its creation, in part because the early code makers contributed to the decentralized architecture of the Internet,[32] this full progressive capacity remains constrained by the existing uneven social relations and so at present while enormous changes are taking place in the lives of people in different places which are also becoming increasingly entwined on a global scale as a consequence of the new technologies, social and spatial divisions are still widening. These social divisions and their impact on people's lives and geographical landscapes are discussed further, mainly by reference to examples from the richer countries, in the next chapter.

## Further reading

M. Castells (2001) *The Internet Galaxy: Reflections on the Internet Business and Society*, Oxford: Oxford University Press.
M. Dodge and R. Kitchen (2000) *Mapping Cyberspace*, Harlow: Addison-Wesley.
P. Norris (2001) *Digital Divide: Civic Engagement, Information Poverty and the Internet Worldwide*, Cambridge: Cambridge University Press.
M. Zook (2004) *The Geography of the Internet Industry*, Oxford: Blackwell.

## Websites

NUA (2002) NUA Internet Surveys: http://www.nua.ie.
The Geography of Cyberspace Directory: http://www.cybergeography.org/geography_of_cyberspace.html.

## Notes

1 These different definitions mean that there is a dearth of internationally comparable data. See OECD (2002a) for a discussion of different definitions. E-commerce can imply the method used to place or receive an order and/or the payment and delivery channels.

2 When I checked to see if this company was still trading, there was just a note to the effect that the domain name was for sale at $50,000.

3 See CNN report January 2002. http://money.cnn.com/2002/01/22/technology/amazon/ (last accessed July 2003).

4 The figures are highest in the Scandinavian countries with 90 per cent of businesses with over 10 employees making B2B communications, but just fewer than 60 per cent placing orders and about 20 per cent receiving orders from other suppliers. By contrast in Italy and Greece while over 70 per cent and 60 per cent of companies use the Internet, less than 10 per cent do so for ordering or receiving orders (OECD 2000 and OECD 2002a).

5 See OECD (2002a).

6 See Humphrey *et al.* (2003).

7 See Mansell (2002).

8 Quah (2003) uses the term 'bitstring' rather than knowledge goods to refer to goods that can be digitized and hence are aspatial to allow for the mixed nature of goods with this property, i.e. they could be high level scientific code or a computer game.

9 These references to Brighton and Hove draw on my own research – see Perrons (2003 and 2004) for further information.

10 By February 2003, 57 per cent of adults who had used the Internet for more than three years had made an on line purchase, compared to only 41 per cent of those who had been using it for more than one but less than three years and 25 per cent of those who had been using it for less than one year (National Statistics 2003b).

11 Typically workers in warehouses pick the shopping from the shelves, it is then delivered by van to the household and the driver then physically carries the shopping in carrier bags to the house.

12 SEWA is both a registered trade union and a movement with the objectives of obtaining full employment and self-reliance for women especially those in the disorganized or informal sector. Their guiding principles are *satya* (truth), *ahimsa* (non-violence), *sarvadharma* (integrating all faiths, all people) and *khadi* (propagation of local employment and self-reliance). To see the full range of their activities see http://www.sewa.org/ (last accessed May 2003).

13 Personal communication from Stephanie Barrientos following a SEWA exposure visit to Gujarat but see also their website for further details: http://www.sewa.org.

14 See Kelkar and Nathan (2002).

15 See Kelkar and Nathan (2002).

16 More specifically 24 April 2002 was a national day of action; while the organizers claim 100 demonstrators in Brighton, with the best will in the world I could only count 30.

17 See Malecki (2002) for a detailed discussion of contemporary Internet architecture.

18 AOL TimeWarner is the largest media group in the world. It includes film making, CNN, record labels and magazines. It was responsible for two of the biggest blockbuster films in 2001 – *Lord of the Rings* and *Harry Potter* – but recorded losses in 2002 mainly because AOL had overvalued TimeWarner which it acquired at the height of the dot.com boom and because of slower than predicted growth in Internet subscriptions and on line advertising.

19 Andrew Leyshon (2003) has referred to these as hi tech gift economies and suggests that they have destabilizing effects on what he terms copyright capitalism. Judging from personal experience however, Napster and now mp3s have simply increased awareness of specialist music and increased sales of CDs. It may have done something towards lowering the prices charged for CDs, especially of older recordings, which reduces monopoly profits rather than destabilizes copyright capitalism.

20  See Table 5.2 for the extent of uneven access between the north east and south east of England.

21  These figures have been obtained from an organization which collates national surveys of varying quality and consistency so are probably more like estimates than reliable data but at least they provide a fairly consistent indication of the extent of growth. See http://www.nua.com/surveys/how_many_online/index. html for the figures for Internet use and http://www.world-gazeteer.com/home. htm for the population figures below (accessed November 2003). See http:// www.worldbank.org/data/wdi2001/pdfs/statesmkts.pdf for a World Bank report on the differential development of the Internet.

22  See also Graham and Marvin (2001) and Zook (2001).

23  See http://www.mdc.com.my from which links to different aspects of this development can be found, including the design details of the Petronas towers – the twin towers that are currently the tallest buildings in the world located in Kuala Lumpur city centre and a virtual recruitment exchange.

24  See http://www.cyberjaya.com.my/.

25  In the UNDP 2001 a Technological Achievement Index (TAI) is calculated which is designed to show the relative position of countries in relation to a number of dimensions relating to a country's ability to both create and to implement the benefits of new technologies. There is a greater association between the HDI and the TAI than with GDP per capita, but it was only possible to calculate the TAI for 77 countries, as the measure is quite complex and data was lacking. In UNDP 2002, this measure was not recalculated though information on some of the components was provided.

26  According to their website http://www.digitaldividenetwork.org/content/ sections/partner_gates.cfm, 'The Bill & Melinda Gates Foundation is dedicated to improving people's lives by sharing advances in health and learning with the global community.' It focuses on health care to the poorest but also has various library initiatives in the US and internationally to enhance access to improve life through information technology.

27  See Kelkar and Nathan (2002).

28  Location quotients are a measure of local specialization or the extent to which a particular type of firm or sector is concentrated in a particular location. Specifically Location quotient = $(RS/RT)/(NS/NT)$ where for example $RS$ = the regional number of firms in sector S, $RT$ = total number of firms in the region, $NS$ = the national number of firms in sector S and $NT$ = the total number of firms in the nation. If the location quotient is greater than 1 then the region specializes in sector S.

29  Malecki (2002) comes to similar conclusions having mapped the geography of the 'backbone networks', i.e. the wires that allow fast transmission of large quantities of data.

30  See OECD (2002a).

31  See Coyle and Quah (2002).

32  Even now there is still much Open Source Code, which is freely distributed by the creators, which enables the Internet to work effectively (Quah 2003).

# 7

# LIVING AND WORKING IN SUPERSTAR REGIONS

One of the mysteries of the new economy is why space, place and distance continue to matter. Paradoxically as the power of communications has increased so has the concentration of high level economic activities in a small number of highly interconnected locations. Correspondingly a new economic landscape has emerged consisting of global cities and global city regions, which collectively could be termed superstar regions, with lower order centres, industrial and agribusiness districts, in between, as well as areas of industrial dereliction and rural regions largely bypassed but not unaffected by the global flows.

Global cities and global city regions house the most dynamic elements of global capitalism and are where the corporate executives, government leaders and high level professional and technical workers, whose decisions and innovations shape the lives of people throughout the world, reside. They are characterized by shining skyscrapers filled with government offices, cultural centres, luxury apartments, hotels and shopping malls, but behind the glittering facades there is a more hidden world of low paid workers in cleaning, catering, security and delivery work, as well as those who are socially excluded. These superstar regions are characterized by extremes of wealth and poverty and provide a microcosm of the fundamental contradictions of global capitalism. At the same time they are typically the largest and wealthiest regions in the countries concerned and so have a wide range of industries and services associated with modern econo-mies as well as the living spaces of their employees. What makes the global cities special and different from metropolitan centres in the past and other regions within their territory is the presence of high level, internationally oriented financial and business services. These may only form a relatively small volume of the total activity but it is these together with the cosmopolitan character of the population that accounts for their specificity and for the high degree of inequality within them.[1]

Considerable efforts have been made to define and measure these centres of affluence. Should they be termed world cities, global cities or global city regions?[2] Which cities or regions merit this status and what is their relative ranking? These measures are important to policy makers as many cities and regions aspire to join this global elite. The most comprehensive measurement effort to date has perhaps been the hierarchical inventory of world cities based

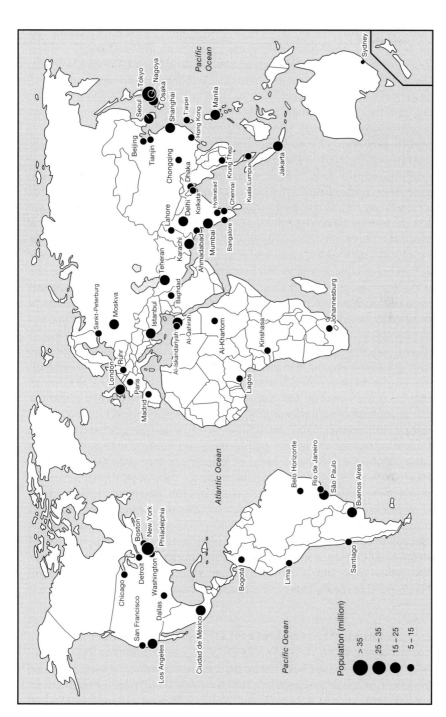

*Figure* 7.1 Global and world cities.

*Source*: Map drawn by Mina Moshkeri.

on the existence of producer service firms with a strong global presence and global connectivity (Taylor *et al*. 2002). Whichever definition is used, however, the same two or three cities, London, New York and Tokyo, appear in the top ranks, with Paris, Frankfurt, Hong Kong, Singapore, Milan, Chicago and Los Angeles following and Seoul, Beijing, Mexico City, Buenos Aires, São Paulo, Sydney, Bangkok, Johannesburg and Kuala Lumpur with its spectacular Petronas twin towers,[3] currently the highest office building anywhere in the world, also making appearances in the top 20–30 depending on the precise measures used (see Figure 7.1 and Hall 2001).

This chapter begins by outlining the processes leading to the formation of global cities and global city regions and the development of social and spatial inequalities within them. This theorization draws on and is set within the perspectives already outlined in Chapters 1 and 3, but also links into the global cities literature. Thus the theoretical concepts of uneven development, the spatial division of labour, the processes leading to social and spatial divisions in the new economy are drawn upon in order to explain the origins, structure and composition of global cities and global city regions. Drawing an analogy with the theory of widening social divisions in the new economy I develop the concept of superstar regions to account for the relatively small number of and internal divisions within these spatial forms. This approach not only encapsulates the processes leading to their formation but also has negative connotations by depicting their elitism and spatial elusiveness. As a consequence it might overcome policy makers' uncritical desire to become a global city, irrespective of their negative side, and also invite investigation and analysis of other urban forms. This concept also avoids some of the difficulties of empirical definitions arising from the dynamic and constantly evolving nature of both economic organizations and contemporary processes of urbanization.

Having outlined the origins and composition of global cities, global city regions and superstar regions, the lives of people living and working in these regions are discussed and consideration is given to whether current patterns are socially sustainable. Particular attention is paid to the question of how both the superstars and their 'servants' manage their work and life. Reference is also made to their increasing ethnic diversity, the inequalities experienced by migrants and to continuing gender inequality. The final section discusses spatial divisions within these regions, paying particular attention to the development of gated communities. Reference is also made to the way that the extent of inequality within these regions also varies between countries, depending on the prevailing welfare regime at the national level, issues developed further in Chapter 8.

## GLOBAL CITIES AND GLOBAL CITY REGIONS

Throughout history there have always been some cities that have had influence beyond their immediate hinterlands, in their national territories or internationally, for example ancient Athens, Rome, Istanbul, Beijing, the medieval Italian city states of Venice, Florence and Genoa, and subsequently London in the era of British colonial dominance. In the context of globalization however increasing attention has been paid to what have been termed world cities by Peter Hall (1966) and John Friedman (1986), information cities and later metropolitan regions by Manuel Castells (1989, 2001) and perhaps most influentially global cities by Saskia Sassen (1991, 2000 and 2001a). Since then the term has been subjected to further academic scrutiny and other terms have been developed, for example the global city region,[4] networked cities and divided cities.[5]

## Global cities

Global cities house the strategic control and command points of the global economy, in particular the coordinating centres of the world's major corporations, the major banks, financial institutions and government. In *The Global City: New York, London, Tokyo* Saskia Sassen (1991) argued that these three cities were similar in that they contained the institutions that play key strategic roles in controlling the increasingly decentralized production activities of large corporations and thereby were analytically distinct from other large cities. In subsequent writing Sassen (2000 and 2001a) has added other cities to this category, specifically Paris, Frankfurt, Zurich, Amsterdam, Los Angeles, Sydney, Hong Kong, Bangkok, Seoul, Taipei, São Paulo, Mexico City and Buenos Aires. However, the three cities she first characterized as global remain significantly larger than the others in terms of their financial dominance and connectivity on a global scale as opposed to being prominent nationally and within their wider geographical regions. Thus Seoul and Taipei are significantly connected within East Asia but less so than Tokyo on a global scale.

For Sassen (2000; 2001a) global cities are the outcome of the asymmetry between the spatial dispersal of production and the continued centralization of control within large corporations. As decentralization becomes more extensive, the task of coordination becomes more complex and corporations subcontract some of the high level producer service functions such as accountancy, law or public relations to specialist firms. These activities benefit from localization and urbanization economies.[6] Being in the information loop is crucial to facilitate innovation and minimize risk and some activities still require a physical presence, especially for non-routine operations. Global cities are therefore where the key strategic and coordination functions take place and consequently have become the 'command points in the organisation of the world economy' (Sassen 2000: 4) as well as being key production sites and marketplaces for the leading firms in finance and producer services. This analysis implicitly incorporates ideas from the new international division of labour theory developed in the 1970s which specified how the vertical division of labour within the firm was being expressed horizontally over geographical space, with the high level or level 1 activities being located within the most developed regions. Global value chain literature could also be drawn upon to explain why these coordinating functions and not the routine production activities are able to appropriate a large share of the value that enables them to pay the high rents demanded in these locations. The new economic geography literature on clustering helps to explain why despite new ICTs many of these high level producer service firms are located in close geographical proximity and the theoretical ideas about social divisions in the new economy can be drawn upon to explain the income differentials between the high paid knowledge workers and the low paid workers who take care of their daily lives.

Producer services are extremely specialized and require highly skilled professional workers. As the work is decidedly pressurized, premiums are paid for people with proven talent, thereby bidding up salary costs and increasing the earnings differential between these and other workers in the locality, especially the low paid service sector workers and the workers in the more flexibly organized manufacturing activities that remain within these centres. As discussed in Chapter 1, the competitive environment also means that work is often project based and employers draw in professionals as and when required on individualized short-term contracts to offset their risks. This insecurity in turn tends towards long working hours as both employees and employers take on projects as and when they are available. Correspondingly, the knowledge workers have little time to manage their own day to day reproduction, which leads to a growing demand for a wide range of personal services. However, because of the inherent economic properties of this work together with its gender and ethnic coding this work is low paid.[7] The social divisions in the new economy are therefore particularly visible in the global cities where the highest paid workers are found and these and low paid workers work and sometimes live in close proximity.[8]

The present trajectory of development is also characterized by increasing spatial divisions between these global cities and other centres outside of the global network. The high level networked service firms need to provide a global service so must have branches or affiliates in other countries, which, Sassen (2001b: 83) suggests, leads to a 'series of transnational networks of cities', and perhaps to 'transnational urban systems' stretching across the globe and including both poor and rich countries, leading to new divisions in the global economy between places in and outside of the global network, also termed 'urban splintering' (Graham and Marvin 2001). This differs from the spatial division of labour in relation to manufacturing as it represents a horizontal rather than vertical division of labour as new branches or partnerships with local firms are established in different regions to draw upon local knowledge of, for example, legal systems and institutional arrangements but at the same time the branches retain the format and reputation of the global players. Thus, new forms of marginality emerge as towns and cities in both rich and poor countries not included in the network lose power and position relative to cities that are connected (Sassen 2000). Yet this does not mean that these cities and regions are totally bypassed by globalization, or 'off the map' as the global city perspective may sometimes imply (Robinson 2002), but rather that they are integrated in different ways, both by the very first international division of labour, through the continued export of agricultural commodities and mineral resources, often on highly unequal terms (see Chapter 3) and in new ways such as agribusiness, decentralized manufacturing and tourism, often attracted by the coexistence of physical remoteness and virtual connectivity through contemporary ICT.[9] Furthermore few, if any, countries are unaffected by globalizing ideologies and the policies of supra national institutions which influence development strategies.

Thus the theories of uneven development discussed in Chapter 3 can be applied and developed to account for the formation and general character of global cities, though the specific forms of each and every global city or region will be different, depending on their history, the level of development of the country as a whole, prevailing macroeconomic policies, welfare regime and political philosophy as well as the precise nature of activities present. In some ways the term global city is therefore inevitably 'fuzzy' (Markusen 1999), because it is an intermediate theoretical concept, that is it has a precise meaning or specificity deriving from the presence of internationally oriented command and control functions, but the explanation of the functions themselves and their locational requirements depends on broader theories of uneven development and their precise development will be contingent on the specificities of the macro context referred to above.

## Global city regions

Global city regions also derive from the ideas of world cities and global cities but are defined in a broader way to encapsulate their significance as both 'essential spatial nodes of the global economy and as distinctive political actors on the world stage' (Scott *et al.* 2001: 11). Empirically these regions are often polycentric and multi-clustered, housing the global cities and the high technology centres, cultural industries and revitalized craft industries, and are similar in this sense to the metropolitan regions of nodes and networks outlined by Manuel Castells (2001). Some centres such as New York, or London within its regional context of south east England, Los Angeles within the context of southern California, Kuala Lumpur, with the multimedia super corridor, and Hsinchu-Taipei in Taiwan[10] are simultaneously global cities as well as centres of high technology industries. Other new economic landscapes are centred mainly on high technology including Rhône Alpes and Sophia Antipolis in France or craft industries such as Emilia Romagna in Italy but are less confined to the triad of North America, Western Europe and Japan, and also include Hyderabad (Cyberabad) and Bangalore in India or Shenzhen in China.[11] Both Castells (2001) and Scott *et al.* (2001) point to the Pearl River Delta area in China stretching between Hong Kong, Shenzhen, Canton, Macau and Zuhai and which contains about 60 million people, as being an extreme example, but more generally argue that such regions contain a number of centres with different activities, finance, culture and high technology as well as new centres at the perimeters or edge cities.[12] Another example would be the Bay Area (San Francisco) which 'consists of a constellation of cities containing 7 million people living in an expanse 60 miles long and 40 miles wide' and if this was linked into the Los Angeles city region (Scott *et al.* 2001)/southern Californian metropolitan region (Castells 2001: 229) which includes Los Angeles and Silicon Valley in Santa Clara County there would be a further 17 million people. Likewise the functional region of London consists of an area of 30 miles by 30 miles with 6.5 million jobs, 40 per cent of them outside Greater London.[13] Global city regions thus represent a complex combination of decentralization and recentralization. As some people and activities have moved away from the central areas new centres have developed and people from rural regions or other countries have moved in and flow between these centres diversifying their ethnic and cultural mix.

The global city region approach has a broad territorial and industrial perspective while the global city approach focuses more on producer services and has a more abstract concept of space. The global city does not have to consist of a highly geographically concentrated central business district (CBD) but can be a 'partially deterritorialized space of centrality', that is, it could consist of digitally connected 'dense strategic nodes spread over a broader region' (Sassen 2001b: 85). Thus it is the interconnections between the activities that demarcates the global city,[14] whereas the global city region, despite its varying internal structure, potentially has clearer, albeit fluid geographical boundaries, and in some accounts the potential of becoming an important political unit.

The analysis of clustering between the global city and global city region perspectives is similar. Both emphasize the significance of external economies in the high level knowledge sectors which offset the decentralizing tendencies of contemporary ICT. Existing skills are enhanced, knowledge workers and firms in related and non-related industries are attracted to the region together creating a vibrant urban environment attracting other firms and workers in a cumulative and reinforcing way. So while there has been an increase in the number of teleworkers, fuelling the new economy myth of the nomadic worker hot-desking from office to office or working on the move with laptops and mobile phones, a large proportion of high level work continues to be based from offices concentrated in a small number of locations. Whether the attack on the World Trade Towers in New York in September 2001 will finally realize the decentralizing potential of new technologies (Davis 2001), predicted since at least the 1970s, remains to be seen. If it does, perhaps the most likely scenario would be the development of partially 'deterritorialized spaces of centrality' (Sassen 2001b: 85) or 'concentrated decentralization' in the form of 'secured citadels' (see Marcuse 2002: 596 and 600 and below) contained within global city regions.[15]

While Castells refers to the people in the southern Californian metropolitan region as 'living, working, consuming and travelling in this territory without boundaries, name or identity' (Castells 2001: 229), somewhat conversely Scott *et al.* (2001: 11) consider global city regions not only as the motors of growth in the contemporary economy but also 'distinctive political actors on the world stage' and the appropriate level for governance in the global economy, as they maintain that the powers of nation states have weakened relative to those of supra and sub-national institutions (the hollowing out thesis – see Chapter 8).

Seeing the region as an entity or subject[16] in this way implies a regional identity and a homogeneity of interests shared by the inhabitants in competition with other regions.[17] Even though disparities are referred to, greater emphasis is placed on the internal collaborative linkages, considered central to the growth dynamics of these 'motors' of the global economy, also said to contribute to the formation of the shared political identity. Correspondingly the global city region approach is implicitly less critical of the system that sustains and reproduces unequal outcomes. Sassen (2000, 2001a) also refers extensively to the processes generating inequality but policy makers who draw on both these bodies of work tend to focus only on the globally oriented high level activities and the dynamic clustering elements and overlook their downsides as they seek to replicate their existence; the presence of at least one such node being considered vital to contemporary modernization and global status.[18] Yet as Ian Gordon (2002) points out, in quantitative terms these activities represent only a small proportion of the total economy of these regions and developing policies that prioritize their interests risks overlooking equally if not more valid claims for support. Similarly, Jennifer Robinson (2002) urges urban theorists to move away from the preoccupation with global cities and devise new concepts drawn from a wider range of contexts in order to encapsulate and validate the diversity of cities and the range of activities found within them.

## Superstar regions

The clustering perspective tends to emphasize micro issues relating to the firms within specific regions, such as economies of scale, the exchange of knowledge and the significance of collaboration and trust between firms and locally based institutions. It also sustains ideas about the region as a harmonious subject with a common identity and set of interests and leads into regional marketing or boosterism. To explain why there are only a small number of global cities or global city regions however it is important to recognize the importance of power, the existence of conflict and the competitive nature of capitalist society. It is the combination of the continuing significance of major corporations, the global reach provided by contemporary ICTs, together with the fact that an increasing number of countries throughout the world, whether by choice or coercion, have become part of the global market economy that accounts for the scale of their growth. Returning to the ideas of Quah (1996) discussed in Chapter 1, the key activities of global cities and global city regions are quint-essentially knowledge based and correspondingly have equivalent properties to knowledge goods: that is, they are weightless, with an almost infinite global reach and in some respects they are infinitely expansible. These properties are clearest in software but even in consultancy or architecture, where each project may seem to be bespoke and will indeed differ in detailed content, the projects still have many common features, allowing companies to realize some economies of scale. Similarly, consumers develop preferences for products and firms of greater renown or 'superstars', reinforcing their cumulative growth as their market share and income will not be constrained by geographical distance. Thus superstar firms materialize in London or New York, as they are believed to be superior to ones in Middlesbrough, Malmö or Mumbai, simply by virtue of being there, irrespective of their actual merits, and they correspondingly capture an increasing share of the market and continue to grow.

The emergence of superstar firms can be seen literally in the architecture and design in world cities. While there is some evidence of different cultures or the vernacular in more minor stylistic features, for example the Muslim influence in the Petronas towers, new urban landscapes such as waterfront redevelopments and shopping centres, office and hotel complexes have similar appearances as a limited number of global architects are involved in their design.[19] Likewise to be taken seriously a firm aspiring to global status has to have a presence in more than one of these locations, which in turn reinforces its national and global standing.[20] Thus the limited number of global cities or city regions can be explained by the continued dominance of large corporations in the global economy; indeed to operate successfully enormous resources are required, which has led to a whole range of mergers and acquisitions, and in turn to the development of business service firms managing these transactions such as the global investment banks. Depending on the overall level of growth, this means that firms elsewhere will inevitably disappear. Large firms build upon knowledge that can be codified at least to some extent, while simultaneously drawing upon a continual stream of learning and innovation developed within smaller firms in their immediate locale, hence the clustering. These regions could then by analogy be termed superstar regions, which is a theoretical concept, one that embodies inequality and has the added advantage of conveying some negative connotations, because by definition not everyone can be a superstar.[21] Only the most crass publicity statements would claim to seek superstar status, while the aspiration to become a global city or global city region superficially at least seems far more positive, modern and progressive despite their negative features.[22]

Superstar regions arise because of their role within both the national and international economy and because they house the global elite. Both the highest and lowest paid workers within their national territories will be found there so earnings differentials and income inequalities will generally be greatest in these regions, the comparative extent of the inequality depending on national fiscal policies. At the same time these regions also contain a wide spectrum of economic activities corresponding to the scale and wealth of the population and likewise a wide range of employees, from electricians and bank clerks to social workers. Thus although these areas are the most polarized, there is also a wide middle, the size of which will depend on the scale of the territorial boundaries used. So just as it would be wrong to assume a harmony of global city-region interests, it would also be misleading to suppose that these regions are in a constant state of internal turmoil and unrest. However, taking these areas as political actors or subjects, in competition with similar regions else-where, risks assuming a unity of interest, and inevitably some interests, usually the business classes, are prioritized. These regions have become very prominent economically but whether they are more connected with each other than with other regions within their own territories or whether they constitute new political entities as maintained in the hollowing out thesis is more questionable and is discussed further in Chapter 8.

## SOCIAL DIVISIONS IN THE
## SUPERSTAR REGIONS

One of the most visible indicators of globalization is transnational migration, most evident in the superstar regions, where both high and low paid migrants concentrate geographically leading to increasing ethnic and cultural diversity. Although the United Nations (2002) estimates that only about 2.9 per cent of the world's population have migrated, the rate of migration has doubled in the last 25 years, and in absolute terms represents 175 million people, the majority of whom have moved from less to more developed regions, in particular to North America (41 million) and Europe (56 million) (see Table 7.1) both of which have positive net annual migration balances (1.4 million and 0.77 million respectively). Asia has a stock of 50 million migrants but an overall annual average migration loss of 1.3 million; the main gainers are Japan, Hong Kong and Singapore, others being the oil states and Israel.

Migrants are attracted by the opportunities in the centres of the global economy. Clearly, not all migrants are ethnic minorities and neither are all ethnic minorities migrants, but past and present migration has increased the ethnic diversity of global cities. In London in 2002 28.4 per cent of the popu-lation belonged to an ethnic minority compared to 7.6 per cent nationally, and in some inner boroughs the white population is the minority.[23] Likewise ethnic minorities are overrepresented in New York and Los Angeles, enhanced by the 'white flight' to the suburbs. In Los Angeles, by 2005, more than 50 per cent of the population will be Latino.[24] Latinos, together with Asians, have become the major source of population growth in the US large cities. In Tokyo, where overall migration is lower, migrants similarly concentrate in central Tokyo, especially around Shinjuku where there is the newly developed Tokyo Metropolitan Government Building, soaring skyscrapers, dazzling department stores and the seedy red light district,[25] again a visible manifestation of the social divisions in the new economy, and, as with other superstar regions, one that attracts high and low paid migrants.

Table 7.1a  Migration – more and less developed regions

| Country or area | Population (000's) | Migrant stock | Migrant stock as % of population | Number of refugees (000's) | Net migration (annual average) | Net migration rate (per 1000 population) | Workers' remittances (billions dollars) | Workers' remittances as % of GDP |
|---|---|---|---|---|---|---|---|---|
| World | 6,056,715 | 174,781 | 2.9 | 15,868 | 0 | 0 | 62,239 | 0.2 |
| More developed regions | 1,191,429 | 104,119 | 8.7 | 3,012 | 2,321 | 2 | 12,535 | 0.1 |
| Less developed regions | 4,865,286 | 70,662 | 1.5 | 12,857 | -2,321 | -0.5 | 49,704 | 0.7 |
| Least developed countries | 667,613 | 10,458 | 1.6 | 3,066 | -306 | -0.5 | — | — |

*Source*: UN 2002; see http://www.un.org/esa/population/ publications/ittmig2002/ittmig 2002.htm.

*Notes*: The more developed regions comprise all regions of Europe and northern America, Australia/New Zealand and Japan.
The less developed regions comprise all regions of Africa, Asia (excluding Japan) and Latin America and the Caribbean and the regions of Melanesia, Micronesia and Polynesia and the countries listed below.
  The least developed countries as defined by the United Nations General Assembly in March 2001 include 49 countries: Afghanistan, Angola, Bangladesh, Benin, Bhutan, Burkina Faso, Burundi, Cambodia, Cape Verde, Central African Republic, Chad, Comoros, Democratic Republic of the Congo, Djibouti, Equatorial Guinea, Eritrea, Ethiopia, Gambia, Guinea, Guinea-Bissau, Haiti, Kiribati, Lao People's Democratic Republic, Lesotho, Liberia, Madagascar, Malawi, Maldives, Mali, Mauritania, Mozambique, Myanmar, Nepal, Niger, Rwanda, Samoa, Sao Tome and Principe, Senegal, Sierra Leone, Solomon Islands, Somalia, Sudan, Togo, Tuvalu, Uganda, United Republic of Tanzania, Vanuatu, Yemen and Zambia.
  Number of refugees: Statutory UN definitions: see UNHCR (2000).

Table 7.1b  Migration – world regions

| Country or area | Population (000's) | Migrant stock | Migrant stock as % of population | Number of refugees (000's) | Net migration (annual average) | Net migration rate (per 1000 population) | Workers' remittances (billions dollars) | Workers' remittances as % of GDP |
|---|---|---|---|---|---|---|---|---|
| Africa | 793,627 | 16,277 | 2.1 | 3627 | -447 | -0.6 | 8755 | 1.6 |
| Asia | 3,672,342 | 49,781 | 1.4 | 9121 | -1311 | -0.4 | 24,205 | 0.3 |
| Europe | 727,304 | 56,100 | 7.7 | 2310 | 769 | 1.1 | 11,854 | 0.1 |
| Latin America and the Caribbean | 518,809 | 5944 | 1.1 | 38 | -494 | -1 | 17,131 | 0.8 |
| Central America | 135,129 | 1070 | 0.8 | 28 | -347 | -2.7 | — | — |
| South America | 345,738 | 3803 | 1.1 | 9 | -75 | -0.2 | — | — |
| Northern America | 314,113 | 40,844 | 13 | 635 | 1394 | 4.6 | — | — |
| Oceania | 30,521 | 5835 | 19.1 | 69 | 90 | 3 | 293 | 0.1 |
| Micronesia | 516 | 116 | 22.6 | 0 | -2 | -4 | — | — |
| Polynesia | 606 | 79 | 13 | 0 | -5 | -8.2 | — | — |

Source: UN 2002; see http://www.un.org/esa/population/publications/ittmig2002/ittmig2002.htm.

Table 7.1c  Migration – selected countries

| Country or area | Population (000's) | Migrant stock | Migrant stock as % of population | Number of refugees (000's) | Net migration (annual average) | Net migration rate (per 1000 population) | Workers' remittances (billion dollars) | Workers' remittances as % of GDP |
|---|---|---|---|---|---|---|---|---|
| India | 1,008,937 | 6271 | 0.6 | 171 | –280 | –0.3 | 9034 | 1.9 |
| Philippines | 75,653 | 160 | 0.2 | 0 | –190 | –2.6 | 125 | 0.2 |
| Thailand | 62,806 | 353 | 0.6 | 105 | –5 | –0.1 | — | — |
| Singapore | 4018 | 1352 | 33.6 | 0 | 74 | 19.6 | — | — |
| United Kingdom | 59,415 | 4029 | 6.8 | 121 | 95 | 1.6 | — | — |
| Germany | 82,017 | 7349 | 9 | 906 | 185 | 2.3 | — | — |
| Mexico | 98,872 | 521 | 0.5 | 18 | –310 | –3.3 | 6572 | 1.1 |
| United States of America | 283,230 | 34,988 | 12.4 | 508 | 1250 | 4.5 | — | — |
| Australia/ New Zealand | 22,916 | 5555 | 24.2 | 63 | 103 | 4.6 | — | — |
| South Africa | 43,309 | 1303 | 3 | 15 | –5 | –0.1 | — | — |
| Gambia | 1303 | 185 | 14.2 | 12 | 11 | 9.1 | — | — |

Source: UN 2002; see http://www.un.org/esa/population/publications/ittmig2002/ittmig2002.htm.

Globalizing bureaucrats and politicians, globalizing professionals, merchants and media moguls as well as corporate executives – in Leslie Sklair's (2002: 160) terms the transnational capitalist class – are attracted by the supra national institutions, government and major corporations. Beyond this global super elite, firms in the superstar regions have the power and resources to select the best students and workers from all over the globe and often at a lower cost than local labour. In Silicon Valley in 1990, for example, one-third of the high level engineers and scientists were foreign born with Taiwanese (16.4 per cent), Indians (11.9 per cent) and Chinese (9 per cent) being the most sizeable groups and made up of children of immigrant families, former employees of US subsidiaries located abroad, former students at universities in the United States and high tech immigrants on special visas. They were also disproportionately male (82.5 per cent) and more qualified on average than the native population. Even so they rarely gained managerial positions in US owned companies, though some became managers in foreign owned companies within the United States or set up their own companies.[26] Others returned to their countries of origin and developed their own companies or worked in the growing IT sectors there. The ability to set up new companies also draws upon social networks in both regions[27] and the deepening social division of labour characteristic of the new economy, in which larger firms look to smaller companies for innovations. Thus connections which started in the 1960s and 1970s as flows of capital from the more to the less developed regions and flows of people in the opposite direction, as people left to study in the United States but rarely returned owing to the lack of job opportunities, have subsequently been displaced by slightly more complementary flows. The Taiwan government for example encouraged Taiwanese IT workers to return in order to upgrade the industry there; see Box 7.1 for Minn Wu's story.[28]

AnnaLee Saxenian and Jinn-Yuh Hsu (2001) find that the social networks built up by Taiwanese scientists and engineers while studying and working in the United States[29] were instrumental in promoting the IT sector in Taiwan, which now has both large-scale producers of electronics goods and companies producing highly innovative products, and overall is ranked third in the world, behind the United States and Japan (Saxenian and Hsu 2001). Initially various organizations were formed by Taiwanese students and workers in the United States for social reasons to overcome their isolation and exclusion but subsequently these became important for finding jobs and venture capital (Saxenian and Hsu 2001). These contacts have been maintained, illustrated by frequent temporary movements between Silicon Valley and Hsinchu-Taipei and provide the necessary trust for exchanging knowledge and ideas that have been important in establishing and upgrading the IT clusters. Some people, referred to as 'astronauts', work in both regions, travelling between them as frequently as twice a month. For although there are high end sectors in Taiwan, contact with Silicon Valley is vital; it remains the leading innovator of processes and products as the United States remains the leading market. Moreover, while they work in both regions their families are generally based in California

*Box 7.1* Miin Wu's story: from Taiwan to Silicon Valley and back again

Miin Wu immigrated to the US in the early 1970s to pursue graduate training in electrical engineering. Like virtually all of his classmates from National Taiwan University, he took advantage of the ample fellowship aid available in the US at the time for poor and talented foreign students. After earning a doctorate from Stanford University in 1976, Wu recognized that there were no opportunities to use his newly acquired skills in economically backward Taiwan and he chose to remain in the US. He worked for more than a decade in senior positions at Silicon Valley-based semiconductor companies including Siliconix and Intel. He also gained entrepreneurial experience as one of the founding members of VLSI Technology.

By the late 1980's, economic conditions in Taiwan had improved dramatically and Wu decided to return home. In 1989 Wu started one of Taiwan's first semiconductor companies, Macronix Co, in the Hsinchu Science-based Industrial Park, with funding from H&Q Asia Pacific. He initially recruited 30 senior engineers, mainly former classmates and friends from Silicon Valley, to return to Taiwan. This team provided Macronix with the specialist technical skills and experience to develop new products and move into new markets quickly. Wu also transferred elements of the Silicon Valley management model to Macronix, including openness, informality and the minimization of hierarchy – all significant departures from traditional Taiwanese corporate models. Macronix went public on the Taiwan stock exchange in 1995 and the following year became the first Taiwanese company to list on NASDAQ. The firm is now the sixth largest semiconductor maker in Taiwan, with over $500 million sales and some 3000 employees.

*Source*: Saxenian and Hsu (2001: 909).

'because of lifestyle advantages' (Saxenian and Hsu 2001: 909). Similar flows of people take place between Silicon Valley and the high tech regions of India: Bangalore and Hyderabad.

Migration contributes to the growth of the already powerful regions, although as indicated above there are reverse flows, which assist upgrading, but richer countries maintain control over the parameters of migration and it is far more difficult for people without marketable skills to migrate. Even IT migrants to the US are treated differently according to their qualifications. Lower qualified migrants are more likely to be involved in 'body shopping', the practice whereby firms employ workers on H1-B[30] visas but loan them on to other firms for specific projects. In this sense the main employer is more like an agency and employees find themselves 'on the bench' between projects, during which time they may or may not be paid, but are certainly on lower rates than when working (Mir *et al*. 2000). See Box 7.2 for the contrasting stories of Indian workers working for US firms in India and the United States respectively. The Indian cases demonstrate how class differences, educational background, the

Box 7.2 Contrasting experiences of Indian high tech workers

Krishnamohan manages a project for the Microsoft India Development Centre, which occupies the ninth floor of the ten-storey state-of-the-art building in the largest 'technopark' in Asia. The infrastructure provided to Microsoft (and others) at the expense of the Indian taxpayers includes a satellite earth station, 3000 direct lines, high-speed fibre optic connectivity to the internet, and an integrated services data network. There is even a dedicated power plant that protects the tenants from inconvenient power outages. Other amenities include air conditioning, a dust-free environment, a post office, a bank, a shopping mall, executive residences, medical care, and a clubhouse. The government has even set up an Indian Institute of Information Technology to provide the skilled programmers needed by Microsoft and the other occupants of the Hi-Tec City (an acronym whose search for a name resulted in 'Hyderabad Information Technology Engineering Consultancy City'!).

(Mir *et al.* 2000: 5–6)

[Sunil Roberts] went to an English-medium school in Hyderabad where he was trained to do well in the entrance examination for the Indian Institute of Technology in Bombay from which he obtained a degree in electrical engineering. While there he also learnt how to make a successful application to US universities and subsequently graduated with a masters in computer science from New York University. He currently works for a major US firm as a telecommunications firmware engineer. He works under the HB-1 visa but the company lawyers are trying to organize a green card for him. He lives in Morristown, a cosmopolitan suburb in New Jersey that has a small Indian population, most employed by high tech firms.

(Mir *et al.* 2000)

Appa Rao trained in his home town of Khamman, a rural district in Andhra Pradesh, first in a local school, and then in a series of small 'technical institutes' in fairly routine programming. In all, he spent a few thousand rupees and a little less than two years for his training. He currently shares a two bedroom apartment with three of his Indian co-workers. Appa Rao is one of several thousand people who are 'bodyshopped' every year in the US on HB-1 visas to work on specific computer projects. Currently he is 'on the bench'. Appa Rao is beholden to his employers in multiple ways – he needs their cooperation to get a green card for permanent residency and permanent employment in order to do so. This dependence opens up the possibility of an extra legal exploitative process.

(Mir *et al.* 2000: 6–7 and 27)

timing of the move and the hiring practices of US companies shape people's lives and their experiences of migration. Following Ajun Appadurai (1996), Ali Mir, Biju Mathew and Raza Mir (2000: 28) point out that migrants 'experience and reproduce a deterritorialized culture' but demonstrate how these experiences vary by social class and caste. For the middle class migrants 'the act of being Indian in the US is inscribed within the knowledge of their security and investment in America. It is mediated by long-term mortgages on homes in white suburbia, a press that labels them as 'model minorities', an embeddedness in local Indian cultural and religious organisations and so on, while for lower skilled workers in the high tech industry and other working class migrants the migration experience is mediated by constant insecurity and anxiety about their status as employees and as migrants. While the skills of Sunil Roberts will probably have protected him from any downturn in the US market, workers such as Appa Rao are much more vulnerable. For although the terms of the HB-1 visas, which were designed to protect US citizens from being undercut, should also prevent migrant workers from excessive exploitation, in reality they are rarely applied.[31]

Migration restrictions are far more onerous for people with skills of low market value but their strength and application varies with the labour market requirements of the richer countries. Italy for example has had a series of moratoriums for illegal migrants over the years, and the United States is also inconsistent in its treatment of illegal migrants. Restrictions were intensified in 2001 in the context of an economic downturn and the attack on the World Trade Center, and the terms migrant, asylum seeker, refugee and terrorist have been conflated in a national sense of xenophobia, but even in these circumstances the crucial role played by migrants in the United States economy limits the extent to which these restrictions are likely to be enforced in relation to all migrants in the long term; rather similar to the migration flows in Europe in the 1960s and 1970s migrants have become a structural necessity to the economy.[32]

Contemporary migrants are also faced with difficult labour market conditions in the new economy: weak labour market regulation, contingent and flexible working and widening wage dispersion, including relative and absolute decline in the lowest decile of earnings in both the United States and the UK. The informal sector has also expanded which makes it easier for illegal migrants to find work, but pay and conditions are hazardous. In the UK a growing proportion of low paid employees are placed by employment agencies and although the vetting of skills is weak, rigorous checks are made on passports and migration status. As Polly Toynbee (2003) comments from her experiences of finding low paid work in the UK:

> I watched the other applicants. Most of them were foreign; many seemed to understand barely a word of English. . . . As I looked on, the only strict rule that seemed to apply was that everyone must have a valid work permit or British citizenship before they even got a chance to fill out the forms. Immigration status was the one check that was rigorously carried out almost everywhere I went.

After the briefest of interviews she was given work at an NHS hospital as she goes on to say:

So that was it. No police check: neither of my referees had been contacted. And I was setting off to work in a hospital. I could be just out of prison for serial unplugging of patients' life support machines. I could have been released from a mental hospital with a particular taste for tormenting elderly patients. So much for security. Forget any noble idea of induction into the noble portals of the NHS. There could hardly be a more lackadaisical way of being dumped into it to turn a quick buck for an agency.

(Toynbee 2003: 36)

The migration status is checked as companies could be fined for hiring illegal migrants, though as indicated in the quote above less consideration is given to the actual skills. The practice of subcontracting formerly public sector work to agencies is widespread in the United States and the United Kingdom and forms part of the efficiency and flexibility agenda. The efficiency gains are however made at the expense of decent wages and living conditions for working people. The minimum wage was designed to give workers a 'living wage' and 'end the scandal of poverty pay' (UNISON 2002) but the composition of the living wage was never defined for UK workers, though there is much less hesitancy in doing so for workers elsewhere. According to the UK government-supported ethical trade initiative (see point 5.1 in Box 9.4, Chapter 9) wages should meet minimum standards and 'in any event wages should always be enough to meet basic needs and to provide some discretionary income'. Several ethnographic studies of trying to live on the minimum wage in the UK and the US indicate that despite working very hard in essential jobs this level is simply not reached, and workers find it difficult to avoid sinking into debt.[33]

Personal service work is inherently labour intensive, so productivity increases are difficult and wage costs form a high proportion of total costs. Market logic would dictate that these jobs disappear from the economy, become available only to the elite who can afford to pay high prices or be provided through the public sector where profit is not the key rationale.[34] Unfortunately within the neo-liberal flexibility and efficiency agenda these jobs have been privatized, deregulated and increasingly provided by large profit seeking organizations. The net result is that 'efficiency gains' made by the firms arise primarily through impoverishing the terms and conditions for employees, terms that are only sustainable through the gender and ethnic coding of work and the fact that it is very difficult for these highly flexible workers to organize effectively to protect themselves. In the UK public sector, care workers secured a pay increase in 2002, following protests against low pay, including a 'mum's army' of novice protesters, but these increases would still be insufficient to provide a living wage and the agreement does not cover agency workers.[35] However there is no reason why market logic should be accepted. Certainly the directors of the corporations supplying these services pay themselves on a par with other directors, so earnings differentials between top management and employees is immense, as high as 200:1, even without allowing for directors' perks such as share options, bonuses, golden helloes and goodbyes (Sikka *et al.* 1999). Migrant women are overrepresented in domestic work throughout Europe; though the specific conditions vary with different welfare regimes between countries.[36]

Living beneath the minimum wage via work in the informal sector is even more precarious and leads to a very low standard of living for migrants who continue to meet their obligations to send remittances home. Contemporary circumstances therefore contrast negatively with the large-scale migration to Western Europe in the Fordist era. Then, despite hard work in less favoured jobs in construction, mines and hospitals as well as some in manufacturing, poor housing and discrimination, migrant workers still benefited from the booming economic conditions and the strength of trade unions, who tried to ensure equal pay if only to prevent migrants from undercutting wages. Second and third generation migrants are also faced with these more difficult conditions with the loss of manufacturing employment from the large metropolitan regions where they mainly live, and correspondingly ethnic minorities are overrepresented in unemployment. There are however wide variations in the experiences of different ethnic minorities in different countries. In the UK, for example, Indians are often on a par with or perform better than the white population, but even so they do less well than would be expected from their level of education and skill, so workplace discrimination remains.[37]

Deregulation and flexibility in labour markets have lowered the incomes of already low paid workers. In the United States the wage gap between migrants and the native population in five major cities successively increased between 1970 and 1990 and the gap was larger between recent migrants and those who had been there for some time, especially in Los Angeles and New York (Clarke 1999). Deregulation also makes it difficult to monitor working conditions especially in the sweatshops which have proliferated in major cities, including London, Los Angeles and New York. In Chinatown, New York, which is very close to the financial district around Wall Street, there is a dense concentration of small and medium-sized enterprises in the clothing industry but city officials distance themselves from the Dickensian working conditions, using ethnicity as the criteria for doing so, as Kay Anderson reports:

> In the underground economies of the ethnic enclaves of Vietnamese, Cuban, Dominican, Central American and Chinese, it is a case of immigrants exploiting immigrants. We can't be expected to protect those who are too docile to come forward.
>
> (NYC Labour Department Official
> cited by Anderson 1998: 269)

She goes on to comment that 'the scripting of immigrants in racial terms and the encoding of Chinatown as an "ethnic" space in the speech and practices of such officials feed into the complex matrix of oppressions currently shaping this district's fortunes.' Thus while migrants often form their own communities or enclaves on the basis of ethnic or cultural identities often out of necessity or self-protection, this can simultaneously alienate the majority population and reinforce racism and exclusionary practices.[38]

Discrimination, unfair treatment, racism and xenophobia are also reinforced by restrictive immigration policies, which emphasize difference between those with and without legitimate claims to residence, thereby creating boundaries between groups, often demonizing those to be excluded. At present distinctions are drawn between asylum seekers, refugees and economic migrants, with the first two having some claim to legitimacy, albeit a restricted one, while the latter are denigrated even though this is completely inconsistent with the prevailing neo-liberal ideology. There is a completely contradictory approach towards the free movement of labour and capital. While lifting of restrictions on the flows of money and goods is considered to be inherently good, individual desire to improve their 'rate of return' through migration is denied. Nevertheless the demand for low wage labour continues, stimulating both illegal migration and the development of new borders.

The EU reflects this contradictory approach. Neo-liberal ideas of free movement operate internally but strong barriers to trade, in particular the import of agricultural commodities and more especially people, exist. Naomi Klein (2002) argues that one of the reasons for EU enlargement is to maintain control over and restrict immigration but to extend the area over which cheap labour can be obtained, by creating new more distant boundaries. Tight immigration controls encourage illegal migration which in turn has generated a highly lucrative criminal activity in people smuggling as well as many tragedies such as in 2000 when 58 people, 54 men and 4 women from Fujian, China, were found dead in a container lorry arriving at the UK port of Dover, despite having paid $20–$30,000 each for the trip. Fujian is a province with a history of emigration. It has benefited from the remittances of successful migrants, who will have had to have worked extremely long hours, given their likely low pay, in order to repay the cost of the trip and send money home. In addition many people are trafficked to more developed countries and regions to work as sex workers.[39] Thus while there is a global elite or international class of symbolic analysts moving freely between the global city regions, such unrestricted access is only possible for the most highly trained; others have to struggle to gain access and once there struggle to survive.

Social divisions on the basis of skill or market value, gender and ethnicity are becoming deeper and ever more complex; there are some highly paid female ethnic minority IT and finance workers, and ethnic majority workers living on or below the minimum wage, thus these divisions take multiple forms. In the new economy, as discussed more abstractly in Chapter 1, these social divisions deepen as a consequence of increasing deregulation and privatization in the more competitive global economy and more specifically from the growth of the more polarized service sector and decline of manufacturing employment. These social divisions are especially evident in the superstar regions where professional and managerial jobs have expanded at one pole and jobs in personal services increased at the other. London, for example, has a higher proportion of residents in the top two occupational groups (which require a university degree or equivalent) than any other region in the UK.[40] Between 1992 and 2000[41] full time employment expanded further in these categories attracting young, highly qualified migrants from elsewhere in the UK and from other countries. Personal and protective services, which includes caterers and care workers, was the only other occupational group where full time employment increased and this category together with sales occupations experienced the largest increases in part time employment. These changes in employment structure, together with the much slower rate of growth of earnings for low paid occupations, have led to a widening of the earnings gap between the top and bottom deciles for both women and men[42] (see Figure 7.2 and GLA 2002). The earnings distribution for men in London is wider than in the rest of the UK and wider for men than women,[43] reflecting both the existence of the highest paid jobs in the country and hence the wider range of overall earnings and women's underrepresentation in these top jobs. Nevertheless, as Figure 7.2 shows, there has been a growing duality between women, as a significant minority of women do work in high paid sectors, if not in the very top jobs, while women overall remain vastly overrepresented in low paid work, especially personal services, itself a cause of the growing feminization of poverty and of the in-work poor. These two phenomena are organically related as the continuing unequal domestic division of labour means that the increase in the number of high paid women workers leads to an increase in the demand for marketed personal services and care.[44]

As discussed in general terms in Chapters 1 and 4 professional and managerial workers are frequently required to work very long hours, and these norms have been internalized by the employees as Sandra Wood, a new mother and an investment analyst in the City of London, reflecting on whether employers recognize the needs of working parents illustrates:

> I mean I used to work probably more like 8.00–7.30. And now I work 8.00–6.00, and you know no one's kind of come up armed with a watch saying that we're going to reduce what we pay you because of that. So I think that's, you know I really appreciate that.

That is she appreciates being allowed to work *only 10 hours* a day which is still in excess of those allowed under the EU Working Time Directive from which employees in this organization have 'voluntarily' opted out. Employers also assist employees to work long hours despite their caring obligations. Thus Sandra's employer has an on site emergency crèche, a designer café, restaurant, dentist, physiotherapist, doctor, nurse and gym which are available to all employees. Nasrin Begum is similarly a new mother and highly qualified professional worker, working in a leading City bank. Nasrin's employer provided fewer facilities but gave her a computer so she could complete work at home if she had to leave 'early':

And I'd said to them, that I was really worried with a baby, how I was going to manage, because I knew I would have to like leave at 5 or 6, I couldn't, like before I had a baby I was working til 7.00, 8.00. 9.00 at night [*having started at 8 in the morning*]. And I could do that, because I didn't have to come home. So they gave me a computer, and so I can work at home, so they have . . . sort have tried to help really.

Similarly to Sandra, Nasrin appreciates her employer's assistance and does not question the long hours culture. Firms generally have equal opportunity and family friendly or work–life balance policies but high level employees perceive that using them will be at the expense of their careers, as Nasrin commented: 'the bottom line is there's more work to be done, than can be done within the 35 hours and if you want to get promoted you have to do the extra.'[45] Such family friendly policies do not always apply to all employees, but where they do they are often not funded and lower paid employees often cannot afford to make use of them.

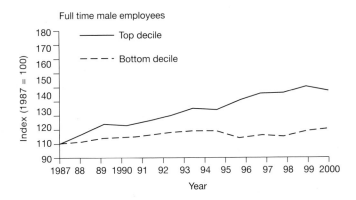

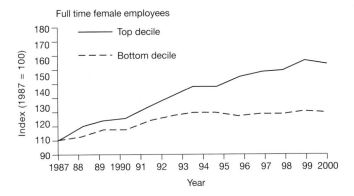

*Figure 7.2* Widening earning differentials in London.

*Source*: Data from GLA (2002).

223

To facilitate the long working hours of people with caring responsibilities, a wide range of personal services and private childcare have developed. As indicated in Chapter 1, these are highly labour intensive with limited scope for productivity increases, in contrast to at least some of the products produced by the knowledge workers themselves. Sandra's household employed a nanny for 11 hours a day 5 days a week, and Nasrin's used a private nursery similarly for about 10 hours a day; in addition both employed cleaners and used the Internet for shopping, which means of course that others, low paid workers, collect their shopping and deliver it to their door. In all cases this expands employment in the low paid caring, retail and personal services sector.

Care workers are generally low paid, though Sandra's household, in the top decile of wage earners, paid their nanny comparatively well, nearly £7 per hour net of taxes[46] which is close to the average weekly earnings of women in London *before* deductions and considerably above the minimum wage. Sandra was still left with between a half and three-quarters of her own salary, reflecting the wide wage dispersion between 'professional' workers and care workers (who might themselves be considered professional if the care of young children was valued differently).

These high earning mothers lead very complicated lives, spending long hours at work and the rest of their time caring for their children or sleeping, but they earn sufficient incomes to enable them to afford housing close to the centre of the city and near their work and to draw on a whole range of marketed services. These illustrations show that women, including ethnic minorities and mothers, can progress within organizations, though statistically most do not; both were unusual in their organizations, in fact the first high level women to become mothers and stay in their jobs. Progress to the highest echelons is therefore rare; caring responsibilities and working times are not the only barriers – prejudice and discrimination operating through male power structures continue to exclude and marginalize women.[47] Despite the new economy, the labour market remains structured by class, gender and ethnicity, with women, ethnic minorities and new migrants overrepresented in this low paid service sector work, on which professional workers depend.

For professional workers with middle range incomes, childcare costs would typically absorb a high percentage of their earnings, but both the short- and long-term gains of working are generally considered worthwhile. Conditions for low qualified lone mothers are particularly difficult. Heterosexual dual person households on lower incomes often manage by one partner, generally the woman, compromising their career in order to manage their work–life balance because they are unwilling to accept the long hours and/or because they are irrational given childcare costs.[48] For other women the lack of accessible and affordable childcare is still a major barrier to employment. The world of a high paid worker, taking taxis when necessary, with a flexible nanny and cleaner at home, able to purchase a whole range of domestic conveniences, is completely different from someone on low pay struggling around on public transport, having to meet nursery and school timetables with no domestic help, circumstances which tend to keep people out of the labour force. Again referring to London, Cara Quinn, younger than the new mothers referred to above, with no qualifications but three children at school or college, lives close by but on a council estate. She has largely been economically inactive, having been unable to reconcile paid work with school timetables, holidays and unpredictable closures as well as the children's illnesses. She currently has a part time job in a local supermarket and her earnings are supplemented by state benefits. Her flat is lovely, but the estate is old; there are lots of drugs, crime and violence and people keep to themselves. She stresses that she has always prioritized her children and would not like them to be looked after by 'strangers' but being a single parent with no qualifications in a context of limited availability of low cost childcare, her choices will have been highly constrained. In other European societies such as Sweden or Germany, her standard of living would probably have been higher, because of the different approaches by the state towards childcare and lone parents.[49]

The above illustrations provide some indications of the complex and mutually constitutive social divisions in the superstar regions, where they are particularly intense simply because these regions contain the highest paid jobs, thus earnings differentials are wider here than in other regions. Contemporary society continues to be hierarchically structured by social class, gender and ethnicity reflecting the significance of capitalist and patriarchal social relations, the colonial legacy and continuing racism and sexism. Given the feminization of employment and the expansion of ethnic minority professional workers, especially in the remaining public sector, it could be however that social class divisions are the most enduring and that in the context of the prevailing neo-liberal agenda they are widening, locally, nationally and internationally.

## SPATIAL DIVISIONS

Social divisions are expressed in the new urban landscapes by spatial segregation with different residential zones or blocks for different income or ethnic groups arising from the competitive bidding for land or market forces, planning or a combination of the two. Individuals have increasing freedom to choose their locations and these lead to spatial sorting by income, abilities, needs and lifestyle preferences as people 'join communities as consumers not as participants' and correspondingly move away as soon as 'better deals come along' (Reich 2001a: 191). In this respect people have increasing freedom to shape their lives and move away from communities they find stifling or unimaginative, noisy or dangerous, etc. but the ability to choose is income dependent; while the rich can switch between areas as their lifestyles and preferences change, the poor have few choices. Just as the low paid service sector workers often cannot afford to buy the services they deliver, so too the city becomes increasingly divided as the poor are excluded from new privatized or gated communities. Some low income neighbourhoods are similarly gated, for example the estate that Cara (referred to above) lived in was protected by several locked gates with entry phones to control access, though the austere prison-like appearance gave the impression that people were being locked in as much as protected.

This spatial sorting by social class leads away from the European model of closely intertwined mixes of land use and social classes to more specialized and privatized environments for those who can pay, while others are left with the diminished quality of services that the state can provide.[50] Clearly the quality of the urban environment depends on the overall balance of public and private expenditure and so will vary between states, and similarly land use patterns vary with the specific histories of cities. In central London, where public housing still exists, the super rich and the extremely poor live in close proximity even though they inhabit rather different worlds. More generally, where the quality of public provision is high then there is less desire to escape to private utopias. Conversely, Leisure Wood, a gated retirement community in Orange County, California with an average age of 75 years, turned itself into a public municipality simply to ensure that the residents' local taxes were spent only on services that they required, such as swimming pools, riding stables and gardening, rather than on schools or services for children, as they were before when administratively part of the wider county (Reich 2001a).

Peter Marcuse, in relation to the United States, distinguishes three forms of segregation (see Box 7.3); citadels (gated communities), enclaves, characterized by the voluntary presence of ethnic minorities or national groups, and ghettos,

---

*Box 7.3* Segregation in the city

*Citadel*
A citadel is a spatially concentrated area in which members of a particular population group, defined by its position of superiority, in power or wealth or status, in relation to its neighbours, congregate as a means of protecting or enhancing that position.

*Enclave*
An enclave is a spatially concentrated area in which members of a particular population or group, self defined by ethnicity or religion or otherwise, congregate as a means of enhancing their economic, social, political and/or cultural development.

*Ghetto*
An outcast ghetto is a ghetto in which ethnicity is combined with class in a spatially concentrated area with residents who are excluded from the mainstream of economic life of the surrounding society, which does not profit significantly from its existence.

*Source*: Marcuse (1997: 238–47).

---

the involuntary concentration of the poor. With some modifications, especially in relation to ghettos, this typification can be generalized. Gated communities have been expanding from 'Los Angeles to Moscow, from Mexico City to Beijing, from Istanbul to Mumbai' (Sklair 2002: 160), enclaves can be found in many cities, and while the connotations of ghettos are highly pejorative an increasing number of cities throughout the world have physically degraded, overcrowded and socially underprovided areas where people, often new to the city, are trapped by income rather than by choice, such as the slums of India, the *favelas* in Brazil, the informal settlements in South Africa, or the outer housing estates in the UK, though in this latter case areas are impoverished further by people moving out rather than by overcrowding.

Gated communities have existed for a long time and in a variety of forms. Sometimes they are created implicitly through urban design, for example by rerouteing traffic away from residential areas to provide peace and quiet, or by constructing rounded bus benches to discourage the homeless. Increasingly they have been developed to provide safety and seclusion for the rich by armed guards or electronic surveillance systems in a wide range of countries. Jo Beall (2002:1) for example argues that Mike Davis's depiction of Los Angeles could equally apply to Johannesburg:

> The carefully manicured lawns of Los Angeles's Westside sprout forests of ominous little signs warning: 'Armed Response!' Even richer neighbourhoods in the canyons and hillsides isolate themselves behind walls guarded by gun-toting private police and state-of-the-art electronic surveillance. . . . We live in 'fortress cities' brutally divided between 'fortified cells' of affluent society and 'places of terror' where the police battle the criminalized poor.
>
> (Davis 1990: 223–4)

Paradoxically while national and local states have been withdrawing from planning and advocating the virtues of the free market, private developers have been constructing highly planned communities with strict regulations which makes it clear that the issue is not between competing systems of the market versus state planning, but one of who is included and excluded from the plan, that is, who is the planning for and over what spatial scale. Implicitly developers have recognized that in land use decisions individual rationality leads to collective irrationality (see Chapter 8). Just as Marx (1976) pointed out in relation to the economy, private firms plan their organizations to the finest degree of detail, but reject overall planning as an infringement of their liberty, so gated communities are planned and managed to ensure a harmonious environment but only for the residents, in the context of a largely unplanned and potentially chaotic city.

Gated communities have been growing especially where social divisions have been widening, for example in Argentina, where an additional incentive has been the declining quality and reliability of public utilities, including water, as a consequence of privatization. Figure 7.3 shows the Lorenz curves for the city of Buenos Aires itself and Greater Buenos Aires. In both cases the distribution of income is highly uneven but more so in the city of Buenos Aires where the top decile have almost 50 per cent of the income in comparison with the lowest decile who have only 0.3 per cent. The difference between the two areas in part reflects the differential structure of employment with the city having a much higher proportion of the population (51 per cent) working in services, compared to 29 per cent in Greater Buenos Aires, and correspondingly a smaller proportion in industry (29 per cent) compared to 45 per cent in the latter (Pirez 2002; see also Ciccolella and Mignaqui 2002). The new, gated communities are largely on the edge of the metropolitan region and are close to but separate from the low income areas, which supply some of the personal services (Pirez 2002). Nordelta is the largest gated community in Argentina covering 1600 hectares and consists of:

> the local campus of a North American University, the Technical Institute of Buenos Aires and sundry elite schools. In the future, it will have services areas, tennis courts, golf courses and football pitches, amongst other sporting facilities. An optical high-speed communications both for Intranet and Internet, with free local calls. An electric train line will be jointly developed by a private company to make access to Buenos Aires faster.
>
> (Pirez 2002: 156)

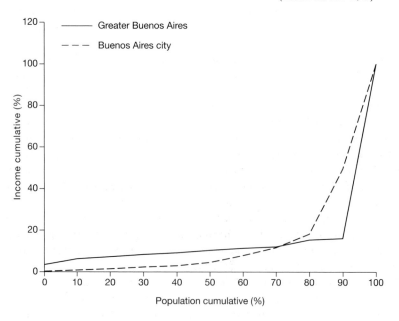

*Figure 7.3* Income distribution in Buenos Aires.

*Source*: Drawn from data in Pirez (2002).

Pedro Pirez (2002: 155) emphasizes the extent of control exercised by the developers in these communities including strict urban zoning, land use controls and building guidelines. Moreover, 'strong behavioural norms are also imposed on purchasers, with rules and regulations on ethics and cohabitation which operates as an admission (or exclusion) policy.' The wide range of services provided means that the communities become almost self-contained and a form of 'private government' emerges within these developments. With increased concerns about security such developments are likely to expand.[51]

In Johannesburg, in contrast to the apartheid era, gated communities are not now entirely white. The growing black middle class has been moving into gated communities as it too has become concerned about safety:

> before it used to be more on the whites you know, because in the townships you got this thing of saying 'the white man has got the money and that's my forefather's money'. Ja if you are a black they never used to bother you. And now because you are staying in the northern suburbs, you are just like a white man. 'You think you are a white man now? Why are you staying in the white suburbs?' So if you are a professional, if you have got a proper job and people know you have got a proper job, then you are also under threat. It's now equal. Whether you are a white man or a black man, as long as you've got money.
>
> (Beall 2002: 6; interview with a young African computer specialist working in a multinational company and living in a townhouse complex)

More generally fear of crime and insecurity was linked to racism in that those who were feared were invariably black, but at the same time the black middle classes were perceived differently and thus divisions were forming on the basis of social difference but a structurally racialized social difference. Forms of closure and self-exclusion have also been developing in socially disadvantaged areas. The hostels traditionally occupied by isiZulu-speaking rural migrants resisted their redevelopment into family housing in order to defend their space in the city[52] though this perhaps fits better with Marcuse's category of enclaves, which have also expanded in the context of globalization as a consequence of the increasing transnational mobility of people.

Sunil Roberts, the IT worker referred to earlier, lives in a small Indian enclave mainly of high tech workers in Morristown; similarly Monterey Park in southern California is known as Little Taipei and there are numerous enclaves around the world, for example the Turkish community in Berlin or in London. Generally, migrants or ethnic minorities choose to join enclaves, for social and economic reasons; they feel less excluded and more likely to find work through contacts. Some enclaves are prosperous, for example Morristown, but others less so. Moreover ethnicity is cross-cut by class and gender, so enclave dwellers benefit to different degrees. The exploitative working conditions in the garment industry in Chinatown, New York were referred to earlier.[53] Nevertheless there is some statistical evidence that suggests that migrants living in enclaves prosper more than migrants elsewhere; conversely black Americans living in ghettos fare worse the more segregated they are (Edin *et al.* 2000; Cutler and Glaeser 1997).

Ghettos are defined by Marcuse (1997) on the basis of class and ethnicity and refer specifically to people who are socially excluded from mainstream society and dependent on state benefits, informal and illegal work for survival. Other writers have used the term marginalized population or underclass (Wilson 1998) in relation to the United States. In Europe the term socially excluded is more widely used. In the United States ghettos have expanded as a result of discrimination in housing markets together with economic restructuring and the loss of manufacturing employment. The black working class and subsequently the black middle classes have moved away, which increases the concentration of poverty. It then becomes difficult to maintain properties, which fall into disrepair and then become occupied by ever poorer groups which in turn increases the racism and stigmatization towards those remaining in a cumulative and vicious cycle. Moreover with the retreat of the working class the poor lose access to the jobs network.[54]

There are contrasting explanations for ghettos, which focus on social disorganization, breakdown and anomie, on individual characteristics and failure and the socialization of people into a culture of dependency fostered by an overgenerous welfare system that undermines individual responsibility.[55] These representations overlook the way that social networks operate within ghettos such that it is not a case that they are disorganized but more that they are organized differently. Moreover they also imply that changing individual lives will lead to a change in a system that produces such inequality.

Even though individualistic explanations based on the supply characteristics of the residents are difficult to substantiate, recent social and regeneration policies are increasingly based on this assumption and directed towards increasing individual employability. These are often well meaning and have some positive impact for the individuals concerned but do little to address the structural processes leading to the problems they are designed to resolve. Thus while the individuals may change and move on and in some cases too the areas will change, other concentrations of poverty are likely to reappear. There is gentrification taking place in Harlem[56] for example and concern now that some of the new businesses moving in, though bringing jobs, are also displacing some of the older smaller-scale enterprises, in part following from Bill Clinton's decision to put his office in the area.

Marcuse's (1997) definition of ghettos refers specifically to the United States but in many countries there are spatially segregated areas of intense deprivation, which have increased in recent years in the context of global restructuring. Even in Cambodia, sometimes thought to be 'off the map',[57] the development of squatter settlements has been linked to the increased integration with the global economy. The spread effects from economic growth in neighbouring countries, the expansion of tourism, and criminal networks, especially in illegal migration and trafficking, have led to the development of 'opulent nightclubs, bars, hotels, and casinos, catering to a new Khmer elite, foreign investors, tourists and aid workers' which has boosted the cost of land and housing and displaced poorer people from the city (Shatkin 1998: 390). Thus the polarizing tendencies of contemporary global capitalism are far reaching and even Phnom Penh has 'taken on an increasingly cosmopolitan feel while emerging social problems related to the growing income gap are increasingly visible' (Shatkin 1998: 390).

## CONCLUSION

Globalization is associated with widening social and spatial divisions. Nowhere is this more evident than in the global cities, global city regions or superstar regions. These regions are pinnacles of wealth and attract people and resources from the rest of the globe. The wealth in these city regions far exceeds that of many nation states, reflected in their dazzling modern architecture, yet they also contain areas of intense poverty and deprivation and arguably display the widest differentials in income and well being found anywhere in the world.

In her original work, Saskia Sassen identified three global cities, London, New York and Tokyo, and, despite the growing number of global cities with increasing global economic integration, these three cities continue to outshine all others in terms of their role as centres of command and control of the global economy, especially in relation to finance. The term global city has very positive connotations; it implies importance, modernity and being at the centre of world affairs and not surprisingly many cities seek this standing, creating some fusion or confusion between the term global city as a status symbol and as an analytical concept. Theoretically the term global city relates to abstract space and consists of the controlling and coordinating activities of the global economy which as Sassen (2001b: 85) points out, could be a 'partially deterritorialized space of centrality', something perhaps more likely following the terrorist attack on the World Trade Center in Manhattan, though Sassen also discusses widening social polarization within global cities and makes reference to other activities found there such as manufacturing sweatshops. Other writers have focused on real physical spaces and used terms such as world cities, networked cities or global city regions to refer to global cities within their regional context and to bring a wider range of cities and modern high technology industrial districts and regions within this category. The broader definition is more inclusive and encompasses the wide range of activities associated with living and working in modern urbanized conglomerations. It includes the analysis of different forms of uneven development within them such as edge cities, cities with borrowed size, gated communities, enclaves and areas of urban decay. Some of these accounts have gone further and considered global city regions to be the new drivers of the global economy and political entities in their own right, issues discussed further in the following chapter.

In this Chapter I used the term superstar region as a spatial analogy to the economic definition of superstars in knowledge/dematerialized sectors, who in the absence of barriers created by distance are able to capture a large share of the market and income thereby creating wider income differentials between themselves and other workers in these sectors relative to others. Superstar regions are the spatial equivalent, in that they form a limited and exclusive group, act as magnets for people and resources and capture a greater share of the world market for their products. This term superstar has the advantage of having negative rather than positive connotations and overcomes the mystification embodied in the term global city which has legitimatized policies promoting the most affluent sectors in the belief that they are the key drivers of the modern economy, vital to competitiveness and economic growth at national and regional levels. Yet in quantitative terms, the distinctive sectors of global cities form only a relatively small proportion of the region's economy, and their impact on the city as a whole as well as the actual effectiveness of the policies to support them is generally unknown. The term superstar by contrast is much less likely to be received with such uncritical acclaim but more importantly it has the advantage of being a theoretical term and one that implicitly embodies the inequality and widening social and spatial divisions which are intrinsic to these regions, both in terms of their relationship to other regions nationally and internationally and in terms of their own internal composition.

Similar to the concept of global city it can be used as a descriptive term to refer to a specific empirical entity, but without breaking this term down into the elements of which it is composed, it would remain a chaotic or fuzzy concept. To be useful, it has to draw upon other theories, in particular the new international division of labour, global value chains and the clustering perspectives, which were elaborated in detail in Chapter 3, and which explain the processes shaping uneven development between firms and regions. It also draws upon the new economy perspective elaborated in Chapter 1 to account for the widening spatial and social divisions, the increasing differentiation between regions and their internal fragmentation, in particular the growing income gaps between knowledge workers and others who supply their reproductive services. Even this approach is not complete however as the tendencies referred to are all market tendencies, and thus rooted at one level in the daily competitive struggle between firms and between firms and their workers over the share of value, and at another level by the specific economic and institutional framework which shapes the precise form taken by the market processes. While many countries now follow frameworks broadly consistent with the neo-liberal agenda, there are still differences between states. Thus inequalities are lower where elements of the social democratic or conservative corporatist regimes remain and in countries where these welfare regimes have not developed in this explicit way there are still differences, for example in Costa Rica the extent of inequality is lower than in Argentina and Brazil, largely because of an active state that has ameliorated the impact of the market. This chapter has paid particular attention to some of the social and spatial inequalities through the experiences of high and low paid workers and by discussing the different living spaces within these superstar regions. It has referred to social divisions by class, ethnicity and gender and tentatively suggested that the divisions by social class and gender seem to be the most enduring, though inflected by ethnicity. The extent of these divisions varies considerably between countries, according to the role played by the state, especially the national state, and also by the actions of voluntary groups and people pressing for change, considered in Chapter 9.

## Further reading

C. Hamnett (2003) *Unequal City: London in the Global Arena*, London: Routledge.
A. Mir, B. Mathew and R. Mir (2000) 'The codes of migration. contours of the global software labor market', *Cultural Dynamics* 12 (1): 5–33.
S. Sassen (ed.) (2002b) *Global Networks, Linked Cities*, London: Routledge.
A. Scott (ed.) (2001) *Global City-Regions*, Oxford: Oxford University Press.

## Notes

1  See Dunford and Fielding (1997) for a discussion of the relative degree of sectoral and occupational specialization in London and the south east region in the 1980s and especially the high concentration of financial services. For information on employment change by sector for the period between 1978 and 2000 see Buck *et al.* (2002).

2  See for example the collection by Scott *et al.* (2001), in particular the chapters by Hall (2001) and Sassen (2001a), and for detailed measurement of global cities the work of Peter Taylor and colleagues at Loughborough University (Taylor *et al.* 2002).

3  Johannesburg and Kuala Lumpur are Gamma world cities on Taylor, Catalano and Walker's (2002) definition, which means that they are global service centres for at least two sectors (from accountancy, advertising, banking and law) and in one of these a major centre.

4  See Scott *et al.* (2001).

5  See Fainstein *et al.* (1992).

6  Localization economies are external economies of scale that derive from proximity to other activities within the sector and urbanization economies arise from proximity to related activities but in different sectors and from the agglomeration itself, for example the presence of a highly skilled labour force.

7  The economic properties of this kind of work are discussed in more detail in Chapters 1 and 4.

8  In the case of London, workers in the internationally oriented financial services sector earn up to a third more than those working for domestically oriented firms.

9  Similar to the export of agricultural commodities, the high value activities associated with tourism usually occur in the richer countries.

10  Hsinchu is a science based industrial park and there is a 50 mile industrial area linking it to Taipei (see Saxenian and Hsu 2001).

11  See Mir, Mathew and Mir 2000; Walcott 2002.

12  See Garreau (1991).

13  See Gordon (2003).

14  In this sense it would be similar to the growth pole concept also conceived in abstract space (see Perroux 1950).

15  Marcuse (2002) provides details of some office decentralization from downtown Manhattan, but so far this has been to areas within the New York metropolitan region; thus Goldman Sachs is moving to Jersey City, just across the River Hudson, but still in a high skyscraper.

16  See also Pirez (2002).

17  Indeed there can be a blurring of boundaries between academic research and the regional boosterism of development agencies.

18  With reference to the multimedia super corridor in Malaysia, discussed in Chapter 6, Tim Bunnell (2002) argues that its development was part of the post-colonial response to past technological and political domination.

19  For example Foster and Partners designed amongst other things Chek Lap Kok Airport in Hong Kong, the new German Parliament in the Reichstag, Berlin, the Great Court for the British Museum, the headquarters for HSBC in Hong Kong and London, Commerzbank headquarters in Frankfurt, the Metro Bilbao, and currently has offices in London, Berlin and Singapore, with over 500 employees worldwide.

20  To be considered worthy in national rankings an international or global reputation is a prerequisite.

21  Ian Gordon (2002) raises similar issues by adapting central place theory to suggest that as the barrier of distance has diminished theoretically the potential for one central place arises, though he suggests that because of horizontal specialization between different activities there will be a number of central places.

22  Jennifer Robinson (2002) suggests that if they had been termed new industrial districts of transnational management and control they would have attracted less attention. She also locates some of the desire to become a global city in the dominance of western urban theory.

23  See Causer and Williams (2002a and 2002b).

24  See Davis (2000).

25  This building was designed by a leading Japanese architect, Tange Kenzō, and has been described as the last great edifice of postmodernism. Close by are three linked towers all topped by glass pyramids and though also influenced by Kenzō contain a whole range of multinational firms and hotels.

26  See Alarcón (1999).

27  See Balasubramanyam and Balasubramanyam 2000; Saxenian and Hsu 2001; Walcott 2002).

28  Although only about 30 per cent of the engineers who studied in the US returned to Taiwan in the 1990s compared to only 10 per cent in the 1970s (Saxenian and Hsu 2001).

29  Taiwan sent the highest number of doctoral students to the US during the 1980s, including entire graduating classes from the elite universities in Taiwan (Saxenian and Hsu 2001).

30  H1-B visas allow people to work temporarily in the United States. Employers have to show that no appropriately trained US workers are able to fill the job, that no US workers will be displaced as a result of the job and that the workers will receive the same terms and conditions as US workers. However these conditions are not monitored effectively or strictly adhered to leading to phrases such as 'techno-bracero' or 'techno-coolies' implying that these technical workers are similar to other low skilled migrant workers as they often end up working for low wages (Mir et al. 2000).

31  See Mir et al. (2000).

32  It is more likely that they will be selectively enforced. At the time of writing the

migrants who are Muslim face particular scrutiny, owing to their perceived links with terrorism.

33 See Ehrenrich (2001), Abrams (2002) and Toynbee (2003).
34 See Baumol (1967), Quah (1996) and Chapter 1.
35 See Chapter 4.
36 See Anderson and Phizacklea (1997), Lutz (2002) and Chapter 4.
37 See Cabinet Office (2003).
38 See Marcuse (1997) and Castles (2000).
39 See also Chapter 4.
40 Look back to Table 5.2.
41 More formally the increases were in the top three highest paid occupational groups, managers and administrators, professional occupations and associate professional and technical occupations (GLA 2002).
42 Earnings data relate to the period 1987 to 2000.
43 The inter-decile ratio being 4.3 for men in London compared to 3.6 nationally. There is no significant difference between the London and UK figures for women (GLA 2002).
44 See also Chapter 4 for a more detailed discussion.
45 The research reported in this section comes from an ESRC financed project on living and labouring in London, carried out by the author in association with Linda McDowell, Kath Ray, Colette Fagan and Kevin Ward and the author's own Leverhulme financed study on Brighton and Hove. These extracts were from interviews carried out by the author in November/December 2002. All names are fictitious.
46 Though her nanny did work 55 hours a week so 15 of these should have been at a premium rate.
47 See Cockburn (1991); McDowell (1997).
48 See Jarvis (2002).
49 See Lewis (1997).
50 Recent plans for the regeneration of urban areas have suggested a greater mixing of land use (see DETR 1999) but this has tended to result in the poor being displaced from attractive central city sites.
51 See Marcuse (2002).
52 See Beall (2002).
53 See Anderson (1998).
54 See Wacquant (1997); Wilson (1998); Sanchez-Jankowski (1999).
55 See Murray (1999).
56 See Smith (1996).
57 See Robinson (2002).

# Part IV

# SHAPING DEVELOPMENT

Looking back over the last half century there have been enormous economic, social and technological changes; people have moved from dependence on basic agricultural goods to a wide range of industrially and digitally produced products, can communicate between each other almost instantaneously across different parts of the globe, work in high tech buildings towering thousands of feet above the ground and have even been into space. These developments and possibilities can be found in at least some places in nearly all countries around the world but while many people share in these developments many also remain excluded and the overall increase in living standards, literacy, health and quality of life serves only to highlight the plight and injustice experienced by the excluded found mainly in poorer countries but also in close proximity to richer people in the wealthiest cities and countries of the world. Further, despite the ending of the cold war, conflicts within and between countries continue, resulting in death and destruction, systematic violence against women and the vast displacements of people – refugees and asylum seekers. Perhaps one of the most tragic consequences of contemporary unequal development in the globally connected world is the rapid expansion of trafficking in women and children.

These changes and patterns of development all result from conscious human action but within constraints over which people have varying degrees of control. More powerful nations, organizations, corporations, groups and individuals will have imposed some of the changes over less powerful ones, while other changes will be the outcome of negotiation between different groups, resistance by the less powerful, or of individually constrained or free choice. So it is important to ask where are the sites of power in the global economy and how can ordinary people shape their lives and locations and protect themselves from the adverse consequences of contemporary developments? These two questions form the central theme of Part IV. Chapter 8 considers the sites of power by exploring the roles and functions of the state theoretically and at different levels and the extent to which these have developed and changed in the context of globalization. Chapter 9 considers how people, including non-governmental organizations, anti-capitalist protesters, regional and social movements, have challenged the more powerful states and institutions and tried to develop more inclusive patterns of development.

Interestingly in the last decade some of the contradictions of liberal market economics have been recognized in official discourse and the state at various levels has taken measures to include some of the demands of protest groups, for example in ethical trading and through empowering people through greater participation in planning and public affairs, although the effects of these developments are often contradictory. The World Bank and the IMF have recently stated that more socially inclusive patterns of development may be more conducive to overall economic growth and prosperity and since the mid-1990s the World Bank has been advocating a form of participatory development within the projects it funds. National and local states have similarly been involving people in the determination of local area regeneration plans. Furthermore, in response to NGO pressure and consumer demands, large corporations have been introducing voluntary ethical codes of conduct which seek to improve the labour standards of people working on goods in their supply chains. These forms of inclusion are not without criticism and contradiction but nevertheless indicate that while liberal market economic philosophies may be dominant there is now at least some official recognition of their contradictions. Part IV and indeed the book concludes by arguing for more participatory democracy in order to increase the influence that people have over the events shaping their lives. It argues further that representation is more likely to be effective if it takes place through existing geographically nested units such as localities, regions and states as people are material beings, living out their lives in real geographically rooted spaces, or places where many of the processes, which may extend across these boundaries, come together and shape their immediate existence, even though they are also engaged in many different networks which extend beyond their immediate locality.

# 8

# SOCIAL REPRODUCTION
# AND THE STATE

The economy operates on a global scale but social reproduction takes place within definable territorial units – the household, cities, regions and the nation state – and while capital may flow ceaselessly around the world, people, even the 'global astronauts' referred to in the previous chapter, have to touch down from time to time to recuperate and recharge their batteries. Individuals generally renew themselves within households and families and overall social reproduction takes place within cities, regions and the nation state. In the current context the stability of families is waning, inequality is widening and there is a growing imbalance between the economy, which increasingly operates on a global scale, and the powers of nation states, which are primarily national, which potentially threatens individual and social sustainability.

To retain power and influence within the new global economy, states have become international in orientation and have transferred some of their powers and responsibilities to supra national institutions. To maintain social reproduction other duties and functions have been handed down to cities and regions. This apparent shifting of responsibilities to the supra and sub-national levels has led to the hollowing out thesis, which suggests that the powers of the nation state have weakened as a consequence. By reviewing some of the functions of the state this chapter suggests that it is more a case of changes in the role of the state and the relative powers of nation states, in particular the rise of the United States as a superstar state with unprecedented economic and political power, than a universal diminution in the powers of nation states, relative to supra or sub-national institutions.

The increasing adherence to a neo-liberal agenda, especially the decline in welfare support, undermines social reproduction, which is further undermined by the impact of neo-liberal policies on household reproduction. In particular deregulation and privatization have led to increasingly flexible and insecure working patterns and falling real incomes especially among low paid workers. There have also been changes in household composition, a tendency towards family fragmentation and more individualized lifestyles. The extent and impact of change varies between countries and between social groups. Thus while states are contributing towards and responding to contemporary changes in similar ways, outcomes differ, reflecting the different positioning of states in the international division of labour, their level of wealth and capacity for choice as well as the differential balance of power between different social groups or classes within national territories. This chapter considers how states have shaped and adapted to the new context and the impact this has had on people in different places and in particular on social sustainability.

To analyse the extent to which the change in the orientation of the state and the rearrangements of powers affect economic and social reproduction the chapter begins by revisiting some traditional theories of the role of the state. It considers how these traditional functions have been affected by globalization and how the state has responded by developing and utilizing supra and sub-national institutions and by making individuals more responsible for their own reproduction. The chapter reviews the roles of supra national, regional and urban institutions and concludes by examining the problems of social reproduction or social sustainability in the new economy. As a prelude it is important to note that by the state I am referring to national governments and supportive institutions such as the police, the military, civil service and the growing range of institutions through which public affairs are administered. By nation state I am referring to the common sense view of the state – or to a 'formally sovereign territorial state presiding over national territory' (Jessop 2002a: 174). By social reproduction I am referring to the processes through which a society sustains itself from day to day and from one generation to the next.

## TRADITIONAL ROLE AND POWER OF THE CAPITALIST NATION STATE

What is meant by the state and what role do states play? The contemporary form of the nation state came into being in the course of the eighteenth and nineteenth centuries in Western Europe when single authorities were able to claim and legitimize sovereignty over defined territorial units. This power was extended transnationally in the colonial era and following liberation/ independence the entire globe is divided into nation states.[1]

Past western political theorists such as Aristotle, and later Thomas Hobbes, John Locke and Jacques Rousseau, all considered that a state or sovereign was necessary to bring about social order and the common good. Most evocatively, Thomas Hobbes (1983 and 1996) writing in 1647 argued that without the state there would be no society, no social progress and no art or culture. In the state of nature, Hobbes maintained that human beings (men) were equal, self-interested and prone to conflict, especially when resources were scarce, so there was a continual 'war of all against all' as people lived in constant fear that others would steal their resources thereby making life 'nasty, brutish and short'.[2] To counter this outcome people would form contracts in which they agreed to respect each other's resources so that they could realize the fruits of their endeavour. Each individual believes, however, that the other is also self-interested and so is continually tempted to renege on the contract. In Hobbes's perspective this uncertainty creates the desire for an all powerful sovereign or state to enforce contracts in the collective interest and by so doing allow individual and social living standards to rise.

Jean-Jacques Rousseau (1968) (writing in 1762) also believed in the idea of a common good or general will, defined as that which rational people believe would be good for everyone. Similarly, this would not come about naturally, as the general will was not equivalent to the sum of individual wills or the will of all. Thus for Hobbes and Rousseau individual rationality or preferences do not necessarily lead to collective welfare, so an all powerful state is necessary to bring the common good into being; that is, effectively, individuals have to give up their freedom to the state in order to secure their freedom to enjoy life and realize universal welfare.[3]

Within this perspective the state is created through consent, but until quite recently it was largely concerned with 'public' issues and correspondingly a very male sphere. Private property was considered sacrosanct, and women and children were considered to be male property. They were largely confined to the home or the private sphere where gender inequality was the norm. Thus the social contract, which recognized that 'all *men* were born equal', was implicitly underwritten by an unequal sexual contract (Pateman 1988) and while the term 'men' is used in its generic sense to refer to all people, in practice it related more to men than to women. Historically, therefore, the state developed as a patriarchal state, which for the most part concerned itself with public issues such as international affairs and the reproduction of the economy. At the local level the state supported day to day reproduction in a limited and largely punitive way through the Poor Laws, which for a brief period at the end of the eighteenth and early nineteenth centuries became quite liberal, but otherwise day to day organization of social reproduction was left to the private sphere of the household.[4] Only later, towards the end of the nineteenth century, did the state assume some responsibility for social existence. The contradictions arising from this separation of spheres are discussed towards the end of the chapter.

In a contrasting vein other writers argued that the state undermines both individual freedom and market efficiency. Adam Smith, writing in 1776, advocated free unfettered markets because: 'It is not from the benevolence of the butcher, the brewer, or the baker, that we expect our dinner, but from their regard to their own interest' (Smith 1976: 18). Thus, according to Smith, self-interest ensures that people produce things that are of value to others; otherwise they would not be able to survive. So according to Smith, in market societies individuals acting in their own self-interest will lead to collective well-being:

> By directing that industry in such a manner[5] as its produce may be of greatest value, he[6] intends only his own gain, and he is in this, as in many other cases, led by an invisible hand to promote an end which was no part of his intention. By pursuing his own interest he

frequently promotes that of the society more effectually than when he really intends to promote it. I have never known much good done by those who affected to trade for the common good.

(Smith 1976: 477–8)

Smith nevertheless argued that the state should provide a framework within which the market could function effectively. The state should therefore enforce contracts, regulate patents to stimulate innovation and develop large-scale public works, such as roads and bridges, to further national wealth. Further the quotation above, often used to justify free markets, comes from a chapter which argues for restraints on certain kinds of imports in order to encourage domestic industry. Thus Adam Smith was not quite the advocate of unfettered free markets as is sometimes believed, indeed much less so than contemporary 'IMF technocrats' (see below). Nevertheless the liberal tradition is associated with a minimalist state, on the one hand because state action is thought to interfere with fundamental freedom or what were seen as natural rights of individuals, and second because it is believed that the state cannot be trusted to act effectively in a market economy, because of its sheer complexity. John Stuart Mill (1989), for example, writing in 1869, when the state was becoming more active in economic and social affairs, argued strongly against the 'tyranny of the majority' and maintained that the only justification for the state to infringe individual freedom was to prevent individuals from causing harm to others, except in the case of individuals who because of age or incapacity were incapable of rational thought.[7] Focusing on economic issues and the complexity of the market, William Stanley Jevons (1911), writing in 1871, argued that interfering with the market was doomed to failure and equivalent to asking a child to dismantle and rebuild a clock. Later, Friedrich von Hayek (1945), whose work provided the intellectual background to Thatcherism, likewise emphasized the complexity of markets and the impossibility of any single authority having the necessary knowledge to orchestrate them effectively. For Hayek, business knowledge was time and space specific and thus while accessible to the immediate entrepreneur was difficult to codify and therefore regulate. Furthermore, the outcome of any action often had unknown and unintended consequences, so the impact of state action or planning was always unpredictable.[8] For Hayek the beauty of markets lay in their simplicity, in that all the information necessary for decision making was summarized in prices. Thus, similarly to Smith, he argued that by allowing individuals to act in their own interests in response to price signals societies would through a process of trial and error gradually evolve towards collectively desirable outcomes.[9] Any errors or inefficiencies would be eliminated through competition so the role of the state was simply to allow this spontaneous order to unfold. This minimalist view of the state underlay the Thatcher/Reagan project and their quest to roll back the state through deregulation and privatization in order to allow markets to function freely and as a consequence raise economic efficiency. It also underpins the contemporary neo-liberal agenda.

Karl Polanyi (1957) however, writing in the mid-twentieth century, was acutely aware of the limitations of market economies and in particular that there was a world of difference between utilizing the market mechanism to allocate some goods and services and subordinating the whole of society to the market. Moreover, in contrast to Smith (and twenty-first century neo-liberals) he argued that there was nothing natural or evolutionary or even 'free' about markets. For him they did not evolve from individual predispositions to 'truck, barter and exchange' (Smith 1976: 20) but rather by conscious and often violent actions by the state, which repeatedly overturned long standing social rights in the process. State violence was frequently used to create markets in labour and land. To create a labour market[10] and provide workers to sustain the industrial revolution in England, people were forcibly evicted from the common lands through enclosure (or what would now be termed privatization) and the Poor Laws were amended to become more draconian, supplying relief only to those prepared to go into the workhouse. Likewise in the British colonies, land and labour markets were created simultaneously by removing indigenous people from the land and forcing them to pay taxes in money that they could only acquire by working on the capitalist farms. The state had to create these markets because land, money and labour are in Polanyi's terms 'fictitious commodities' because although they are essential to a market society, they cannot be produced by it in the same way as true commodities, such as apples and pears. Marx (1976) illustrates the fictitious nature of labour as a commodity in his discussion of how a labour force was created in Australia, then under British colonial rule (see Box 8.1). For Polanyi, subjecting labour and land to the market mechanism means that the very substance of society is subordinated to the economy, something which is now happening on a global scale.

Polanyi (1957: 3) points out however that a pure market society only lasted for about 30 years in England in the first half of the nineteenth century, before it became unsustainable, that is the society that was created was a very 'stark utopia' and was in danger of being unable to reproduce itself. Thus, 'inevitably society took measures to protect itself' and introduced a series of ad hoc regulations relating to working time, industrial safety, health and education. These measures demanded by workers and supported by an enlightened bourgeoisie were the precursors to the contemporary welfare state. In the mid-twentieth century these measures were combined with Keynesian macro-economic management, and strengthened again during the era of Fordism (see Chapter 5) when it became increasingly accepted that the market could not be relied upon to produce optimal quantities of goods such as health, education or housing which are associated with social as well as private benefits.[11] Public utilities were also regulated or nationalized in order to ensure efficient provision and universal coverage. Altogether these measures contributed to rising levels of state expenditure and to the unprecedented expansion of western economies in the third quartile of the twentieth century. While Polanyi focuses on England, similar developments were taking place in other European countries.

---

*Box 8.1* Creating a free wage labour force

Marx is commenting on Edward Gibbon Wakefield who developed a theory of colonization and was involved in establishing a permanent settlement in New Zealand in 1839. The problem he encountered was obtaining a labour force, because land was plentiful so people found they could set up as independent producers, that is they had independent access to the means of production. He describes Wakefield's problems in the following way:

> First of all, Wakefield discovered that, in the colonies, property in money, means of subsistence, machines and other means of production does not as yet stamp man as a capitalist if the essential component to these things is missing: the wage-labourer, the other man, who is compelled to sell himself of his own free will. He discovered that capital is not a thing, but a social relation between persons which is mediated through things. A Mr Peel, he complains, took with him from England to the Swan River district of Western Australia means of subsistence and of production to the amount of £50,000. This Mr Peel even had the foresight to bring besides, 3,000 persons of the working class, men, women and children. Once he arrived at his destination, 'Mr Peel was left without a servant to make his bed or fetch him water from the river'. Unhappy Mr Peel, who provided for everything except the export of English relations of production to Swan River!
>
> (Marx 1976: 932–3)

---

There are parallels between Polanyi's account of nineteenth century England and the contemporary revival of neo-liberalism in the late twentieth century both in terms of the strong advocation of free market policies and, to a lesser extent, with the subsequent reaction to the adverse outcomes.[12] The United States and the UK have embraced neo-liberalism following the collapse of Fordism, and despite important differences between European states it is also evident in the form taken by increased economic integration in the European Union. In addition neo-liberalism has been imposed on states seeking assistance from supra national institutions: in response to the economic crisis in Latin America in the 1980s and 1990s, in the East Asian crisis in the mid-1990s and following the collapse of the Soviet system in Russia and Eastern Europe. Manuel Castells (1997: 269) attributes the fervour with which 'IMF technocrats' enforced their policies and removed all 'remnants of political controls over market forces' in Africa, Latin America and Russia in the early 1990s to a 'deep seated, honest, ideological commitment to market economics as the only way to build a new society, rather than anything to do with "capitalist domination".[13] Irrespective of individual motivations, their policies nevertheless secured the interests of the internal and external dominant social classes, or in the case of the transition countries the emergent capitalist class, and in this respect the Marxist analysis of the state has a more powerful and differentiated understanding of social interest. Contemporary writers in the Marxist tradition have developed a sophisticated understanding of the mechanisms through which different interests compete for power, control and hegemony within the state.[14]

Recognizing society as a capitalist society, founded on class relations rather than being natural or preordained, Marx and Engels (1983: 16) argued that the state was not above and beyond civil society acting in the interests of the common good but a class state that would act in the interests of the dominant social class. Hence they argued that 'the executive of the modern state is but a committee for managing the common affairs of the whole bourgeoisie'. In their view therefore the state would reproduce rather than challenge the inherent inequalities in capitalist society. This does not mean however that the working classes will never benefit from state action. Referring back to Chapter 1 and consistent with the ideas of Hobbes and Rousseau, individuals (including capitalists) acting according to their immediate self-interest will not necessarily secure their long-term goals. For example, while individual capitalists may prefer lower taxes and lower wages they also gain from well paid workers who provide a healthy labour force and a market for their products. Likewise, as Marx pointed out, there are differences within the capitalist class. For example, in his day, there was a conflict between the owners of large-scale machine based industry who supported workers, demands for reductions in the length of the working day while smaller producers, reliant on labour intensive methods, maintained that such cuts would eliminate all profit.[15] Thus in this perspective the state steps in and tries to act as sovereign in the longer-term interests of the dominant fractions within the capitalist classes, which may differ from both individual capitalists' self-interest and from the 'general will' or the interests of everyone, that is, all social classes, assuming such a general interest could be defined.[16] Accordingly the role of the state in capitalist society has consolidated around a number of functions, which aim to ensure the reproduction of *capitalist* society, though the scale and manner of provision differs between states according to the level of development, geopolitical position, the internal balance of class forces or class fractions and their differential abilities to compete or collaborate to press their claims.[17] Just as there is a struggle over the social product, there is also a struggle for power within the state, so specific outcomes vary over time and across space leading to different varieties of capitalism, as illustrated by different welfare regimes, or pathways discussed in Chapter 1 and referred to again at the end of this chapter. Moreover, because the state tries to act in this way it does not mean that it necessarily succeeds. As Hayek (1945) pointed out actions have unintended consequences and as Marx (1976) always emphasized change takes place through a dialectical process of action and reaction. Even so it is important to identify these basic functions in order to understand why and how the form and functions of the state have changed with the growing discontinuity between global economic space and national political space.

The first task of the nation state is to secure its borders and maintain internal peace and to do so the state has armed services and a near monopoly over legitimate violence.[18] Once this basic task has been fulfilled the key role of the state is to secure the general conditions for economic and social reproduction, day to day and between generations, which involves the kinds of tasks identified by Adam Smith: namely, to guarantee property rights, to enforce contracts through the development of legal and monetary systems, to support innovation through patents, to manage international trade, to provide large-scale public works and a taxation system to finance these activities. During the twentieth century the state expanded its role in regulating the economy, in particular to manage effective demand and avoid crises of underinvestment or overproduction by ensuring effective demand for the output of the mass production associated with the introduction of Fordism.[19] The state also has to ensure the availability of a labour force of an appropriate size, and thus manages migration and regulates unemployment by introducing special measures such as training schemes in times of labour surplus and ways of coercing people to work in times of shortage. Additionally there is an array of goods and services ranging from street lighting, transport, water, health and education to low income housing and care, that are vital to the effective functioning of a modern economy, but where market imperfections are so endemic that they would be unlikely to be produced competitively, efficiently or in sufficient quantities by private capital. These traditional functions of the state relate mainly to the public sphere because until recently[20] the reproduction of the labour force took place largely within the private sphere of the home, through unpaid female labour. With the increasing feminization of the labour force, discussed in Part II, there is growing concern about social as well as economic sustainability.[21] Thus states are becoming involved in childcare. In the EU for example there is a Directive on Parental Leave and states are asked to monitor childcare provision. The rationale for state action arises therefore because the market cannot even guarantee economic and social reproduction of itself, i.e. a market society, let alone what might be construed as a more desirable social system.

Land use and location are particularly clear examples of where individual rationality leads to collective irrationality. This arises in part because, following Polanyi (1957), land is not a true commodity. The supply of land is not limitless and uses cannot change instantly. Moreover in land transactions two goods are purchased simultaneously in one transaction, a physical site, i.e. the land itself, and its relative location. Sometimes land prices reflect existing uses or relative locations but discontinuities can arise, so states generally have some form of land use regulations or zoning to ensure compatibility between uses to maintain property values. Without any controls, for example, a mobile phone base station could locate in an area of luxury housing. It would probably depress house prices, thereby rendering the purchasing decisions of existing residents irrational. With the tendency towards deregulation, states have relaxed some of these land use controls, in some cases resulting in developers making their own rules to ensure compatibility, for example in the gated communities (see Chapter 7) but the city as a whole can be left with sub-optimal outcomes.

Harold Hotelling (1929) illustrated the divergence between individual and socially optimum outcomes by his classic example of the ice cream sellers. Under certain conditions, two ice cream sellers on a linear beach would gravitate towards the middle rather than space themselves at the quartiles, which would be the socially optimal solution from the consumers' perspective (see Figure 8.1). This example illustrates the more general tendency for market solutions to differ from the socially optimum. The media and political parties also tend to drift towards the middle ground to maximize sales or support; they can discount marginal preferences because they are unlikely to defect to the opposition. Private decisions also tend to be myopic and prevent the realization of more socially desirable outcomes in the longer term. A more general illustration of the divergence between private and socially optimal outcomes arising from individuals acting in their own interests is summarized by the prisoner's dilemma (see Box 8.2). These examples endorse the idea that the socially desirable outcome, even from the perspective of the capitalist classes, does not always arise from the summation of individual decisions.

The market is even less reliable in relation to longer-term changes. As pointed out in Chapter 3, historically, during most periods of economic transition, new growth trajectories have depended on significant increases in state involvement in economic management and modernization.[22] Certainly, the economies that grew most rapidly towards the end of the twentieth century, the 'four tigers' and other NICs in South East Asia, were associated with significant degrees of state involvement and the subsequent crisis in that region can be attributed more to their increased openness rather than to the state (see Box 8.3).[23] Some nation states keep tighter degrees of control than others, thus while Russia and Eastern Europe have embraced free market ideology wholeheartedly, the Chinese state retains control over the framework within which it has allowed a market

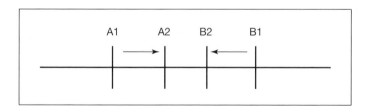

*Figure 8.1* Individual rationality and collective irrationality: the ice cream sellers.

*Note:* The starting-points for the two ice cream sellers are A1 and B1, which is the social optimum position. The consumers are evenly distributed along the beach. To increase their sales A will encroach on B's territory and likewise B will encroach on A's territory and this process will continue until the two sellers end up side by side in the middle of the beach, which is the private solution but sub-optimum from the social perspective. Neither need fear a loss of sales from the people at the ends of the beach as their demand is assumed to be inelastic, i.e. they will still buy ice cream even though they have to travel further.

*Box 8.2* The prisoner's dilemma

| Prisoner 1 | | Prisoner 2 | |
| --- | --- | --- | --- |
| | | Confess | Not confess |
| Confess | | p, p | p, P |
| Not confess | | P, p | F, F |

The prisoner's dilemma is a classic puzzle and illustrates that individuals acting in their own self-interest does not lead to a collectively rational outcome.

The situation envisaged is of two prisoners who have been caught for a crime they did commit. If they confess, their punishment will be less than if they remain silent but are then found guilty. A conviction however depends on their confessions, so if they do not confess there is a chance they will go free. The prisoners are held in separate cells so neither knows what the other will do. They therefore have a dilemma, because each knows that if they cooperate and keep silent then they will both go free (F). But if they do not confess but their partner in crime does, then they will be punished more harshly (P) than if they do confess (p). Acting in their individual self-interest and in a situation of uncertainty, where they cannot trust their partner because they know that there are marginal gains from confessing, the probability is that they will both confess and both end up in prison. This means that they both avoid the harshest sentence but fail to achieve the optimum outcome, which would be to go free.

This puzzle has been used to illustrate a range of issues which are also referred to sometimes as the free rider problem from the development of urban dereliction to the failure of private companies to engage in training policies, where the collectively desirable outcome fails to come about in the absence of some overriding authority or the state to bring it into being.

society to develop and in relation to Hong Kong, sees itself as a one nation, two system society. Thus somewhat ironically, when the Chinese state, still formally run by the communist party, issued a hundred year bond in 1996 it did so through one of the major capitalist corporations, J P Morgan in New York, rather than dealing with the US state, which illustrates how states too have become players in the global market.[24]

While there is a certain amount of consensus that the state should intervene in cases of market failure, it nevertheless remains a class state hence there is less consensus over what constitutes market failure, about how the state should respond to it, and more generally over the areas of life with which the state should become involved. Even in land use decisions, while the state may use zoning to secure property values, it does not generally act to affect the socially

*Box 8.3* Global financial instability: the Asian financial crisis

The Asian financial crisis was triggered by the devaluation of the Thai baht in July 1997, which had a domino effect on other countries in the region – Malaysia, Indonesia, South Korea, Taiwan, Hong Kong and the Philippines – and implications for many other countries as far apart as Brazil and Russia and even for north east England.

The crisis was effectively caused by financial speculation, which led to a devaluation of the currency but has had significant effects on the real economy including dramatic falls in GDP per capita as well as longer-term implications for the Asian model of development. Prior to the crisis these countries had strong real economies, high rates of growth, low inflation and low levels of government borrowing.

*Background*

1. The Asian model of development* was based on controlled relations with the world economy and close relations between the state, industry and the banks. Industrial investment was financed largely from the comparatively very high levels of household savings via the banks rather than via equity. Although banks lend on a long-term basis, firms are vulnerable to shocks which disturb their revenue stream as debt is repaid on a fixed basis not according to profits, as in the case of equity. For this reason banks, firms and the government worked closely together to minimize shocks.
2. Exchange rates were largely fixed and the currencies were pegged to the dollar.
3. In early to mid-1990s there was a change to the Asian model of development. Capital markets were liberalized and domestic controls loosened, stimulated by internal and external profit seekers and IMF pressure. There was competition between Thailand, Malaysia, Singapore and Hong Kong to become the regional financial centre and this entailed liberalization. The development of a financial services sector was seen as a source of high level employment and economic growth. Liberalization generated vast inflows of international capital (between 1994 and 1996 there was an increase from $47 billion to $96 billion to South Korea, Indonesia, Malaysia, Thailand and the Philippines) arising from excess liquidity in world markets in the context of low rates of growth in Japan and Western Europe. These funds could be borrowed more cheaply than domestic capital leading to a large amount of foreign debt that had to be repaid in dollars. Thus there were high levels of domestic and foreign debt, which was sustainable as long as the industries remained competitive and repayments could be made.

continued . . .

*Immediate causes of the crisis*

1. With the fixed exchange rate system foreign borrowing had to be matched by expanding the domestic money supply, leading to inflation.
2. The currencies appreciated relative to the Japanese yen and Chinese yuan which squeezed exports and increased import prices – leading to current account deficits of between 4% and 8% of GDP.
3. Declining profitability in the manufacturing export sector, domestic inflation, and the continuing propensity to save led to high levels of investment in property, which was used as security against future borrowing. Property speculation increased; in Bangkok property prices rose at 40% per year in the early 1990s.

*Thailand crash*

1. The property boom collapsed, falling prices meant loans could not be repaid, many companies went bankrupt and the stock market crashed in 1996.
2. Foreign investors saw falling property prices and became concerned about the ability of domestic borrowers to repay loans in foreign currency.
3. The slowdown in economic growth led to concern about the possible devaluation of the baht – foreign and domestic companies tried to sell the baht for dollars, which further reduced its value, making devaluation more likely. In July 1997 the baht was floated breaking the link with the dollar meaning that a much greater volume of real resources, measured in baht, had to be used to repay loans leading to many bankruptcies.
4. An IMF rescue package was implemented that involved freezing financial companies and in effect made the panic worse.

*From Thailand to the Asian crisis*

1. By the end of 1997 other countries had devalued; Indonesia floated its currency and Taiwan devalued. The term Asian financial crisis was first used, leading to pressure on the Hong Kong and South Korean currencies. As all these countries are competitors, investors feared a competitive devaluation.
2. A net inflow of $93 billion in 1996 turned into a net outflow of $12 billion in 1997, which is equivalent to 11% of the pre-crisis GDP.
3. IMF conditionality packages were introduced, based on open economies, free capital markets and no government subsidies, thereby undermining the Asian model of development.

*Effects on the real economy*

1. There was an increase in foreign ownership of assets in South East Asia.
2. Domestic consumption and investment fell and unemployment rose, millions lost their savings and their jobs, and the burden of loss was not equally shared.

3. Production was restructured as firms found it difficult to finance investment. There was an expansion of exports to repay debt – South Korea for example expanded memory chip production and sold at low prices out-competing chip production in the newly constructed plants in north east England, leading to the closure of two major plants there, Fujitsu and Siemens, the latter having only opened in the previous year (see Chapter 5).

*After the crisis*

1. The long-term effects on the countries are varied; the stronger economies such as South Korea have largely recovered but severe problems remain, especially for the poorest groups in the Philippines and Indonesia.
2. The Asian model of development has to some extent been undermined.

*Sources*: Singh (1999a); Wade (1998).

*Note*: *Governments took a pragmatic view about integration with the world economy, integrating only when it clearly suited their interests, being more interested in exporting than importing and in science and technology more than finance or multinational investment. The extent of this collaboration varies between the countries, being higher in Japan, South Korea and Taiwan than Indonesia and Thailand where links are looser (Singh 1999a; Wade 1998).

desirable outcome illustrated in Figure 8.1.[25] That is, in practice the state does not act as a neutral arbiter enacting the general will, in Rousseau's terms, but to secure the interests of the dominant classes, while making sure that the pursuit of their individual self-interest does not undermine overall social stability, and hence their long-term interest, though these intentions are not necessarily realized. There is correspondingly a tension between how much of the state's resources go towards sustaining capitalist accumulation and how much towards social rights or social sustainability, the resolution of which varies between countries depending on the balance of power between different class fractions and social groups. In the global economy the new global elite has increasingly emphasized the former and urged states to deregulate and privatize their economies and cut back social rights and social expenditure, especially during the 1990s (see Box 3.1 on structural adjustment). During the early years of the third millennium there are some indications that the contradictions of market economies are being recognized and the neo-liberal state is beginning to construct its own regulatory mechanisms and modes of governance at different spatial scales to contain some of the adverse consequences of unfettered market relations in the global economy, points developed later on in the chapter. First though the changing role of the state in the context of globalization is discussed.[26]

# GLOBALIZATION AND THE STATE

The revolution in contemporary information and communication technologies provides the physical foundations for a global economy, and free market economists, following Adam Smith, argue that as long as the factors of production are allowed to flow freely efficiency gains will arise from the increase in the division of labour that this geographical extension of the market potentially allows. This section considers how nation states have been affected by and in turn affect the process of globalization and in particular how they control their economic, social and geographical space in this new global context or whether given their capitalist nature they actually seek such control.

Some writers have questioned the contemporary powers of the nation state, arguing that they have weakened with the development of globalization. Keniche Ohmae (1995) for example argues that 'we now live in a borderless world in which the nation state has become a fiction and where politicians have lost all effective power'. Likewise it has also been suggested that:

> Even the most dominant nation states should no longer be thought of as supreme and sovereign authorities, either outside or even within their own borders. The United States does not, and indeed no nation state can today form the centre of an imperialist project.
>
> (Hardt and Negri 2001: xi–xiv)

Such views imply that past nation states were comparatively very powerful, which may not have been the case, especially as the strongest forms of the nation state, fascism in the early to mid-twentieth century and state socialism in the latter part of the twentieth century, were both defeated.[27] Further although no nation state can be immune from terrorism from within or without, the United States now has economic, political and military power on a scale unrivalled in history. This perspective also tends to prioritize the role of economics over politics, rather than taking a political economy perspective that explores their inextricably intertwined nature. In particular it overlooks how the state has acted and continues to act in the interests not of a generic capitalist class or even a transnational capitalist class but its own dominant capitalist class, which could be national, international or global in orientation. The interests of this class will vary and from time to time and may correspond with the interests of capitalist and elite classes elsewhere in the globe, but not invariably so. Within this latter perspective the role and powers of the nation state may have changed but are not necessarily weakened. Perhaps as Manuel Castells points out:

> State control over space and time is increasingly bypassed by global flows of capital, goods and services, technology, communication, and information. . . . the nation state seems to be losing its power, although, and this is essential, not its influence.
>
> (Castells 1997: 243)

Thus it is important to analyse how the nation states that characterize contemporary societies have exercised their influence by adapting their role to match the contemporary global context. As David Held suggests, 'economic globalization by no means necessarily translates into a diminution of state power; rather, it is transforming the conditions under which state power is exercised' (Held *et al*. 1999: 441).

Correspondingly there have been changes in the functions and form of the nation state and a differentiation in power, as some states, especially the United States of America, have increased their power while that of other states has declined.[28] Moreover rather than just responding to the new global context the state has been an active agent in the process of globalization, changing its form, the scale on which it operates and its actions, in order to implement and accommodate the global capitalist economic system, albeit to different degrees depending on its relative global power and the internal balance of different social groups.

Global capital still requires supportive state institutions to operate effectively, in particular to provide infrastructure and enforce contracts. Indeed the state has played a key role in promoting the global economy by providing infrastructure, liberalizing trade and capital movements, privatizing its assets – even allowing foreign firms to buy up and run basic utilities – and changing accountancy systems to facilitate global transactions.[29] Even though international private systems have developed to resolve international legal disputes and provide credit ratings to legitimate potential global players, the state remains the arbiter of last resort.[30] Internally states have reduced expenditure on social issues, promoted flexible labour markets, and increasingly transferred control over monetary policy to central banks, rendering what in reality is a political choice to a technical concern. In this last case, states have voluntarily surrendered some power, but in the other cases the state has continued to exercise its power or influence but in a different way.

Thus states have adapted their strategies to accommodate and promote the interests of global capital, in accordance with the relative power of the globally oriented capitalist class within their own territory. Just as Polanyi (1957) argued that the state established market society in the nineteenth century, the contemporary state has been equally instrumental in establishing a framework for global capitalism in the twentieth and early twenty-first centuries. Thus it is not a question of the state's powers diminishing but more a question of how its role has changed to meet the requirements of the new dominant social groups, specifically the nationally based global elite which has been very effective in seizing hegemony within individual nation states and within supra national institutions, and having done so has encouraged the adoption of neo-liberal policies with scant regard to their unequal outcomes or implications for social reproduction, as Jim Glassman argues:

> Post-colonial imperialism has successfully created a 'counter-nationalist' intelligentsia, which though still needing to some degree to have national credentials in order to serve within the nation state

has none the less taken on board most of the perspectives of its international mentors. . . . This counter-nationalist intelligentsia is in effect a 'transnational kernel' within the technocracy and a crucial player in the process of alliance formation which promotes capitalist internationalization.

(Glassman 1999: 686)

Some 40 per cent of the top 400 Thai civil servants had masters or PhD degrees from western universities.[31] There is also an international cadre of young professionals with control over large volumes of capital who trade virtually in the new financial products such as options, futures and derivatives which have increased the speed and scale of transactions across the globe and unsettled many economies.[32] The increase in virtual transactions raises the danger that both companies and states will lose track of what is happening. Following some spectacular crashes, such as the Enron affair, and the destabilizing effect of financial speculation, as in the Asian (1996/7), Russian (1998) and Argentinian (2001/2) financial crises, both states and companies are beginning to seek greater control.[33] Even the World Bank now accepts that some controls over short-term speculative flows are necessary to protect national economies. These controls however fall far short of the Tobin[34] tax supported by many NGOs including War on Want, who highlight the potential gains from a 0.1 per cent tax on speculative flows, currently in the order of $1 trillion a day (see Figure 8.2).[35] Thus both supra national institutions and NGOs are seeking

*Figure 8.2* Demonstrating for the Tobin tax.
*Source*: War on Want/Ben Blackall.

some control, making the issue not one of 'whether but how to manage the global economy' (Weiss 1999: 140). Or as Ulrich Beck (2000b: 102) points out: 'the first wave of national deregulation makes necessary a second wave of transnational regulation'. This is similar to Polanyi's (1957) comments in relation to the establishment of market society at the national level. More generally, given the way that the boundaries of the space economy extend beyond the powers of the nation state a wide variety of supra national institutions have become instrumental in governing the global economy.[36] In this respect it is useful to distinguish between government – control exercised by the nation state, through formal (usually elected) political parties – and governance – control exercised by a variety of public and private institutions that have been established at different spatial scales.[37] The role of these institutions and their relationship with nation states is discussed below.

## Supra national institutions

Just as Hobbes argued that individuals surrender their freedom to the state in order to survive, nation states have surrendered some of their autonomy to supra national institutions to secure overall economic and social reproduction. Thus as Allen Scott suggests:

> The contemporary economy is one of incipient globalization in which national economies are evolving from a condition in which they are less like billiard balls in holistic interaction than they are permeable entities in various states of amalgam with one another.
> (Scott 1998: 25–6)

Michael Hardt and Antonio Negri (2001) have taken this idea a bit further and argued that supra national institutions have led to the development of a new form of sovereignty, that they term Empire, which consists of both national and supra national institutions and:

> has no centre of power and does not rely on fixed boundaries or barriers. It is a decentred and deterritorializing apparatus of rule that progressively incorporates the entire globe within its open expanding frontiers. Empire manages hybrid identities, flexible hierarchies, and plural exchanges through modulating networks of command. The distinct national colours of the imperialist map of the world have merged and blended in the imperial global rainbow.
> (Hardt and Negri 2001: xii)

The analogies of merging billiard balls and especially the imperial global rainbow imply some sort of cooperation between equal powers and while in practical terms a common neo-liberal agenda is followed, this is not invariably the result of free choice. Moreover it is still enacted differently by different nation states, depending in part on their current global positioning, in terms of the division of labour, extent of debt and geopolitical location, but also on the internal balance of class forces. Moreover as the discussion below indicates the powers of the nation states are differentiated within these supra national institutions. In many ways however these institutions have also become a medium through which the United States secures an apparent global consensus and correspondingly legitimation for its own strategies. If this support is not forthcoming, as in the case of the war against Iraq (2003), then it acts alone or with close allies.

Many supra national organizations were set up immediately after the Second World War to establish greater political and economic stability.[38] These include the United Nations (UN), the International Monetary Fund (IMF), the World Bank, the World Trade Organization (WTO) and the International Labour Organization (ILO).[39] These organizations have humanitarian goals to make the world economy run more effectively, that is smoothly, peacefully and in ways that lead to increasing economic prosperity (see Box 8.4). Thus, the IMF provides loans to countries experiencing sudden financial shocks, as in the 1997–8 Asian financial crisis,[40] or environmental disasters, for example to Kenya to cope with a severe drought in 2000. The World Bank also finances a wide range of development programmes. The WTO was set up more recently in 1995 but is the new name for the General Agreement on Tariffs and Trade similarly set up in the 1940s and is designed to promote the free trade of goods and more recently services on a global scale. These organizations have been in existence for 50 years, yet despite their stated intentions, in the last 25 years the world has become increasingly unequal and experienced an almost continual state of war.

One explanation for the imbalance between the humanitarian goals which perhaps approximate the 'global will' (making an analogy with Rousseau's concept of the general will) and the unequal outcomes arises from the unequal distribution of power within the supra national institutions, so that in practice their policies reflect partial interests, and, in particular, the interests of the dominant states. In the World Bank, power is related to a country's economic standing: there are 5 permanent executive directors – from France, Germany, Japan, the UK and the US – and the remaining 19 directors rotate from the 184 member countries.[41] Thus the lending countries that determine what is done are located in areas where the bank does not operate (Monbiot 2002a). Similarly in the IMF voting rights are positively related to GDP and in the United Nations five countries, the United States, Russia, France, Britain and China, are permanent members of the security council and have the power of veto over resolutions.[42] Although the primary purpose of the United Nations is to maintain international peace and security (see Box 8.4), in the recent past it has increasingly been used to legitimize war, ostensibly for humanitarian reasons, but in a selective way and in accordance with the material interests of the permanent members, especially the United States.[43] In 2003 the UN

*Box 8.4* Stated aim and role of supra national institutions

*United Nations (UN)*
The main purposes of the United Nations 'are to maintain international peace and security; to develop friendly relations among nations; to cooperate in solving international economic, social, cultural and humanitarian problems and in promoting respect for human rights and fundamental freedoms; and to be a centre for harmonizing the actions of nations in attaining these ends' (United Nations 2003).

*The International Labour Organization (ILO)\**
The ILO became a specialized agency of the UN in 1946 and 'seeks the promotion of social justice and internationally recognized human and labour rights. . . . It formulates minimum standards of basic labour rights: freedom of association, the right to organize, collective bargaining, abolition of forced labour, equality of opportunity and treatment, and other standards regulating conditions across the entire spectrum of work related issues, (ILO 2003b).

*The International Monetary Fund (IMF)*
Is 'an international organization of 183 member countries, established to promote international monetary cooperation, exchange stability, and orderly exchange arrangements; to foster economic growth and high levels of employment; and to provide temporary financial assistance to countries to help ease balance of payments adjustment' (IMF 2003).

*The World Trade Organization (WTO)*
Was formed in 1995 from the General Agreement on Tariffs and Trade (GATT) which was set up after the Second World War and is 'the only international organization dealing with the global rules of trade between nations. Its main function is to ensure that trade flows as smoothly, predictably and freely as possible' (WTO 2003).

*The World Bank*
'The World Bank Group is one of the world's largest sources of development assistance. In fiscal year 2002, the institution provided more than US$19.5 billion in loans to its client countries. It works in more than 100 developing economies with the primary focus of helping the poorest people and the poorest countries. The World Bank helps developing countries fight poverty and establish economic growth that is stable, sustainable, and equitable' (World Bank 2003b).

*Note*: *The ILO was originally founded in 1919 and is the only surviving major creation of the Treaty of Versailles which brought the League of Nations into being.

withstood US pressure and declined to pass a resolution to legitimate the US/UK war against Iraq, who then proceeded to conduct the war illegally, thereby illustrating that the global institutions only have the power that powerful nation states choose to bequeath to them.

The WTO differs from the others in that decisions are generally reached by consensus and where differences arise voting takes place but on the basis of one country one vote.[44] Furthermore if countries fail to comply with decisions, sanctions are taken by member states, not the WTO, which has no direct powers to do so, which may be one reason why richer countries can continue to subsidize their own producers while forcing free trade on other countries. Both the IMF and World Bank, which do have sanctions, generally act in accordance with WTO rules. The governing body of the ILO also has 10 out of the 28 permanent members from the main industrialized countries, the others being rotated every three years. Policy is determined by the annual conference, at which workers and employers are represented and the determination of Conventions and Recommendations is made on the basis of a two-thirds majority. These then have to be ratified through national legislation, though there is no compulsion on member states to do so.[45]

The more general inequality in power means that while the powers of some nations are superseded, others are able to exercise their influence more fully, but through the guise of apparently impartial supra national organizations, which raises the question of whether these institutions are world or global institutions in the sense of acting in the 'global general will' or whether they simply reflect the existing unequal balance of power between and indeed within states. Parallel to the way that nation states generally claim to be acting in the national interest when in reality they reflect the interests of the most powerful groups, so these institutions have come to reflect a singular neo-liberal ideology, consistent with the dominant class fractions in both rich and some of the poorer countries, thereby questioning the relationship between democratic and economic control in the global economy. For example, nation states retain authority in the sense that they do not have to apply for loans, but once accepted, they are governed by the conditions attached, which does undermine democratic control, because although debtor nations can vote to change the government, if the country is already locked into debt, it cannot vote for a fundamental change of economic policy.[46] In this respect it will be interesting to see whether the left wing government of Brazil, which assumed power at the beginning of 2003, is able to introduce and sustain genuinely different policies given the IMF loans incurred by its predecessors, issues discussed further in Chapters 9 and 10.[47]

There are however some important differences between these supra national institutions. The IMF is more reflective of the neo-liberal ideology and emphasizes stabilization programmes, specifically using monetary instruments to lower debt and restore economic health on the assumption that these strategies will promote economic stability from which social well being will follow automatically, though there is little evidence to support such a view. Such policies typically involve reducing public expenditure, switching public expenditure from non-productive to productive uses and transferring labour from non-tradable to tradable sectors of the economy, all of which prioritize the economy over society and undermine social reproduction. The World Bank, on the other hand, advocates development more than growth and so is more prepared to allow public expenditure to be committed to investments in productive and social infrastructure.[48] So although in each case the richer countries have a greater voice in the policies of these institutions, there are some differences in orientation which affect different classes in society differently, financial interests being more aligned with the IMF and industrial and public sector professional workers being more aligned with the World Bank. Thus there are fractional class interests within and between countries that cut across the national differences.

The United Nations also directly finances a range of humanitarian projects from child poverty and the trafficking of people to international war crimes and moreover has a number of covenants related to social and human rights. In contrast to the policies of the IMF or World Bank however there is little or no enforcement of the UN's humanitarian policies. Nevertheless, the international standards and measures discussed in detail in Chapter 2, especially the Human Development Index, provide benchmarks against which countries can assess their performance and which can be used as political leverage. In 2000, government leaders expressed concern about the current levels of inequality and 'recognized their collective responsibility to uphold the principles of human dignity, equality and equity at the global level'. They set eight development goals to be achieved by 2015 – the millennium goals (see Box 8.5 and UNDP 2002:16) – but they have no means of enforcement; this remains the task of the individual member states.

The International Labour Organization (ILO) was set up in 1919 in response to unhealthy and unsafe working conditions, fears of widespread industrial unrest and, even more, to concerns that without standards there would be a competitive descent to the lowest levels, placing countries with better conditions at a competitive disadvantage.[49] While the ILO has developed many standards and publishes information and analyses of worldwide working conditions, it has very little power with which to enforce standards, though as with the HDI it can provide useful benchmarks around which trade unions and NGOs can bargain.

There are also a growing number of regional trade blocs, which in some ways promote the global economy and in others maintain boundaries around its extent. Different types of trading blocs – export processing zones, free trade areas, customs unions, common markets and economic unions – are associated with different degrees of economic and political integration (see Figure 8.3). Export processing zones (see Chapter 4) are areas where states have effectively denationalized parts of their territory, by suspending normal customs duties and regulations to create free market conditions for global capital, and absolve firms of many social responsibilities. In principle free trade areas simply promote free trade between member states; customs unions in addition to free internal trade have common trade policies towards external states; common markets

---

*Box 8.5* The millennium development goals

At the UN General Assembly in 2000 world leaders set eight goals for development to be achieved by 2015. Most of the goals are defined quantitatively to allow progress to be monitored.

1. Eradicate extreme poverty and hunger.
2. Achieve universal primary education.
3. Promote gender equality and empower women.
4. Reduce child mortality.
5. Improve maternal health.
6. Combat HIV/Aids, malaria and other diseases.
7. Ensure environmental sustainability.
8. Develop a global partnership for development.

The UNDP 2002 report analyses progress towards these goals and finds that while countries have made progress it is unlikely that the poorest countries will reach them. Having aspirational goals and monitoring progress towards their attainment is valuable for assessing the relative performance of different countries, but it is also important to understand the processes generating the problems being tackled.

*Source*: See UNDP (2002).

---

| | Free trade area | Customs union | Common market | Economic union |
|---|:---:|:---:|:---:|:---:|
| Removal of trade restrictions | ✔ | ✔ | ✔ | ✔ |
| Common external trade policy | | ✔ | ✔ | ✔ |
| Free internal movement of the factors of production | | | ✔ | ✔ |
| Harmonization of economic policies | | | | ✔ |

*Figure 8.3* Forms of Economic Integration

*Note*: ✔ indicates the presence of the policy so the economic union represents the deepest form of economic integration have free internal trade, common trade policies and allow free movement of the factors of production; and economic unions have all of these characteristics plus a harmonization of economic policy. In practice real trade blocs have an agreed mixture of these elements. Examples of predominantly free trade areas include NAFTA (the North American Free Trade Association (consisting of the US, Canada and Mexico);[50] MERCOSUR/MERCOSOL (Argentina, Brazil, Uruguay, Paraguay); and AFTA/ASEAN (Cambodia, Laos, Indonesia, Myanmar, Singapore, Vietnam, Brunei, Malaysia, Philippines and Thailand).

Economic unions such as the European Union[51] have the strongest level of integration; for example 12 of the 15 member states have a common currency, the euro, and there is a central bank that makes decisions over interest rates that can have implications for the 'real economy' in different localities. The EU also has different levels of regulation in relation to economic and social policy. Directives are mandatory and Recommendations are discretionary. In some policy spheres, voting has to be unanimous while in others, a majority carries decisions. In the latter case the population of member states weights votes. Directives appear to be a strong form of supra nationalism but if their provenance is examined it becomes clear that they build from members' existing practices rather than being imposed from above, so the end result is that their content embodies the practices of the lowest common denominator, i.e. the state with the least onerous/least progressive conditions. In the case of parental leave for example a consensus was reached in 1984 based on the existing practices in member states but opposed by the UK; in 1994 the Directive was ratified without UK support, which eventually was given in 1997, but even so states had until the following year to implement the Directive within their own national laws.[52] Thus some issues have been removed from direct political control at the member state level, though the supra national policy still reflects their voice and the exit option always remains.

The extent to which the different member states can realize their objectives within these organizations is also influenced by power differences. Within NAFTA, for example, product markets are open so goods can be freely exchanged but restrictions on labour markets remain. Thus the more competitively priced and often subsidized products in the US can undercut Mexican producers, who are not free to migrate to the US, an issue that is discussed further in Chapter 9. Moreover while Canada has a free hand in passing progressive social legislation, in economic terms the hegemony of US interests prevails.

In addition to the formal institutions discussed above there are a variety of ad hoc institutions or gatherings of states that come together from time to time. Various groupings of countries, the G7, G8 or G20[53] or smaller or larger variants, form to discuss economic stability, growth and sustainability. For example the G8 meeting at Kananaskis, Canada in June 2002 invited various African leaders to discuss the New Partnership for Africa's Development (NEPAD), devised within Africa by Africans, which like the other supra national organizations discussed above has worthy goals of eliminating poverty, increasing education, reducing child mortality and dealing with debt, in this case through public–private partnerships. Although the colonial background, problems of debt and the workings of the international economic system are referred to, the plan focuses on governance – on problems of mismanagement and poor leadership within Africa – the resolution of which may be a necessary condition for change but unlikely to be a sufficient one. Without some control over the economic forces that have shaped their development and continue to do so, the worthy objectives are unlikely to be achieved. Interestingly Brazil has suggested that poorer countries should set up their own 'G' meetings to discuss issues of common concern and not wait to be invited to the meetings of richer states (Lula da Silva 2003).

The world economic forum[54] is a wider meeting of world leaders likewise designed to 'improve the state of the world'. It is this group together with the WTO that have been the focus of non-aligned international protest, the anti-capitalist and anti-globalization social movements which might be more appropriately termed movements for global justice. These movements have led to the alternative world social forum (WSF), which by contrast is an entirely open forum for those concerned with the adverse effects of neo-liberalism and globalization especially its lack of accountability, though its growing prominence has attracted appearances by world leaders.[55] These movements are discussed further in Chapters 9 and 10.

Some nation states, especially indebted poorer countries, have little power in relation to supra national institutions, while other states have increased their power. In complete contrast to the quotation cited earlier from Hardt and Negri (2001), the United States has become the dominant superpower of the new millennium. Not only does the US exercise power singly and through the supra national institutions, it has sustained its high levels of economic growth partly through its colossal trade deficit and thus its implicit borrowing from other, often poorer, countries. The US debt is equivalent to all the debt of the heavily indebted poor countries put together and equal to over 70 per cent of US GDP. However, this debt sustains the wealth and high living standards of the majority of US citizens, in contrast to the austerity policies imposed by the IMF and endured by the populations of other debtor countries. This debt has arisen in part because of the role played by the dollar in international trade (see Box 8.6).

Debt is emphasized as one of the key problems currently facing many poorer countries, but these problems arise more from the way in which the debt has to be repaid in dollars, rather than in their own currencies or other currencies with which they trade, than from debt per se. Payment in dollars traps these countries into devoting resources to products that can be traded on world markets and as the terms of trade have turned against primary products, prices are low and falling, so more and more of these goods have to be produced to meet the debt repayments, which makes it difficult to accumulate funds for invest-ment necessary to achieve other goals.[56] The US also gains because returns on foreign assets held in the United States are lower than the returns on assets held by the US in foreign markets, which both reflects and further boosts its economic power. Countries and individuals are prepared to invest in low

---

*Box 8.6* The gold standard, the dollar and trade deficits

In the past international trade was regulated by the gold standard, which was a self-correcting mechanism through which trade balances were maintained.

The money supply in any country was linked to its gold reserves. If a country had a trade surplus it would receive gold from debtor countries. This gold would allow it to expand its supply of money or credit. As a consequence prices were expected to rise, exports would become comparatively more expensive, and imports comparatively cheap, so in time the trade deficit would be restored. If a country had a trade deficit the reverse would happen, that is, the supply of money would be reduced, prices would fall, imports would become comparatively expensive, and exports comparatively cheap, so again the trade deficit would be eliminated. The gold standard was suspended during the First World War, revived during the 1920s but collapsed in the recession of the 1930s.

The Bretton Woods negotiations of 1944 established the dollar as equivalent to gold and countries linked their currencies to the dollar which in turn was pegged to gold ($35 per oz). The system was expected to work in the same way as the gold standard but be more flexible because international settlements would be made in dollars rather than gold.

The convertibility into gold was ended in 1971 and this has effectively ended the stabilization mechanism and allowed the US to build up a trade deficit, as it can simply print dollars irrespective of its trade balance or gold reserves. This tendency to increased liquidity is reinforced because other countries keep dollars in order to settle their international accounts and countries with trade surpluses accrue dollars which they do not simply hold in bank vaults, but invest in the United States, which increases the supply of money there and allows the US to finance its deficit.

Thus the break with gold has created an unstable system, as there is no self-correcting mechanism – some countries can go on building up trade surpluses, while the US can sustain a deficit. The corresponding increase in liquidity and credit brings instability and a tendency to economic booms and busts as too much money is chasing too few investment opportunities, which in turn facilitates phenomena such as the dot.com boom or the property boom in Thailand, linked to the mid-1990s Asian financial crisis (see Duncan 2003).

---

yielding returns in the US, because they earn dollars, necessary to settle international accounts, and because they are more stable. Figure 8.4 shows that the US gains disproportionately from the resources it borrows from other countries and from the resources it lends to other countries as a consequence of these differential rates of return. Thus far from there being a deterritorialization and decentring of capitalism, it seems that economic resources are accumulating ever more strongly within the United States. The US is a true superpower whose might has increased with globalization and the development of supra national institutions over which it exercises control or ignores depending on its own

**a** Assets/liabilities as percentage of GDP

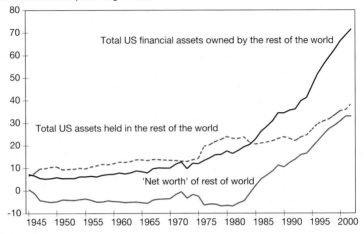

**b** Exports/imports/trade balance as percentage of GDP

*Figure 8.4* The debt of the United States. (a) Transfer of investment to the US; (b) The US trade deficit.

*Source*: (a) Data from Federal Reserve System (2003) (b) US Department of Commerce. Bureau of Economic Analysis (2003) National income and product accounts, Washington DC: Bureau of Economic Analysis.

*Note*: Figure 8.4 demonstrates the way that funds are being transferred from the rest of the world to the United States. Figure 8.4a illustrates that the rest of the world owns assets in the US equivalent to 70% of the annual GDP of the US while the US holds assets in the rest of the world equivalent to around 35% of the annual GDP of the US. Thus on balance there is a transfer of funds to the US. Figure 8.4b illustrates the trade deficit that the US incurs with the rest of the world which widened significantly in the late 1990s as the import of goods and services rose while exports fell.

immediate concerns; it respects neither the rulings of the UN nor the strictures of the IMF.

Recognizing the power of the US super state, together with the way that different nations have accommodated and responded to globalizing tendencies in different ways, questions the validity of one side of the hollowing out thesis, namely that nation states have lost power to supra national institutions. In the post-Iraq situation in 2003 this looks a rather strange thesis and perhaps derives from placing too much emphasis on the form of the state and questions of governance, rather than on the role that contemporary states play. Clearly, given the global context, states cooperate in areas such as policing, migration and to some extent in economic and environmental affairs but in so doing they each seek to preserve what they take to be their national interest, and the extent to which they are successful depends on their relative power as a nation state. Before turning to questions of social reproduction, which takes place at the sub-national level, the role of sub-national states, cities and regions is briefly considered in order to examine the other side of the hollowing out thesis – that nation states have lost power to regional institutions.

## Regions and city institutions

There are three main sets of arguments to suggest that sub-national, regional or city region institutions[57] have increased their powers relative to nation states: the paradoxical increase in the significance of the region as a key economic territory in the global context; the advance of neo-liberal supply side policies and the growth of regional consciousness or identity, which in some instances has led to devolution; and greater political, economic and social autonomy for some regions. While there is substance to each of these claims, discussed below, the main argument in this section is that while there has been some fragmentation of nation states leading to new nation states, for example in the former Soviet Union and Yugoslavia, and some devolution of responsibility to regions on a wider scale such as in Spain or the UK, the powers of nation states are rarely undermined from below. There have been some changes in functions and responsibilities but largely at the discretion of the nation state.

Despite the development of a new economy and dematerialized knowledge, firms and organizations retain a material presence. Moreover knowledge based goods and services are often highly clustered, owing to the significance of spatially dependent economic, political and cultural externalities and untraded interdependencies, as discussed in Chapters 3 and 7, which have increased the significance of cities and city regions where they are located. Place has therefore become a significant factor in economic competitiveness, and there is an association between strong regional economies and supportive regional institutions, though the direction of causality is unclear. Establishing regional institutions and raising regional institutional capacities may not therefore be a sufficient condition for generating regional development, especially as these institutions will be in competition with one another.

While local connectivity is important, economies of scale remain, and modern ICTs facilitate greater material and virtual connectivity and enable regions to market their products globally; thus there are a limited number of superstar regions throughout the world. It has been argued that these regions establish connections with comparable regions in other countries leading to transnational urban and regional networks, which bypass the nation state. Alliances have developed between regions with related interests, such as the Four Motors Region of Baden-Württemberg, Rhône Alpes, Lombardy and Catalonia. In the European Union these networks have been stimulated through EU financial and research initiatives, which are invariably conditional on a cross-national composition. A Committee of Regions was also established under the Maastricht Treaty in 1993, in order to give local and regional authorities a direct voice in the formulation of policy making at the supra national level. Elsewhere, Allen Scott (1998) refers to the way that the Malaysian Prime Minister visited California to promote Malaysia's multimedia super corridor (see Chapter 6) directly to high technology and media business circles, rather than going through official US channels, to support the idea of the growing significance of the region and connections between regions in the contemporary global economy.[58] Furthermore he argues that regional intuitions may emerge because people's well being is tied more directly to the: 'fate of the city and its region rather than to the state, [so] some aspects of citizenship may begin to be associated once more, as they were in earlier historical periods with city regions not with states' (Scott et al. 2001: 27).

Castells (1997: 273) also suggests that metropolitan regions have become more important because: 'national governments in the information age are too small to handle global forces, yet too big to manage people's lives.' While Scott (1998) recognizes that the nation state still retains the most power and that people may identify more closely with city regions, he also suggests that regional institutions are emerging to manage regional economic dilemmas arising from competition between regions in a global economy in which nation states no longer assist regions of economic disadvantage. This interpretation links to the second reason for the significance of regional institutions.

With the demise of Fordism and Keynesian macroeconomic demand management and the widespread adoption of neo-liberal economic policies, regions become the sites at which the supply side policies such as raising competitiveness, employability and training are implemented. In Europe there are numerous economic development agencies at different local and regional scales competing for inward investment, each of which promotes its own area as 'the place to be'.[59] In the competitive global context, therefore, place marketing is widespread and some regions have developed 'the capacity to act entrepreneurially' and build global advantage by enhancing place specific international competitiveness and developing 'institutional and organizational features that can sustain a flow of innovations' (Jessop 2002a: 189). Many more regions have perhaps become entrepreneurial simply because there appears to be no alternative. It has been claimed, for example, that 'by 2008 the north east [of England] will be the most creative region in Europe'[60] a claim echoed, no doubt, by countless other regeneration managers and place promoters, just as Brighton and Hove claims to be the cyber capital of Europe and many cities compete to become the European City of Culture.

Underlying some of these strategies are partnerships between the public and private sectors, but to what extent these represent new forms of local representation and control and to what extent a forum through which the business interest is given an extra voice is unclear. What they do illustrate however is how the role of the state has changed from being a managerial one in the era of Fordism to an entrepreneurial one today.[61] Not all regions can be the most creative, the cyber capital or the capital of culture. There will inevitably be winners and losers, but the ideology of individual competitiveness is pervasive. Indeed there is a close correspondence between academic concepts of clusters, learning and knowledge regions and the regional boosterism of development agencies, both of which are aligned with the neo-liberal business agenda to promote winners rather than to secure objectives such as regional well being or balanced regional development within the country as a whole. In this competitive context, the idea of networks of cities and regions acting collaboratively and somehow against the nation state sits rather oddly with the idea of regions and localities in fierce competition with each other for inward investment. The chief beneficiaries of this zero sum game will undoubtedly be the multinational companies, who gain from the incentives and whose ties to the region are inevitably weakened by the continual presence of offers of funding from competitive regions.[62]

One of the reasons why cities and city regions have become entrepreneurial[63] and aim to create appropriate socio-economic infrastructures for the contemporary economy, such as the quayside developments in north east England or developments in Baltimore in the US (Harvey 2000), is because central funding for redistribution policies has been scaled back and targeted only towards areas of intense deprivation. So their only option is to subscribe to the trickle down philosophy, i.e. that social benefits will flow from business support, despite the lack of evidence. Social policies are only introduced in extreme cases and where deprivation threatens to undermine competitiveness and, as discussed in the final section on social sustainability, are similarly supply side oriented. Nevertheless the promotional policies conform with rather than oppose the neo-liberal agenda so it is not clear how the competitive policies of the regions undermine the nation state.

The significance of the regional agenda within the European Union can also be overstated. The EU has a hybrid nature. On the one hand it pursues neo-liberal economic policies in the belief that they will raise the competitiveness of Europe relative to the US and Japan. On the other it retains ideas about the significance of social cohesion, reflecting the continental European corporatist, social market philosophy and the recognition that unless all member states benefit from increased integration, the integrity of the Union would be threatened. The existence of conflicts between these two agendas is rarely discussed. Moreover, in addition to the overall stabilization programmes, which challenge cohesion, some of the substantive policies, for example in agriculture and research and development, run directly counter to the cohesion objectives.[64]

In relation to cohesion, the European Regional Development Fund was strengthened in 1987 with the formation of the single market, and cohesion policies to address the specific problems faced by the least developed countries, Ireland, Greece, Portugal and Spain, were introduced with the Maastricht Treaty on political and economic union in 1993. The future strength of these policies is currently in doubt with EU enlargement, specifically the incorporation of 10 new states from Eastern and Southern Europe in 2004, which with the exceptions of Slovenia and Cyprus all have lower levels of GDP per capita than the existing member states, and all are below the EU average. Moreover in relation to the original four cohesion states while there has been cohesion at the national level, regional differences within these countries have widened. Even with the strengthening of regional and cohesion policies only around one-third of the EU budget and less than 1 per cent of the combined GDP of the member states is allocated to lowering regional inequality. So, although there is an overwhelming array of documents, policies and discursive forums in relation to regional affairs, commitment in terms of real resources remains quite low. The allocation of regional funding takes place through the nation states and at the national and supra national levels the stabilization and competitiveness agendas associated with neo-liberalism dominate. Thus the concept of a Europe of Regions in which the sub-national regions have autonomy and engage directly with the supra national EU is more fantasy than reality except at the discursive level.

The contradiction between competitiveness and cohesion within EU policies is however more general and particularly reflected in global cities or superstar regions, which in spite of or perhaps because of their dynamic economies are also characterized by a polarized social structure. Correspondingly they find themselves in the Janus like situation of simultaneously engaging in 'beauty contests', that is marketing their attractive features to attract international capital, while highlighting their downside in 'ugly sisters contests' for social funds[65] (Jessop 2002a: 192). The significance of competition has therefore been extended and regions have to compete for state welfare support from the national, and in the case of the EU, the supra national state. Even so the form of this assistance is supply side oriented. Thus regions have been individualized in the sense that they have to compete for inward investment and for welfare support and what is more the dominant way of dealing with economic and social disadvantage is similarly individualized and based on enhancing the supply side characteristics of individuals within the regions, discussed further in the final section.

The third main reason for the increasing significance of regions arises from the increasing attention given to regional consciousness and regional identity (see also Chapter 9). These sentiments often derive from the presence of populations with a common cultural, ethnic and linguistic heritage that have nationalist aspirations, such as the Catalonia or Basque regions in Spain. In Canada the Québécois see themselves as a distinct nation and in these and other cases demands for devolution derive from a sense of both cultural and economic injustice. Not only are the populations being culturally oppressed but typically there is also a sense that the regions are economically disadvantaged and that gains would arise from greater autonomy over their resources. Economic disadvantage was a significant factor in the demands for devolution in the UK from Scotland and Wales or north east England. Conversely, in richer regions, there is a sense that their resources are unfairly taken to finance other areas within the national territory, for example the Lega Nord in Italy. These movements have sometimes successfully pressed their nationalist claims, leading to the formation of new nation states such as in the break up of the former Yugoslavia, while other movements are content with or settle for greater autonomy within the nation state. Belgium for example has become a federal state to recognize the distinctiveness of the different regions, Wallonia, Flanders and the Brussels region, and Spain has introduced autonomous regions. In the EU, national states without a strong regional tradition have created regions in order to benefit from EU funding, as in Ireland and Greece,[66] though in the case of Ireland, at least, these relate to the historic provinces.

Regionalism, decentralization and devolution are 'in fashion' and there has been a growth in the number of regional institutions especially from the 1970s onwards. The discussion here has focused mainly on Europe but similar trends have been identified on a wider scale.[67] However, it is more questionable whether the new regional institutions undermine the nation state or whether nation states have devolved responsibility for issues that they either can no longer or choose not to manage. While a common trend for devolution can be identified in the underlying processes, the form, power, resources and extent of popular legitimation for the devolved institutions vary. One key difference is whether the new institutions are established through popular demand or whether they have been established for administrative purposes and effectively act as a local arm of the nation state. In general, however, the powers of new regional institutions are quite limited, as is the extent of resources over which they have control.

In Italy, for example, although devolution expanded in the period from the 1970s, nationally controlled programmes accounted for 92 per cent of regional expenditure.[68] Likewise the new regional authorities in the UK have limited resources and similar to the Italian case much of the budget is assigned in advance to national government programmes. At the metropolitan level there is a similar story. While London now has a mayor, the funding and powers are restricted. Moreover evidence of support for this level of devolution and whether people perceive that real power is exercised at this level is varied. Where there are long standing conflicts between the regions and the state, autonomy at the regional level is often popular, but elsewhere, for example in England, there has been some resistance to and scepticism about the establishment of new tiers of government. One reason for this may be because the state at all levels has been hijacked by the interests of the globally oriented capitalist class and the traditional functions of the state in terms of supplying public goods and ensuring living standards are no longer guaranteed. In this sense the neo-liberal ideology of individualization and individual responsibility for welfare has come full circle. In other words people have internalized the philosophy to the extent that they no longer see the value of collective action and expenditure through the state. This cynicism is reflected in low voter turnouts and in the case of Hartlepool, in north east England, even though there is quite a strong regionalist movement, a person campaigning in fancy dress as the football team's mascot, H'Angus the monkey, was elected as mayor; several cities, including Brighton and Hove, voted not to have an elected mayor.

To substantiate the argument that regions are displacing the powers of nation states it would be necessary to show that the new regional bodies have the necessary resources and control to enact independent programmes and are recognized as legitimate by the population they administer. In the case of Western Europe and North America there is little evidence to suggest that this has happened. Apart from Catalonia in Spain and the fully constituted sub-national authorities such as the states in the United States, the German *Länder* or the Canadian provinces, which predate these discussions, there is little evidence to support these claims.[69] Thus it is perhaps more a case that with the widespread endorsement of neo-liberal policies nation states have to varying degrees abrogated their responsibility for social reproduction and deflected it to the regional and local levels, though without adequate resourcing.

## SOCIAL SUSTAINABILITY IN THE NEW ECONOMY

Poorer people in rich and poor countries have experienced real reductions in their living standards in the last two decades as a consequence of changes in the organization of work, reductions in state welfare expenditures and falling public sector employment. This section considers how these changes in the orientation of the state, together with changes in the organization of work and in household composition, have affected day to day social reproduction. The focus is on the United States and the European Union; first though the problem of maintaining a livelihood in this new context for people in poorer countries is briefly considered.

Poorer countries have been forced to open their economies to international trade, thus many of the industries established under import substitution policies have disappeared and the terms and conditions of employment in export oriented factories are sometimes inferior.[70] New employment is often precarious and increasingly in the informal sector (see Chapter 4). Primary producers have experienced deteriorating terms of trade and it has become harder to sustain a livelihood in agriculture. Thus it has become more difficult to acquire money income and at a time when charges have been introduced for what were in the past publicly provided and free services, in particular health and education. Social reproduction on a day to day basis has correspondingly become increasingly hazardous. The task of filling this deficit has often fallen to women, whose work burdens have increased as a consequence (see Box 8.7 for a discussion of Ghana). In some cases these problems are compounded by the fragmentation of households and the increase in the number of women raising children without male support.[71] This fragmentation is often exacerbated when people move away to find work because although the remittances help to sustain household incomes,[72] social and kinship ties that in the past sustained households and communities through difficult times are broken.[73]

Western countries have similarly liberalized their economies and cut back on social spending, making individuals increasingly responsible for their own

---

*Box 8.7* The impact of neo-liberal policies on Ghana

Ghana, once the star of structural adjustment policies, is now receiving aid under the Highly Indebted Poor Country Initiative (Donkor 2002: 226). Jobs were lost in the public sector, in education, the civil services and state enterprises and the only new employment was predominantly low paid work in hotels, financial services, insurance and private medicine. Furthermore incomes in agriculture have fallen because the terms of trade have switched towards the export crops internally, yet externally the terms of trade have moved against these crops. Thus while more people are encouraged to produce for export the rewards for doing so have declined significantly, though to a lower degree than in the domestic sector, thereby increasing internal inequality. Low interest credit to the agricultural sector was also withdrawn as being inconsistent with a liberal economy and rural poverty was further intensified by the liberalization of food imports which lowered prices for domestic production therefore making it extremely difficult to earn a living in agriculture. Furthermore charges are made for all services, so for example treatment in a state hospital costs 10,000 Ghanaian cedi (80 pence) when the minimum wage is only just over 50p a day (Donkor 2002). These changes impact particularly on women as they have the task of managing social reproduction.

Given these circumstances, many families encourage at least one person to migrate and then live on the remittances but while their mode of existence has been undermined by increased openness they face a much less open world when it comes to migration. Migrants try all sorts of ways of getting into richer countries, travelling on tourist visas and staying on. They often work in the informal sector, living very meagrely with friends and relatives in cramped conditions and as cheaply as possible in order to send money home. Although many will be working for the minimum wage or less, its purchasing power in Ghana is still significant.

welfare. The decline in state welfare and traditional systems of support such as trade unions has similarly occurred at a time when the intensification of competition has rendered labour market conditions more precarious. With contemporary ICTs and global financial markets investors can easily switch their funds, making firms vulnerable to changes in share prices. Firms are consequently very sensitive to their current performance and adjust quickly to any changes. Many of these costs of heightened competition are transferred to employees through increases in flexible working practices, increased monitoring of individual performances, the individualization of the wage relationship, as well as empowerment strategies, which devolve responsibility to employees. As Pierre Bourdieu (1998) points out, while employees are in reality 'simple wage labourers in relations of strong hierarchical dependence' they are nevertheless 'held responsible for their sales, their products, their branch, their store, etc. as though they were independent contractors', which leads to a high pressure competitive working environment. All of these conditions converge to undermine collective solidarities and:

> in this way, a Darwinian world emerges – *it is the struggle of all against all* at all levels of the hierarchy, which finds support through everyone clinging to their job and organization under conditions of insecurity, suffering, and stress.
>
> (Bourdieu 1998; my emphasis)

These comments from Bourdieu echo the ideas of Hobbes cited earlier and imply a weakening of social solidarity and a return to the state of nature in terms of people's behaviour towards one another in the workplace. Bourdieu's remarks also suggest that people are driven to work long hours because they internalize the competitive pressures. Other prominent social theorists also refer to these tendencies towards increasingly insecure and individualized working relations.[74] Ulrich Beck (2000a) refers to the Brazilianization of the west and Richard Sennett (1998) to the Corrosion of Character but empirical evidence is rather mixed.[75] It is important to recognize that academia is not immune from globalization and English has become a very dominant language to the extent that concepts developed in one context are rapidly transmitted to another and assumed to mean the same and yet the underlying substance could be different. Moreover the impact of similar working practices can vary and be viewed differently by different people depending exactly on how they are implemented and in particular whether the employees have any control over the practices being introduced.

At one level it seems as though the European Union with its corporate conservative and social market tradition is also subscribing to the neo-liberal and flexibility agenda, illustrated by the four pillars in the EU's Employment Strategy: employability, adaptability, entrepreneurship and equal opportunities. These pillars are potentially contradictory, and in practice are open to different interpretations. Adaptability for example could involve continual training and reskilling to cope with structural changes in the economy or it could mean flexible and precarious terms and conditions of employment to allow the workforce to be adapted to variations in demand.[76] Thus the same broad policy framework can be implemented differently within member states. Indeed quite different traditions remain between the members of the European Union in terms of the relations between the state and society; in this case whether working practices are imposed or introduced through agreement between the social partners. More generally different welfare regimes and regulation theory remain leading to quite different levels of overall welfare. The most extreme differences are between countries following a social democratic tradition, as in the Nordic countries, and those that have increasingly embraced neo-liberalism, such as the UK.

The EU has been concerned about its comparatively high levels of unemployment and relatively slower levels of economic growth than the United States and has considered ways of trying to raise competitiveness while maintaining social cohesion. It is generally assumed that everyone benefits to some degree from economic growth reflected in rising levels of GDP. However, when figures are examined more closely the meaning and desirability of growth is more questionable. In the last 15 years in the United States overall GDP per capita has risen and is higher than in any country in the world except Luxembourg. A large part of the gains however has been absorbed by the top 20 per cent and an even larger proportion by the top 5 per cent. A Congressional Budget Office study using income tax data (cited by Krugman 2002a) found that between 1979 and 1997 the after tax income of the top 1 per cent increased by 157 per cent while families in the middle of the distribution only gained by 10 per cent. Comparing this performance with Sweden, Krugman finds that while the overall USA GDP per capita is higher, families at the median share very similar living standards. Lower down the distribution however the welfare levels of the Swedish families are far higher. Thus the poorest 10 per cent in Sweden have incomes that are 60 per cent higher than equivalent families in the US (Krugman 2002a: 8). This is only one measure but a range of other social statistics such as gender inequality, the welfare of lone parents and child poverty (see Figure 5.1 which shows that only 2.6 per cent of children live in poverty in Sweden compared to 22.4 in the United States and 19.8 in Britain) also demonstrate that Sweden is a more inclusive society. Inequality is also racialized in the US with nearly a quarter of the blacks as compared to an eighth of the whites living in poverty.

Sweden and other social democratic societies are generally portrayed as having highly interventionist states, and undeniably there are active welfare policies and high levels of taxation. But state action is also required to sustain the high levels of inequality in the US both internally and on the world stage. The huge levels of wealth are maintained through the political and economic supremacy of the United States backed by military force while internally the US has a very authoritarian welfare to work policy and one of the highest prison populations of all western countries; with 1 in 142 of its population in prison in mid-2002, the total prison population topped 2 million for the first time (BJS 2003). These very high figures cast a new gloss on the relatively low levels of unemployment found in the United States that the EU is trying to emulate. Similar to the poverty figures black people are more likely to be incarcerated, with one in eight black men compared to 1 in 63 white men being in prison; black men have a 1 in 3 probability of being in prison at some point in their life (BJS 2003). While young black men are disproportionately in prison young black women are overrepresented in low waged employment, which they have to accept as part of the welfare to work policies. Thus while one side of American life reflects a gilded age,[77] the other reflects an authoritarian and punitive one with a 'carceral' state.[78]

While the neo-liberal state in practice is prepared to be draconian in terms of its direct dealings with the poor, it appears to take a more laissez faire stance with respect to unemployment. Unemployment is attributed either to the restrictive practices of trade unions that have boosted wages above market levels or to the characteristics of individuals that make them unemployable. Correspondingly unemployment is tackled by removing any remaining restrictive practices and by introducing supply side policies to raise individual employability. In general these amount to very superficial training, often little more than basic work discipline. People on the training schemes welcome the opportunity of free access to information and the personal support they are given by the workers in the centres while they search for work but the schemes do nothing to increase the number of jobs in different locations,[79] and neither do they enhance the career prospects of trainees, except in a superficial way. The jobs that are obtained are often precarious so people cycle through low paid work, employment schemes and benefits which makes the existence of a jobs deficit clear.[80] By contrast in the corporatist conservative countries of France and Germany training schemes and job qualifications tend to be more formal which gives people some chance of moving forward. Furthermore in France working hours have been reduced to 35 a week in order to try and increase the number of jobs.[81] Similar differences in approach are found in policies designed to tackle social exclusion in areas of deprivation. Currently the state in many countries has responded to areas of intense deprivation through ad hoc locally based initiatives, though the European Union has a more comprehensive strategy to address social inclusion. A comparative European study found that although other countries lagged behind Britain in terms of designing area based programmes, their more regulated labour markets and greater support for the welfare state limited the severity of social exclusion in the first instance.[82] In particular the authors pointed out that inequality had not increased as much as in Britain during the 1990s so they concluded that: 'Area-based programmes are valuable ways of addressing the problem of social exclusion once it has emerged. But in responding to social exclusion, the wider comparative lesson is that prevention, rather than cure, is the more intelligent strategy' (Parkinson 1998: 4).

Ironically as some people are unemployed others work extremely long hours or at least claim so to do. This situation also arises from the heightening of competition as explained in Chapter 1. Long hours threaten social sustainability, as Martin Carnoy (2000) has argued in relation to OECD countries:

> With increased competition in the globalized economy and the rapidly rising capacity to use 'world time' to enhance productivity, the very best workers are now those who never sleep, never consume, never have children, and never spend time socialising outside of work.
>
> (Carnoy 2000: 143)

Carnoy's concern is that this pressurized economy threatens social sustainability and in particular that parents no longer have time to spend with their children. Ulrich Beck (2000) similarly argues that traditional structures have declined leading to individualization in both work and the family with increasing opportunities but also risks:

> The old forms of work routine and discipline are in decline with the emergence of flexible work hours, pluralized underemployment and the decentralization of work sites. . . . The traditional family is being displaced by a 'post-family' or a 'new negotiated provisional family composed of multiple relationships'.
>
> (Beck 2002: 202–3)

Throughout the EU and the US the marriage rate is falling, and the divorce rate and the proportion of lone parents are rising although the levels differ between different countries. The UK has the highest divorce rate and the highest proportion of isolated lone parents, that is those not included within a wider family, in the EU[83] as well as highly flexible working practices. The discussions about social sustainability are driven by a number of related themes. One is the concern with social reproduction – actually reproducing children physically given the declining fertility rates and social sustainability in the sense of ensuring that people have adequate time and resources to maintain themselves and look after their children. A second theme is concerned with the ageing of the population in western societies and the problems of financing the pensions and care of the much larger generation of future pensioners. A further theme related to the sustainability of the traditional family as an institution and this concern stretches across the globe with the feminization of employment. In detailed qualitative research in Costa Rica, where there has been an increase in the number of female-headed households, Sylvia Chant (2002: 132) found that perceptions of the changes were differentiated by gender. While men perceived that there was a crisis in the family, she suggests that their concerns relate more to the decline of the patriarchal family, which has occurred partly as a consequence of women's increased involvement in paid employment. Women however considered that the family, which they take to mean their extended network of blood relatives, was as strong as ever. Although they identified various practical problems arising from paid work, they generally considered it in a positive light as it provided a way of getting ahead and providing for their children. Chant also found optimism among the younger people who saw the changes not so much as a crisis of the family but 'family breakthrough or betterment'.

With the ascendancy of neo-liberal ideologies, the boundaries of the state are changing and the contradictions of market society identified by Polanyi (1957) seem to have been forgotten. Globalization has taken this trend further and not only are basic facilities such as water, transport and power supplied through the market, they are even supplied by international firms. During the summer of 2003 for example there were four major power failures in Italy, North America, parts of London and even in Scandinavia.[84] The question is how and why have these elites been successful in pressing their claims despite their contradictions. Neo-liberalism is a powerful ideology and appeals to people's self-interest. It implies that free markets are somehow a natural and inevitable state of affairs in which individual endeavour will be rewarded and perhaps because of this the poor accept growing inequalities because they think they have a chance of becoming rich themselves as society appears to be freer and more open. Moreover, there have been real increases in wealth for most states in the aggregate even though overall the rate of growth has been lower than in the Fordist era. Furthermore because state policies operate at a more abstract level, managing the money supply for example rather than managing enterprises, the connections between state policies and outcomes are less obvious. Unless there is explicit gender, ethnic and regional budgeting for example the connections between abstract macroeconomic policies such as interest rate changes and the well being of different groups or regions in society remains unclear. Yet these differentiated effects can be quite profound, for example rises in interest rates in the US leading to increases in unemployment have twice the effect on the black population.[85] Likewise policies favouring private rather than public transport favour men more than women as do tax cuts, men being higher earners and women in general being higher consumers of state services.

These trends in terms of the increase in flexible working and work in the informal sector have been increasing on a global scale. There have also been comparable changes in household structures, which combine to create unease about social reproduction. Referring back to Polanyi's analysis of the state in the nineteenth century, it was the contradictions between the market economy and social reproduction that led the state to take action in areas of health, education and housing to guarantee social reproduction. In the global context, however, the reproduction of the national population is less significant as labour can be drawn upon from anywhere. Thus increasingly social change will have to rely much more on social protests (including cross-national or global protests) and social morality rather than capitalist self-interest except at a global level. There are supra national institutions such as the ILO or UN, discussed earlier in the chapter, that might fulfil this role, but so far these still reflect the uneven power of their constituents, the member states rather than the 'global will'.

## CONCLUSION

This chapter has examined theoretically and empirically the role of the contemporary state and its changing character and form. Ideas about the ideal role of the state have been contested over time, and during the last half century there have been two distinct variants: a more active and interventionist state concerned with social welfare in the 1950s to 1970s, and the contemporary neo-liberal state in the context of a global economy, that emphasizes free markets and individual responsibility. In both of these perspectives however a state or sovereign, that is some collective institution, is considered necessary to provide a framework for capitalist society, to maintain law and order, to manage international relations and to act in cases of market failure. What constitutes market failure however varies so there are debates about when it is appropriate for the state to take action. As societies and economies have become more integrated with globalization some of these traditional functions of the nation state are managed at the supra national level.

The main argument in this chapter is that despite globalization, nation states retain considerable power and influence over the organization of economic and social life. Some powers have been delegated to supra national institutions and others to sub-national institutions but the powers of these institutions rest on the powers granted to them by nation states. Thus the key issue is that different states have different degrees of power and in the early years of the new millennium the United States has become a super state with unprecedented economic and political control over the rest of the world.

One of the ironies of the contemporary era is that just at the moment when more nation states are formally democratic, many determinations or influences on economic and social life lie outside the nation state. Nevertheless nation states retain some powers to organize their internal affairs differently and there are quite distinct welfare regimes that can be identified analytically and in practice, which suggests that nation states remain a worthy object of political struggle for those seeking change. The most extreme form of neo-liberalism is found in the United States where writers have referred to the development of a 'carceral' or 'penal' state.[86] Even in the US there are some progressive social programmes run by the state or state supported voluntary organizations and even more so within the European Union, where despite increased economic integration nation states still have some autonomy to follow different trajectories. These differences in state practice together with the formal democratic structure suggest that working within the state, while at the same time challenging its direction, is still a worthwhile task for those seeking social change – or to revive the slogan of the 1980s to act simultaneously 'in and against the state'.[87]

## Further reading

*Antipode* special issue on the neo-liberal state, July 2002, Volume 34, Issue 3.
D. Harvey (2000) *Spaces of Hope: Towards a Critical Geography*, Edinburgh: Edinburgh University Press.
B. Jessop (2002a) *The Future of the Capitalist State*, Cambridge: Polity Press.
K. Polanyi (1957) *The Great Transformation: The Political and Economic Origins of Our Time*, Boston: Beacon Press.
H. Wainwright (2003) *Reclaim the State: Experiments in Popular Democracy*, London: Verso.

## Websites

International Labour Organization (ILO): http://www.ilo.org.
United Nations: http://www.un.org.
World Bank: http://www.worldbank.org.

## Notes

1 See Mann (2000).
2 Hobbes (1983) (originally 1647) Preface to *De Cive*. These ideas are developed more fully in *Leviathan* (originally 1651; Hobbes 1996).
3 Interestingly for Hobbes if the sovereign/state threatened the self-preservation of individuals then the contract between the people and the state was effectively broken, as self-preservation is the fundamental rationale for the state. Rousseau by contrast thought the state had the right to eliminate people who challenged the attainment of the common good.
4 This occurred primarily through the Speenhamland system passed in 1795, which built on the Elizabethan Poor Laws but it was more liberal because it was paid to able-bodied people. It was only payable within the parish to which people belonged and so undermined labour mobility. It was abolished with the new Poor Laws passed in 1834 (see Polanyi 1957).
5 Smith (1976) is actually referring to individuals who employ capital in support of domestic industry in a chapter which is justifying restraints on particular imports.
6 Note here that the use of the male pronoun reflects the language of the time which also reflects and informs understanding; that is, there is fusion between the he as a generic pronoun to refer to any individual and he being an intentionally gender specific pronoun.
7 Mill (1984) endorsed the prevailing racist views of his time considering other races as 'children' requiring leadership, though he had more progressive views on women's equality.
8 See Wainwright (2003).
9 Hayek (1945) noted that there was a growing consensus about the role of markets and that even Leon Trotsky was in favour of them, but there is an important difference between using markets as part of an allocation procedure and managing the whole of society as a market.
10 A free wage labour market is a proletariat or population whose only means of existence is the sale of his or her labour.

11 See Barr (1998) for an introduction to the economics of the welfare state.

12 See Wacquant (1999).

13 See Stiglitz (2002) and Chapter 9 for a rather different interpretation.

14 See Jessop (2002a).

15 See Marx (1976: Chapter 16).

16 In this particular case reductions in the working day may have closely approximated the general will.

17 See Jessop (2002a) for a detailed discussion of the nature of the capitalist state.

18 In contemporary times there has been a rapid expansion of private security services, for example in the gated communities, and in many societies parents are still allowed to use some violence against their children and husbands against their wives so the capitalist state is also in many ways a patriarchal state.

19 David Harvey (2003) argues that war is one way in which surplus production is absorbed. As Marx pointed out, in other kinds of societies surplus production would be regarded as a bounty, something to be celebrated, but in capitalism it is a problem because capitalists will be left without profit (see Chapter 1).

20 By recently I am referring to the last two decades though the rate and extent of the feminization of the labour force varies between states.

21 See Carnoy (2000).

22 See for example List (1909) and Freeman (1987).

23 See Wade (1990), Amsden (1993) and Singh (1999a and b).

24 Indeed a large proportion of deals that take place between buildings in Manhattan are actually international, sometimes not even involving US firms, and for this to be possible new financial institutions have developed (Sassen 1999).

25 See Lojkine (1976).

26 See Wacquant (1999) and Peck and Tickell (2002).

27 See Mann (2000).

28 See Peck (2001).

29 The liberalization of services and public utilities has occurred as a consequence of GATS, the General Agreement on Trade in Services, implemented in the mid-1990s but extended to public utilities in 2002. See http://www.wto.org/english/tratop_e/serv_e/gatsqa_e.htm (last accessed July 2003).

30 See Sassen (1999).

31 See Glassman (1999) and Sklair (2000).

32 See Sassen (1999).

33 Enron recorded the largest bankruptcy in US corporate history in December 2001. It was a medium-sized energy company that diversified into the knowledge economy and expanded dramatically in the 1990s partly through a complex web of offshore partnerships and holdings that inflated its visible profits and growth and disguised debt. The structure of the company became so complex that it was difficult to monitor, though its auditor Arthur Andersen was prosecuted for corruption. The company was exposed when the e-bubble burst. New rules for corporate trading have been established.

34 James Tobin, a Nobel prize winner for economics, proposed a tax on financial speculation in the 1970s.

35 See War on Want (2003).

36 The same could be said of environmental risks (see Beck 1992).

37 See Peck and Tickell (1996 and 2002).

38 These institutions were established following the Bretton Woods conference in 1944.

39 The WTO was formed later than the others in 1995 but evolved from GATT, the General Agreement on Tariffs and Trade, which was founded in the immediate aftermath of the Second World War. The websites of these organizations provide detailed information about their role and mission as well as detailed accounts of

specific projects. They also provide a valuable source of data. It is important to remember however that organizations' websites usually present a positive self-image with little or no critical reflection. The addresses are http://www.un.org, http:// www.worldbank.org, http://www.wto.org and http://www.imf.org/ (last accessed February 2003).

40 For example the IMF pledged $21 billion in 1997 to assist Korea reform its economy, restructure its financial and corporate sectors, and to recover from recession. Within four years, Korea was able to repay the loans and, at the same time, rebuild its reserves IMF (2001).

41 See World Bank (2003b).

42 The UN security council consists of five permanent members and ten other members elected for a period of two years.

43 See Chomsky (2003).

44 See http://www.wto.org/english/thewto_e/whatis_e/tif_e/org1_e.htm (last accessed July 2003).

45 For further details of the ILO constitution see http://www.ilo.org/public/english/about/iloconst.htm#a7 (last accessed July 2003).

46 See Monbiot (2002b).

47 At the time of writing, six months after he came into office in January 2003, he remains widely popular. He has maintained the neo-liberal economic policies, but has also tried to build a wide consensus for reforms, including pensions to forge greater equality between high and low paid workers.

48 See Chapter 10 for a brief discussion of the World Bank's opposition to an air traffic control system for Tanzania as it conflicted with the poverty reduction strategy.

49 International Labour Organization: http://www.ilo.org/. For more details on the history and role of the ILO see: http://www.ilo.org/public/english/about/history.htm.

50 The NAFTA website is http://www.nafta-customs.org/ (accessed February 2003).

51 For information about the EU and its activities see http://europa.eu.int/index_en.htm# and for a brief history of the European Union see http://europa.eu.int/abc/history/index_en.htm.

52 See Wilkinson (1997).

53 The G7 are France, the United States, Britain, Germany, Japan, Italy and Canada. The G8 = G7 plus Russia; the G20 = G8 + Argentina, Australia, Brazil, China, India, Mexico, Saudi Arabia, South Africa, South Korea, and Turkey with two spaces reserved for Indonesia and Malaysia or Thailand, similarly with the stated objectives of achieving stable and sustainable world growth that benefits all.

54 See http://www.weforum.org/.

55 Naomi Klein (2003) points out that the presidents of Venezuela and Brazil both attended the forum in Porto Alegre in 2003, and asks rather sardonically, 'how on earth did a gathering that was supposed to be a showcase for new grassroots movements become a celebration of men with a penchant for three-hour speeches?' The WSF remains a grassroots movement, correspondingly there are many WSF websites; to access them use the Google search engine, entering World Social Forum and the different websites will appear.

56 See Wade (2001).

57 In this discussion the term region is understood as a sub-national territory, that in some instances may extend across national frontiers. These regional entities can be defined in different ways, as functional economic regions, historic or ethnic regions, administrative or planning regions and political regions with elected representatives. In some instances regions defined on these different criteria will overlap, in others not (see Keating and Loughlin (1997) for a fuller discussion).

58 See Chapter 6 for a fuller discussion of the multimedia super corridor.

59  A slogan or logo used by Brighton and Hove in 2000, subsequently becoming 'Where Else'.
60  See Chapter 5.
61  See Harvey (1989; 2000).
62  See Murray (1991).
63  See Harvey (1989 and 2000).
64  See Perrons (2000b).
65  Though this behaviour is not unique to global cities. Brighton and Hove in south east England portrays itself as a city of culture, a new media hub and as deserving of social support to counter deprivation (see Perrons 2003).
66  See Keating and Loughlin (1997).
67  See Rodríguez-Pose and Gill (2003).
68  Dunford and Greco (forthcoming) citing Piattoni (1994).
69  See Keating and Loughlin (1997).
70  See Cravey (1997).
71  Female-headed households are not inevitably the poorest of the poor (see Chant 1997).
72  See World Bank (2003a).
73  See González de la Rocha (2003).
74  See Beck (2000a) and Sennett (1998).
75  See Doogan (2001).
76  The existence of contradictions between these four pillars, for example adaptability and equal opportunities, does not seem to have been fully recognized (see Perrons 1999b). In 2003 the European Employment Strategy was revised but these contradictions remain.
77  See Krugman (2002a).
78  See Wacquant (1999) and Peck (2003).
79  See Turok and Webster (1998).
80  See Skyers (2003).
81  See Fagnani (2002).
82  See Parkinson (1998).
83  See Chambaz (2001).
84  See Brough (2003).
85  See Peck (2003).
86  See Peck (2003) and Wacquant (1999).
87  See L-EWRG (1980) and Wainwright (2003).

# 9

# GLOBALIZATION, PARTICIPATION AND EMPOWERMENT

How do individuals, organizations and social movements gain influence over their immediate living and working environments or places within the global economy and in what circumstances do those in power allow and indeed encourage decentralization of decision making and greater involvement? This chapter addresses these questions by examining how people have tried to influence the trajectory of development through social movements within nation states, regions and localities and as individuals through different kinds of association such as NGOs or protest movements. This differentiated analysis of how people influence and respond to processes affecting their environments illustrates that human beings are active agents who are capable of shaping developments, something that can be overlooked if globalization is understood as some kind of inalienable process. Some movements are more successful than others but the ways in which all individuals and groups exert and sometimes secure their demands are rarely harmonious.

Societies are a curious mixture of conflict, struggle, cooperation, compromise and resolution. This is no surprise as individuals are filled with anxiety and internal struggle over their own preferences, except where circumstances are so dire that they preclude any choice. Thus in social groups, the uncertainty, range of possibilities and potential conflicts multiply, probably in a geometric progression. Conflict is therefore endemic but ways of resolving conflicts are neither predetermined nor inevitably violent. The majority of people live in some kind of equilibrium most of the time, having accepted or adapted their preferences,[1] internalized prevailing social norms[2] and roles, or become reconciled to their existing situations even though these can sometimes be life threatening. Consequently it is the episodes of conflict and contestation that generate new individual and social trajectories. Having considered the role of the state in managing conflicts and social reproduction in Chapter 8, this chapter examines how people have challenged local and nation states, supra national institutions and multinational organizations and how these institutions have responded to protests, in some cases, by widening the boundaries of participation in decision making.

The first section examines two different regionalist or nationalist campaigns, one that has been relatively successful, the Catalan nationalist movement in Spain, and one that is still struggling to secure its objectives, the Zapatista movement in the Chiapas, Mexico. The second section discusses three very different illustrations of widening participation in decision making at the local level: in a participatory project concerned with rain fed farming in eastern India; an area regeneration scheme in a deprived area of inner London; and the development of participatory budgets in Brazil. The final section considers the extent to which NGOs and other activists have been able to influence the activities of transnational corporations through fair trade and ethical trading initiatives.

These movements and organizations differ widely in political orientation but raise similar and fundamental questions about where control over territorial space, living and working conditions resides. They also reflect the demand for and the increased tendency for people to become more involved in decisions that affect their daily lives. The main argument in this chapter, which links these different cases, is that the key issues are not so much who is included in the decision making processes, but over what issues they are able to exercise their preferences and further, that it is less significant who is in control of particular territories than how that control is exercised. Reference is also made to the anti-globalization or anti-capitalist protests, or more positively the movement for global justice which is in some ways an umbrella movement for protests against supra national institutions and the neo-liberal project and encapsulates many of the issues covered in the case studies below. For example, the Zapatistas support and are widely supported by this movement and Port Alegre, where participatory budgets first evolved, held the 2002 meeting of the World Social Forum, where alternative ways of living were discussed in addition to protests against the adverse consequences of globalization. Many NGOs are also part of the anti-capitalist alliance and participate in its activities. This movement is of relatively recent origin, often dated to the protests against the World Trade meeting in Seattle in November 1999, and has a varying composition of social movements, including some of those discussed below with longer individual provenances. It is not yet clear whether this collective movement will become a major force for change or whether it will fade away in the context of the more divided world following the war against Iraq, issues discussed further in the concluding chapter. What is more certain, however, is that the elements of which it is composed or new organizations will emerge to address existing and new injustices, as struggle and conflicts are an inherent feature of social life, especially in dynamic societies and those that exclude billions from sharing in their wealth.

# NATIONALIST, REGIONALIST AND LOCAL
## SOCIAL MOVEMENTS

Within nation states people at regional and local levels have tried to play a greater role in shaping their fortunes, sometimes in response to adverse regional and local consequences arising from relations between the nation state and the global economy, as in the case of the Chiapas region of Mexico, or because there is a sense that regional economic, political and cultural interests are being undermined or suppressed by the nation state, for example in the Basque region of Spain or in north east England. Movements for greater autonomy also arise in richer regions, which sense that their fortunes are undermined by unification with poorer regions, such as the Lega Nord in Italy, or simply as a consequence of national borders being out of synchrony with contemporary national/regional identities as in parts of Northern Ireland, Catalonia or in war torn regions of Africa, where national boundaries were imposed by colonial powers without any sensitivity to local attachments, or through some combination of these processes. At the local level, different kinds of social movements have also sought to influence the trajectory of development. These include: NIMBY (not in my back yard) movements that resist developments considered threatening to their current neighbourhood lifestyle such as hostels for asylum seekers or motorways; more radical protest groups that try to protect local firms, farms, villages[3] and cooperatives from the demands of large-scale, sometimes multinational commercial, retail and property developers; and groups trying to establish their identity in particular localities, such as gays and lesbians, religious groups and ethnic minorities.

These regional and local movements vary widely in their aims and political complexions but share a common sense of injustice in the status quo and the aim of establishing greater control over their own living spaces, that is the struggles are territorially based, even though they may be challenging processes whose origins lie beyond their localities. In this respect, the main differences between them lie in the relative degrees of power they possess, the strength of internal solidarity within the region or locality and the degree of resistance they are likely to face. Thus while these movements seek increased control over their immediate environment or place, the extent to which this desire fundamentally challenges existing power relations or the global socio-economic order varies enormously. Furthermore only rarely will there be total unanimity for their aims; people within localities generally remain divided by social class, gender, ethnicity, age and cultural outlook and even though particular social movements may capture the local popular and intellectual imagination from time to time, it is likely that there will be dissenters, some of whose wishes may be repressed by those advancing the cause. Arundhati Roy (2001), for example, argues that the movement opposed to the construction of dams in the central and western states of India, the Narmada Bachao Andolan (NBA), 'is a fantastic example of people linking hands across caste and class. It is the biggest, finest, most magnificent resistance movement since the independence struggle.'

Though as Roy herself points out it has also been alleged to be a middle class movement that prioritizes the environment over the needs of the poor. Recognizing that differences exist within as well as between territories, irrespective of spatial scale, and relatedly that similarities exist across territories, helps to explain both the existence of non-territorial struggles and the search for a set of universal human rights and capabilities at a more abstract level, around which there could be more general agreement. These rights could provide a framework within which local conflicts could be resolved without repressing the wishes of within group minorities, issues discussed later in the chapter.

The section begins by briefly discussing the Zapatista movement in Chiapas, one of Mexico's poorest regions, partly because this movement has been very effective in using contemporary technologies to make its case known and partly because it represents a very clear illustration of the conflict between self-determination and autonomy on the one hand and pressures for global integration on the other. It then turns to the struggle for autonomy in Catalonia in north east Spain, the country's richest region on the basis of GDP per capita, which has been more concerned with establishing cultural rather than economic autonomy, at least in the more recent past. In both cases the history and politics are highly complex. This section draws on them to highlight different ways in which people in different regions of the world experience and react to more powerful national and global forces and to indicate that the resolution of a territorial struggle does not automatically lead to a disappearance of other social conflicts. Indeed, the converse could be more likely, in that it is precisely when nationalist or territorial demands have been resolved that other social divisions come to the fore.

## Chiapas and the Zapatista movement in Mexico

The Chiapas borders Guatemala in the south east of Mexico, and although rich in natural resources, is one of the country's poorest regions in terms of GDP per capita (see Figure 9.1). A high proportion of the population are indigenous Amerindians, including Mayans. The economy is based on agriculture, especially corn growing, organized around the *ejido* or communal lands that were established after the Mexican revolution, led by Emiliano Zapata in 1910–17. When Mexico joined the North American Free Trade Association (NAFTA) along with the US and Canada in 1994, the constitution was amended to abolish these communal land rights and to open the economy to market forces. This sparked a 12 day uprising and subsequent ongoing guerrilla activity by the Zapatista movement (named after the early revolutionary leader). The Zapatistas were concerned that increased marketization would intensify social polarization within Mexico, further impoverish the already poor and make it increasingly difficult to follow an autonomous, more ecologically sensitive development trajectory. In particular they were concerned that corn produced in the USA would undercut production in the Chiapas, undermine the coherence of the regional economy and displace the indigenous population from their land.

Following the initial uprising, the struggle in the Chiapas has largely been peaceful despite the extensive presence of government paramilitaries within the region, possibly because the Zapatistas have established and maintained international awareness of their struggle on the Internet, which probably constrains government repression. One illustration of peaceful protest was the 2000 mile march, the Zapatour, from Chiapas to Mexico City in 2001. This march took place shortly after the World Economic Forum meeting in Cancun, a resort on the Caribbean coast (see Figure 9.1). Ironically, tourist information for Cancun highlights excursions to historic Mayan cultural sites including the coastal city of Tulum[4] (see Figure 9.2), suggesting perhaps that when fossilized and commodified, traditional ways of life are compatible with globalization even though their lived reality is construed as an impediment. A further reason lies in the change of government in 2000, which promised a swift and peaceful end to the dispute.

The new government has placed greater emphasis on dialogue but remains strongly committed to further global integration illustrated by the Plan Puebla-Panama (PPP), which covers 14 southern states of Mexico, as well as the other 7 countries of Central America.[5] Indeed the government regards the PPP as a far more effective solution to regional discontent than autonomous development. The plan aims to narrow regional inequalities within Mexico, through improved infrastructure and expansion of maquiladoras, currently concentrated in the states bordering the US and around Mexico City (see Figure 9.1).[6] The Plan also recognizes the significance of the region's natural resources, and aims to maintain the biodiversity and protect the rights of indigenous people, who would be consulted as part of the development process (Global Exchange 2002).

Opponents argue, however, that the plan prioritizes the interests of banks (the Inter American Bank (the regional arm of the World Bank) and the Central American Bank of Economic Integration are both sponsors) and the major corporations. These organizations are attracted by the strategic location of the region, between the Atlantic and Pacific and therefore between the two key trading areas for NAFTA, Europe and Asia. The area is also attractive to the US government because it provides a direct corridor to Venezuelan oil and a means of extending free trade into Central America. Correspondingly, in the opponents' view, the main outcomes of the Plan will be the construction of mega-highways or 'dry canals' to supplement the Panama canal, hydroelectric dams to provide power for the rest of Mexico, and agribusiness to supply foreign markets which will result in 'bio piracy' rather than biodiversity. Furthermore, they argue that the promised maquiladoras will only provide exploitative employment for women and young people. More generally the central objection to the plan is that it prioritizes external interests rather than promotes autonomous development based around the use of local resources and conforming with the wishes of the indigenous population.[7] The Zapatistas have declared that

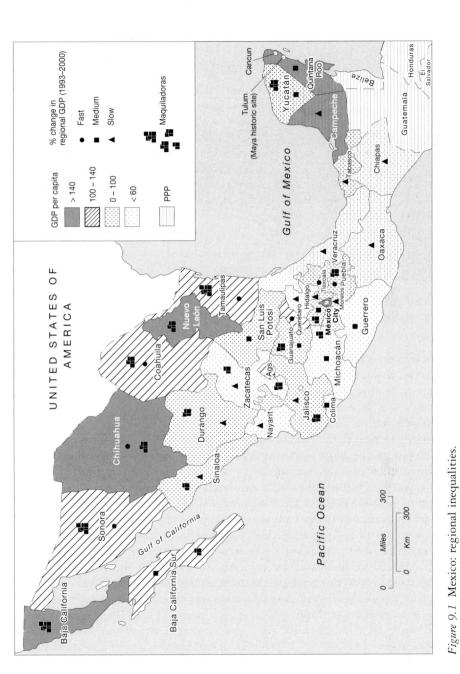

*Figure 9.1* Mexico: regional inequalities.

*Source*: Map drawn by Mina Moshkeri.

*Note*: For data aources see note 6.

*Figure 9.2* Tulum: a Mayan city.
*Source*: Photograph by Silvia Posocco.

they will not allow the PPP in the municipalities they control and continue to argue for an alternative more inclusive, ecologically sensitive strategy.

Formally, there is some agreement about the desired outcomes for lowering poverty and for protecting resources, but there are diametrically opposed understandings of how these outcomes might be attained. The Mexican government, in contrast to the Brazilian government elected in 2002, does not seem to recognize the huge and unsustainable rises in poverty and inequality associated with neo-liberal development strategies elsewhere. For example per capita income in many South American countries, including Brazil and Mexico, is barely higher than it was in 1980 and given that inequality has increased, a significant proportion of the population have experienced an absolute as well as relative decline of welfare.[8] Opponents may however be overlooking some of the more positive aspects of global integration, for example although the maquiladoras are criticized for their exploitative and sometimes hazardous working conditions, there are some advantages, especially for women, who make up just over half of all employees and at times consider factory work as respite from heavy domestic work – 'I yearn to get to work to sit down and rest' (Denman 2002: 6, citing an interview with Patricia, a maquiladora worker (see Box 9.1)). This sort of comment is reported in studies of female workers throughout the globe, even though the absolute circumstances are profoundly different, which reflects the arduous nature and uneven distribution of domestic work.

---

Box 9.1 A typical day in the life of a maquiladora worker

Patricia usually wakes up at seven, then she gets the kids ready for school, cleans the house, makes meals, washes clothes, asks the neighbour to stop the gas truck when it goes by in the afternoon and if she is lucky takes an hour or two to nap and make up for the lost sleep. It didn't bother her so much at the beginning when she started working the night shift, she kind of liked going to work at four thirty in the afternoon and getting back at 2 or 3. Things were usually calmer on the night shift although the flirting and coqueteo, which goes on is more intense than during the day shift. If she stays to do overtime, which she badly needs to do to increase her wages, she can get home by four in the morning, in time to sleep three hours before she has to get the kids up. Her brother and sister-in-law moved into the shack next door when she finally built her concrete brick one room house and they look after the kids at night until she gets home. Her 8-year-old daughter takes care of the smaller daughter and son and feeds them when they get back from school. Patricia works hard to make ends meet, she makes tamales and tortillas to sell, she takes beauty products to work to sell during breaks, she sells 'hielitos' (confectionery) during the summer and often sells cans of soda when she has energy to carry them up the unpaved road, 6 blocks from the bus stop. Somehow she finds time to make small gifts, like she did last Mothers Day for all 18 workers on the production line she coordinates.

Source: Denman (2002: 7).

---

Despite this gruelling life Patricia would never think of 'going back to the village she came from as she has her own house and a job and says it is better to be alone [referring to a partner] than with bad company' (Denman 2002: 7). Furthermore the presence of maquiladoras has been associated with higher levels of regional GDP and higher levels of growth in the last decade (see Figure 9.1), although during the 2001–2 recession some 230,000 or 17 per cent of the jobs were lost through employment contraction and movement to lower cost locations, such as China.

Alternative forms of development based on fair trade for organic coffee have also occurred in the Chiapas, which as indicated below allows people to interact with the global economy while largely conserving traditional ways of life. Lessons could perhaps also be learned from Bolivia where in 2002 original inhabitants including representatives of trade unions and peasant groups gained almost one-third of the seats in the Congress. 'It's a sort of peaceful, democratic, Zapatista revolution' (Vacaflor cited by Rocha 2002: 11). In this way the original or first people are able to exercise a voice in the formulation of plans and projects rather than simply being asked to contribute to their implementation once key decisions have been taken, a point developed below in the section on participation.

289

While the Zapatista movement may have become a cause célèbre for the anti-globalization movement their argument that neo-liberalism will intensify existing inequalities is also supported by Joseph Stiglitz, the ex-chief economist at the World Bank,[9] not necessarily because of free trade itself, but more because of the way that free trade principles are unevenly applied. In this case, while US products are given free access to Mexican markets, Mexican migration to the USA is restricted. Consequently the freedom of the poor and relatively powerless to improve their futures is constrained, while the free movement of capital tends to undermine the abilities of poor people to make a living where they are.

Immigration restrictions are often supported by people, including past migrants, already in the USA, anxious that large-scale migration will undermine their way of life but less concerned that their government's policies will undermine the ability of other people to retain their lifestyles.[10] However, the extent to which restrictions apply vary largely in accordance with the economic needs of the wealthier states. Proposition 187 for example which prevented the children of illegal immigrants from receiving education or health services in California was withdrawn in the booming economic conditions at the end of the 1990s, while following the economic downturn in late 2001/2002, partly linked with the terrorist attack on the World Trade Center, migration is again being curtailed. Nevertheless, both legal and illegal migration takes place, as the border is long and difficult to patrol effectively. Once in the US the undocumented workers experience disadvantageous terms and conditions of employment but contribute to the growth of the US economy.

More generally, less developed countries have been forced to open their markets while developed countries continue to subsidize their own agriculture and restrict immigration and imports from less developed countries. As Stiglitz notes, it is not surprising that the WTO formed the focus of the protests against globalization because it symbolizes the hypocrisy of the already rich countries:

> While these countries preached – and forced – the opening of the markets in the developing countries to their industrial products, they continued to keep their markets closed to the products of the developing countries, such as textiles and agriculture. While they preached that developing countries should not subsidize their industries they continued to provide billions in subsidies to their own farmers, making it impossible for the developing countries to compete. While they preached the virtues of competitive markets, the United States was quick to push for global cartels in steel and aluminium when its domestic industries seemed threatened by imports.
>
> (Stiglitz 2002: 245)

He goes on to point out that these inequities are beginning to be acknowledged in that in the November 2001 Doha negotiations the richer countries 'agreed to discuss' these inequities. He comments somewhat sardonically that 'just to discuss redressing some of these imbalances was viewed as a conclusion' (Stiglitz 2002: 245). This latter point is very similar to the idea that participation is itself a solution to the problem of development or regeneration, when much of this participation, as will be shown later in the chapter, is purely discursive.

While the Zapatista movement in the Chiapas consists largely of peasants and seeks more autonomy from the national state and the world economy, in Catalonia, discussed below, the struggle was led primarily by the bourgeoisie, and was simultaneously for more independence from the national state but for more integration into the wider world economy too in order to foster economic growth.

## Catalonia[11]

Catalonia was an autonomous nation, with a small empire, until the fifteenth century when it became part of Spain following a royal marriage. Following a long struggle for independence it secured the status of an autonomous region in the late twentieth century, which seems to have partially satisfied the demands for political autonomy, in contrast to the Basque region where despite a similar status, ETA (Euskadi Ta Askatasuna) continues its terrorist practices.[12]

The relations between Catalonia and the Spanish state were always tense, with a failed secessionist struggle in the eighteenth century followed by intermittent independence campaigns which resulted in some autonomy in the early 1930s, only to be reversed during the authoritarian Franco era, from 1940 until 1975, when even the use of Catalan, the national language, was prohibited. The repression failed to quell the quest for autonomy and the Catalan culture and language survived through strong neighbourhood associations, a committed intelligentsia (Garcia-Ramon and Albet 2000) and through religious and sporting events. The Nou Camp, football ground of FC Barcelona, was a particularly symbolic site where Catalan could be spoken, contributing perhaps to the continuing rivalry between FC Barcelona and Real Madrid.[13] This suppression reinforced the strong sense of regional or national identity and in part explains the solidarity around the redevelopment of the region and reconstruction of Barcelona, when democracy was eventually restored in 1976. The most visible aspects of the way that the national culture lives on are the widespread use of Catalan, which became an official language alongside Castilian (Spanish) in 1983, and more symbolically in the Sardana, a traditional dance performed every Sunday at midday outside the Cathedrale Santa Maria in Barcelona in an apparently autonomous way.

Catalonia always had a relatively strong economy; indeed this was one source of frustration for the Catalan bourgeoisie in the late nineteenth and early twentieth centuries, whose ambitions were constrained by the more traditional, agriculturally and commercially based Spanish state.[14] Nevertheless the first socialist government of the main city, Barcelona, inherited a decaying urban area with high levels of unemployment. By drawing upon the new sense of liberation and desire for change the city and region embarked on wholesale physical and cultural reconstruction which transformed Barcelona from a grey industrial city in the 1980s to a modern metropolitan city in the 1990s, and now to a global knowledge based city in the new millennium.

The programme began modestly by renewing some of the old industrial spaces and turning them into parks and museums to create much needed open public space in the different neighbourhoods of the city; this increased popular support for the wider project. For example the Parc de l'Estacio del Nord was built around the abandoned station, converting the station itself into a multi sports centre and the abandoned shunting yards into a national theatre. In the open space there is a multi-toned blue and white dune-shaped sculpture by Beverly Pepper, illustrating how public art formed an important component in the city's renewal. A further example is the Parc de l'Espanya Industrial, built between 1982 and 1985 from an old textile factory. This now has a boating lake and sculptures by Catalan artists and while the overall design has been quite controversial, some people likening it to a concentration camp, it is nevertheless widely used by the local population. These comparatively modest developments were followed by more spectacular projects, especially associated with the Olympic Games held in 1992 and the Universal Forum of Cultures in 2004.

The Olympic Games involved four new sporting sites and major infrastructure projects in addition to general renewal. The interests of ordinary Barcelonans were respected by tunnelling large parts of la Ronda, the motorway connecting the sites, and creating parks on the space above. Housing for the additional security forces was subsequently transferred to student accommodation. The modern cultural image was fashioned by using key architects, indeed the global superstars. Santiago Calatrava designed one of the telecommunications towers (at the Monjuic site) and the other was designed by Norman Foster (at Tibidabo) and Arata Isozaki designed the Palau Sant Jordi. Indeed, Barcelona won the Royal Institute of British Architects Gold Medal for architecture in 1999 – the first time the award was made to a city rather than to an individual architect. Besides the stunning architecture the most lasting achievement with arguably the widest social impact was the clearance of derelict industrial sites which opened the city to the sea and provided a major resource for local people as well as a tourist attraction. The Olympic Games were used to promote Barcelona and Catalonia as a modern global city region, with a Mediterranean specificity by building on its natural advantages – the sun and the sea, together with its historic and modern culture (Moragas Spà et al. 1995).

One of the reasons why the Olympic Games were held in Barcelona was because the Catalan Nationalist Party held the balance of power in the national parliament and they were widely supported in the region partly as a means of asserting the autonomy and global significance of Barcelona and Catalonia in comparison to Madrid and the rest of Spain (though there are also rivalries between the city and the region).[15] Charismatic figures such as Jordi Pujol, the President of the autonomous Catalonian regional government, and the two mayors of Barcelona, first Pasqual Magaral (1982–1997) and now Juan Clos, as well as highly committed intellectuals seized the Games as an opportunity to gain national and international finance, utilize the long tradition of public sector led planning and use their new political power to redevelop the area in what they considered to be the best interests of the city region.

Some of the contradictions associated with this redevelopment are now becoming apparent, for example the housing in the Olympic Village, intended as mixed housing, has largely become housing for the elite. Further the Museu d'Art Contemporania and the Centre de Cultura Contemporania, designed by Richard Meier, were built on the former Casa di Caritat – the poor house in the old city, El Ravel, very close to La Ramblas and adjacent to some of the poorest housing in Barcelona occupied by migrants mainly from northern Africa. So far this development has not yet displaced the local population, as gentrification in the immediate environment has been rather limited. Unemployment amongst young people was high in the 1980s and 1990s and consequently they tend to remain in the parental home. Moreover the cost of renovating the traditional housing in the old city would be extremely high. However this particular redevelopment has done little for the existing residents except that it provides some open space which they, or at least their children, use for skateboarding when the museum closes. The contradictions from the redevelopment are perhaps even more evident in the second major reconstruction project, the Universal Forum of Cultures.

Following the Olympic success, the Universal Forum of Cultures, organized in association with UNESCO, will be held in 2004, broadly within the Poblenou (see Figure 9.3), which was one of the main industrial areas of the city. This event differs from past major events or spectacles because of its core themes – cultural diversity, sustainable development and world peace – and because it has a virtual as well as a physical presence. On the web, Forum 2004 provides regular updates on the progress of the programme, an opportunity for discussing various global political and cultural issues, such as world debt, asylum seekers and military interventions in civil wars, and also a site for posting information about related conferences and events. There was for example regular coverage of the African Caravan for Peace and Solidarity which travelled from Cape Town to Dar Es Salaam and staged cultural events and meetings about governance, health and education, the economy, decentralization and regional integration, culture and art on the way. Further, its real material existence (between April and September 2004) involving conferences, exhibitions and cultural events is to be staged within an ecological park.

There are however similarities with the Olympic Games and more generally with other spectacular redevelopments elsewhere.[16] In particular, the venue has provided a focal point for the major reconstruction of a coastal industrial wasteland that will further transform the structure of Barcelona by extending the Avenida Diagonal, one of the main thoroughfares running through the central business district to the sea, thereby fulfilling the plans of the nineteenth century engineer, Ildefons Cerda (see Figure 9.3). The Forum itself has a convention centre, hotel and leisure complex including a marina and an aquatic zoo, which won an EU prize for its sustainable energy programme. More generally, however, private commercial concerns have dominated. The Diagonal Mar centre is very much based on the American model of mixed shopping, housing (consisting of 'residential towers'), offices, hotels, a convention centre, cinema complex and a public park and is clearly aimed at the tourists and the young, upwardly socially mobile, as well as the new global elites. Not far away is La Sagrera, which will be a terminus for the high speed trans European railway but also a site for a park and housing. A further aspect of the development focuses on knowledge based industry known as 22@[17] whose design has been 'inspired by New York's high tech Silicon Alley but in a very Mediterranean way' (Ajuntament de Barcelona 2002). Many old factories in the area have been converted into spaces for small high tech firms and artists to live and work. Undoubtedly high tech workers will be attracted, Barcelona being one of the most culturally sophisticated cities in Europe with a Mediterranean climate and close to the sea. However this style of redevelopment has been criticized for doing little for the current resident population of former industrial workers, who have to adjust to the presence of clubs and bars associated with the incoming younger population and endure living through the massive reconstruction currently taking place (Gdaniec 2000).

Thus this redevelopment also contains many contradictions. While the themes of the Forum are about peace and diversity and establishing harmony and sustainability in the context of globalization, the tangible outcomes are likely to be much more conventional and similar to publicly supported redevelopments taking place elsewhere, as global players acting in commercial ways have played a major role in the redevelopment and it is less clear that the diversity within Barcelona has been respected. In particular, as indicated above it is not clear how the interests of the traditional working class residents of Poblenou are being met as well as those of the culturally more diverse migrant population that was rehoused in La Mina, on the other side of the River Besos. La Mina was built as a response to the development of shanty towns during the period of large-scale migration to Barcelona in the 1960s and 1970s, but is now an area of considerable deprivation, low literacy and high drug abuse.[18] Plans similar to those of other areas of deprivation have been made to revitalize this area, to counter social exclusion and more specifically in this case to enable the population to take part in the Universal Forum of Cultures, though it is not yet clear how effective these will be. Local people have suggested that the plans proposed for the physical redevelopment of the area, in particular the creation of more open space, will not be sufficient to deal with the depth of the social problems found there. As the mayor of Barcelona is clearly aware of the problems of contemporary globalization on a world scale, arguing that 'liberalization is causing havoc to the social protection networks in many third world countries, and many cities are finding that they have to pick up the pieces as that is where

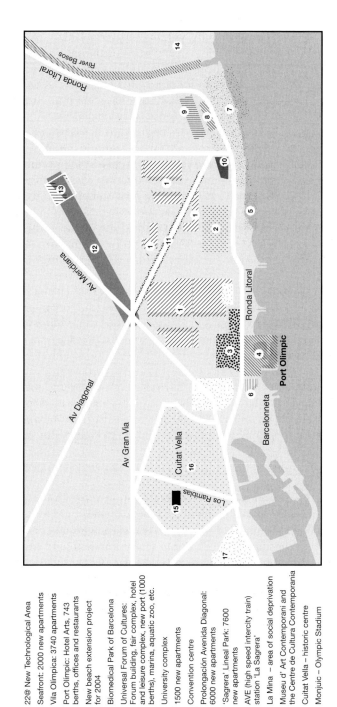

1 22@ New Technological Area
2 Seafront: 2000 new apartments
3 Vila Olimpica: 3740 apartments
4 Port Olimpic: Hotel Arts, 743 berths, offices and restaurants
5 New beach extension project for 2004
6 Biomedical Park of Barcelona
7 Universal Forum of Cultures: Forum building, fair complex, hotel and leisure complex, new port (1000 berths), marina, aquatic zoo, etc.
8 University complex
9 1500 new apartments
10 Convention centre
11 Prolongación Avenida Diagonal: 6000 new apartments
12 'Sagrera' Lineal Park: 7600 new apartments
13 AVE (high speed intercity train) station 'La Sagrera'
14 La Mina – area of social deprivation
15 Museu d' Art Contemporani and the Centre de Cultura Contemporania
16 Cuitat Vella – historic centre
17 Monjuic – Olympic Stadium

*Figure 9.3* Barcelona and the Universal Forum of Culture.

*Source:* Adapted from Diagonal mar (2002).

people are moving to' (Clos 2002), it is to be hoped that people living in this small area on the fringes of his own city, will not be overlooked.

The redevelopment of Barcelona is in many ways a stunning illustration of what can be achieved through public sector led redevelopment. It is associated with a lot of civic pride and the majority of people living there as well as its many visitors will welcome this transformation. The ease with which it has been achieved probably reflects the strong sense of national identity and widespread desire to put Catalonia and Barcelona on the world stage. Indeed Barcelona has become a symbolic icon, and a model for other cities to follow, something which risks overlooking the particular set of historic circumstances that at least enabled the first round of reconstruction to take place relatively harmoniously (Garcia-Ramon and Albet 2000). As a consequence, the mayor has become a global spokesperson, speaking at the United Nations in 2001, the first mayor ever to do so, and somewhat paradoxically at both the World Economic Forum in New York and the World Social Forum in Porto Alegre in 2002, as well as trying to empower cities within the EU decision making apparatus.[19]

Castells (1997) suggests that Catalonia provides an illustration of a nation without a state and so reflects the meaning of identity within the information age. Catalans have their own language, institutions and some autonomy over their economic and political affairs, but remain part of Spain and part of Europe. More specifically he suggests that:

> By not searching for a new state but fighting to preserve their nation, Catalans may have come full circle to their origins as people of borderless trade, cultural/linguistic identity, and flexible government institutions, all features that seem to characterise the information age.
>
> (Castells 1997: 50)

More generally the new global space provides opportunities for regions to engage directly with the global economy and form alliances with other cities and regions. Catalonia, for example, has formed alliances with other regions in southern Europe in the context of European Union funding, and the mayor of Barcelona has become active on the global stage. Consequently, formal secession that perhaps might only be attained through violent means may be unnecessary in the information age, providing that they are only seeking to manage territory and not seeking to undermine the capitalist global system. At the same time this example also illustrates that territorially based struggles overcome or perhaps suspend other social divisions but by no means all, as not all interests are met within a capitalist society. Indeed within Barcelona the remaining social divisions are perhaps more exposed, because solidarity around the Catalan national identity has waned as the objectives of the bourgeoisie have largely been obtained, in contrast to the Basque region, which by comparison remains economically deprived and actions by ETA continue. Nevertheless, the de-linking of identity with territory perhaps marks the way forward by allowing struggles around other social divisions such as social class, gender and ethnicity to be addressed rather than subordinated to a common territorial objective.

Recognition of the widening contemporary social divisions more generally has led local and supra national agencies to encourage greater public participation in planning and redevelopment. Though similar to the Catalan case and in contrast to the Chiapas the range of issues over which people are permitted to exercise their preferences is subscribed within the neo-liberal agenda.

## PARTICIPATION AND EMPOWERMENT

In response to growing social divisions, rising inequality and the apparent failure of development policies, international institutions, governments and NGOs have attempted to widen participation at the local level in order to draw on local knowledge, to make plans more appropriate to local needs and to empower people by giving them some influence in the decisions affecting their future.[20] These tendencies can be observed at different spatial scales and across the globe. Indeed in the 1990s participation has become 'frenetic' (Gujit and Shah 1998: 3), almost a mantra for redressing inequalities, and in many cases a requirement of funding organizations. Institutions from the World Bank to local authorities in places as far apart as eastern India, inner London and Belo Horizonte in Brazil have been widening participation with varying degrees of effectiveness.[21] Thus while economic power continues to concentrate in the most wealthy firms and countries and as states at all spatial scales are increasingly confining their role to supporting private capital, people, as individuals, are being asked to participate in development decisions. This section highlights four key issues: why is wider participation considered important; how representative is participation; does participation lead to empowerment; and finally does participation lead to any changes in the material allocation and distribution of resources.

### Widening participation

At the global level there is growing concern that institutions such as the IMF, World Bank and WTO, founded to assist countries in crisis and maintain global stability, have narrowed their mission to promoting capital market liberalization, securing stability of financial markets and advancing free trade. Joseph Stiglitz (2002: 216) argues that their key actors 'genuinely believe that the agenda they are pursuing is in the general interest',[22] hence the fervour with which structural adjustment policies have been and continue to be imposed on countries needing assistance, despite growing evidence of their adverse consequences. As Paul Krugman (2002b) comments: 'we promised them a rose garden, but even before this latest crisis too many people got nothing but thorns.' However, the uneven impact of policies makes it difficult to believe that the 'key actors' are only misguided by their ideology. Market liberalization, specifically the opening of markets to 'free trade', has meant that many people in less developed areas are no longer able to make a living producing and selling agricultural products, as illustrated in the Chiapas above, as their products are undercut by cheaper (sometimes subsidized) products from elsewhere. In Jamaica even the production of local staples is no longer economically viable, given cheap imports from the US. Thus the fields lie idle, many people are unemployed and debts rise as the country becomes dependent on imports and still has to pay off debts incurred by the infrastructure built to attract foreign direct investment. These people have effectively been economically disempowered by global integration and it is hard to accept that the IMF and World Bank are simply unaware of these adverse consequences. If as Stiglitz suggests they genuinely believe that market liberalization is in the general interest and are not simply prioritizing the interests of the large-scale producers and corporations in the richer countries, why were they not campaigning against the subsidies given to producers in developed countries? Nevertheless as he points out representation in the IMF is highly uneven:

> The IMF's actions affect the lives and livelihoods of billions throughout the developing world; yet they have little say in its actions. The workers who are thrown out of jobs as a result of the IMF programs have no seat at the table; while the bankers, who insist on getting repaid, are well represented through the finance ministers and central bank governors.
>
> (Stiglitz 2002: 225)

Widening participation is correspondingly considered necessary to increase awareness of these adverse consequences and while so far there has been little change in the IMF, the World Bank has been widening participation in concrete development projects as well as overall macroeconomic policies from the mid-1990s. In 1998, James Wolfensohn, the president, stated that:

> Participation matters – not only as a means of improving development effectiveness, as we know from our recent studies – but also as the key to long-term sustainability and leverage. We must never stop reminding ourselves that it's up to the government and its people to decide what their priorities should be. We must never stop reminding ourselves that we cannot and should not impose development by fiat from above – or from abroad.
>
> (Wolfensohn 1998, cited by Aycrigg 1998)

The World Bank (1996) produced a *Participation Source Book* which provided guidelines for and examples of participatory development, understood as 'the process through which stakeholders influence and share control over development initiatives, and the decisions and resources that affect them' in all stages of the project from defining the problem to implementing the solution (OED 2000: 1); see Box 9.2.

In developing its guidelines the World Bank drew extensively on the ideas of Participatory Rural Appraisal (PRA), advanced by Robert Chambers (1983) and designed to draw upon local knowledge to ensure that projects more appropriate to people's needs could be quickly implemented rather than having to wait for extensive surveys carried out by external experts. Correspondingly, agents or motivators trained to listen to and learn from local people are placed in communities to encourage participation and to widen participation by using innovative forms of communication so that for example the non-literate can be included in the discussions and development of local plans. There have been similar concerns with widening participation and including the socially excluded in richer countries. In the UK extended community involvement has been promoted in regeneration strategies in inner cities and declining industrial regions often through the development of partnerships between stakeholders. Partnerships were designed to give voice to local businesses and the community, especially ethnic minorities, and to make the delivery of policies more effective.[23] There was also concern that local government, especially when controlled by the political left, was unrepresentative of local opinion.

This focus on participation assumes that people wish to become involved in decision making at this level and that such involvement will lead to more inclusionary and harmonious outcomes which reflect the community's interest. However evidence that this degree of involvement is desired, that it is representative or conducive to greater social inclusion except in a formal discursive sense is rather mixed. Furthermore, there is an underlying assumption that it is the lack of knowledge about community preferences, or as Stiglitz (2002) suggested above their lack of presence at the negotiating table, that has prevented their interests from being considered and by implication if they were,

---

*Box 9.2* Different levels of participation used to assess the World Bank's projects

| Low level | High level |
|---|---|
| (1) Information sharing – one way communication | (3) Collaboration – shared control over decisions and resources |
| (2) Consultation – two way communication | (4) Empowerment – transfer of control over decisions and resources |

*Source*: Adapted from OED (2000).

a more inclusive outcome would be obtained. There is therefore a rather simplistic assumption that wider participation would lead to the emergence of a singular community interest or the general will. Correspondingly power differences and material conflicts between individuals or groups either within or between different communities, or between community proposals and the interests of fund providers or the broader economic policies advanced by global institutions and economically powerful nations, are rarely addressed.

## Representation and empowerment

There is a small but growing literature on the effectiveness of widening participation. Two key problems have been identified: representation, i.e. the extent to which those involved in the participatory process are representative of the population and second, whether such participation leads to empowerment, likewise a concept with different meanings. Even the Operations Evaluation Department (OED) of the World Bank, which has probably carried out the most extensive evaluation of the effectiveness of participation, is critical of the achievements made so far in projects financed by the World Bank.

The OED (2000) evaluated a random sample of 189 projects with an electronic survey and focus group discussions and assessed schemes on the basis of different levels of participation (see Box 9.2). Overall it found that participation had increased from 41 per cent of the projects in 1992 to 70 per cent in 1998, falling back to 67 per cent in 2000, but that most participation was low level, taking the form of one way communication and listening in the design and implementation stage, rather than in project identification or evaluation. Participation at this stage of a project allows only a very limited form of involvement and, as the OED (2000) recognized, primary stakeholders might consider it as a way of gaining their support for a project rather than a means through which they would be able to shape the project or check it was being implemented according to their priorities. In this respect there are parallels with the idea that participation itself has become a 'new tyranny' (see Cooke and Kothari 2001), that it has become something that is imposed on people in order to appear to be more inclusive, but in reality only secures local legitimation for plans effectively determined elsewhere.

Moreover the OED found that participation was far more frequent in 'people' related projects such as agriculture and health rather than in finance or in adjustment lending, which clearly have as important if not so immediately obvious connections with welfare, and only limited sections of the population were consulted. The 'powerful members of the community dominated the participatory process, and effective participation of women, the poor, and other excluded groups proved limited and elusive' (OED 2000: 2). They found some instances of good practice where people were genuinely empowered, which they define as a 'transfer of control over decisions and resources' but in general 'best practice examples were islands of success in an ocean of participatory needs'

(OED 2000: 2). This limited participation was attributed to the tight time-tables and lack of resources but also to resistance by governments who actually implement the projects and although the World Bank's finance gives them some leverage it does not ensure complete control.[24]

There are many parallels between the OED's findings, which relate to a wide range of projects in poorer countries, and those in the growing number of participation schemes in the UK, elsewhere in Europe and the US.[25] There are so many partnerships in UK regeneration schemes there is a danger of partnership fatigue, low levels of community interest and activist 'burnout'. Some authorities keep registers of potential 'volunteers' in order to meet the deadlines for participation set by the funding agencies, so the same volunteers participate time and time again, raising questions about who and what these partnerships actually represent and risking 'confusing the self-serving advertising of corporate leaders with the real possibilities of a vibrant civil society' (Forester 1998: 214). Thus extending governance could paradoxically weaken local democracy by prioritizing the interests of those who have time or inclination to be involved.

Rather alarmingly, one of the OED's (2000) recommendations was to engage in more *capacity building* to extend participation to a wider range of people, something also strongly recommended in the UK.[26] Indeed capacity building became very fashionable in the early years of the twenty-first century and the theme of many conferences. However, emphasis is often placed on raising people's capacity to engage in the participation process, that is educating local people in the language of the development planners, rather than equipping them with useful skills such as managing budgets. In a deprived area of inner London, Sophia Skyers reports that residents found that consultants followed a set formula rather than responded to their expressed needs and so they simply stopped going to the meetings despite the consultants' efforts to persuade them:

> They tried everything to get people, they changed the times of the session, they did them in the evenings, they put childcare on, they put food on, and one day they were due to have 26 people, they had 3 and I think at that point they [the consultants] decided that this [capacity building] is not right for this group of people.
>
> (Programme manager inner London,
> interviewed by Skyers 2003)

Moreover the same people were involved in successive programmes, as one resident commented: 'I mean, I've been capacity built under three different schemes and quite frankly, my view on capacity building is that if you employ a consultant, the money will go to the consultants' (local resident in Inner London interviewed by Skyers 2003).

The second reason why local people are wary of capacity building schemes is that to them the nature of their problems, often poverty, and the solution, more material resources not increased capacity, are rather obvious. In a study of community initiatives in former coalfield areas, representatives stated that they were not 'incapacitated' but needed funds to realize their capacities (Bennett, Beynon and Hudson, 2000).[27] Or an inner London resident reflecting on area regeneration schemes commented that:

> I think the crucial issue is poverty. Poverty is the key. Poverty is the number one reason for the state of things as they are at the moment. If we get over poverty, if people have got money in their pocket, they've got something constructive to do with their time, they're spending their money, you know, that's where you get social inclusion rather than exclusion.
>
> (Tenant representative and partnership board member, cited in Perrons and Skyers 2003: 281)

Correspondingly, local people remained sceptical and had little confidence that the capacity building programmes would enable them to overcome power-lessness and gain greater control over the circumstances of their lives (see level 4 in Box 9.2).[28]

Parallels can be found in a study of a participatory project in eastern India (Orissa, Bihar and West Bengal) for rain fed farming funded by the Overseas Development Agency of the UK carried out by Cecile Jackson (1997). The project was intended to overcome criticisms that the Green Revolution demanded inputs that were beyond the means of poorer farmers. If villagers, including women farmers, 'catalysed by village-based project staff' became involved in all stages of project design then it was thought that solutions based on low cost technologies suited to local circumstances would be found. The research was based largely on the field diaries kept by village motivators sent to live with the villagers for the duration of the project (three to five years) and encourage their participation. The villagers initially found it difficult to place the village motivators or understand their purpose and considered that they were of little value unless they could bring what they perceived to be tangible benefits, such as fertilizers or a means of accessing water. 'Farmers said, "if you want to develop, give me a well, otherwise nothing can be done"' (Jackson 1997: 240). This reflects the long standing belief in India of the centrality of water to well being[29] and as Jackson comments: 'One can understand the puzzlement of those who, after clearly articulating what they saw as problems [lack of irrigation], are asked to play games with tamarind seeds to discover what the problems are.'

In this case there is a clear conflict in that a project designed to be parti-cipatory did not in practice respond to the expressed needs (water and fertilizers) of the participants, who in this case may have been the richer farmers, despite attempts to include the most marginalized.[30] As the project continued, however, the villagers became attuned to the project's discourse and re-expressed their demands within its vocabulary, for example by arguing that crop-spraying equipment would 'help in group action . . . a sprayer can play a vital role in establishing harmony' (Jackson 1997: 241). This example clearly indicates that people can try and adapt the terms of reference of a project to their own demands but not that these demands will be met, as the resource implications were not only beyond what the project funders would provide but also in conflict with the intention to assist the most marginalized, who given the prevailing distribution of resources, globally and within their region, would be unable to sustain a project based on fertilizers and irrigation. However it may also indicate people's unwillingness to accept what are perceived to be lower forms of development, something that has also become apparent in 'third sector' or social projects elsewhere, which provide valuable means of securing immediate survival needs but few chances of moving beyond this level.

Recognizing diversity and looking at ways of analysing problems and designing solutions in inclusionary ways is clearly a necessary condition for moving towards a fairer society. In particular the inclusion of marginalized groups, such as women or ethnic minorities, in the decision making processes may lead to projects that are sensitive to currently hidden needs. The process of participation may also lead to increases in self-esteem and in some cases individual career possibilities. However, it is important to recognize that discursive inclusion alone is unlikely to be sufficient to empower people economically or to overcome material exclusion. Moreover, key questions about power differences and structural economic inequalities that often lie at the heart of various forms of disadvantage are often excluded from the agenda. In the UK increased participation at the community level often occurs in parallel with a centralization of effective control as well as real cuts in expenditure which directly affect jobs, health and welfare services. Therefore although local groups may be formally recognized, sometimes as part of a 'celebrating diversity' strategy, which critical black and ethnic minorities have referred to as the "Steel Band, Sari and Samosa" approach,[31] the main agenda, in terms of resource allocation and the determination of overall priorities, is decided elsewhere in regional, national and global institutions. Local participation consequently becomes either a means of legitimating national policies and/or a means of making local people responsible for their fate when in reality they have little control over the processes generating the context in which decisions are made. Likewise, while the participatory poverty assessment programmes introduced by the World Bank have raised the issue of poverty, participation is restricted to process and consultation, and issues affecting the underlying determinations of well being such as 'the distribution of assets, income and power across ethnicity, class, gender and caste' (Francis 2001) and indeed the broader macroeconomic context of market liberalization and global integration, 'are off limits' (Stiglitz 2002: 234).

There is a further problem in that communities are rarely so homogenous that a singular interest can emerge from participatory discussions. Communities are generally multiple so within community differences are also important. Some writers drawing upon Jürgen Habermas have argued rather optimistically that consensus can arise through 'dialogical constructions' just as our identities, preferences and beliefs are established in everyday life and are formed through discussions (Healey 1996: 219). Further, even if it is possible to achieve a discursive resolution through 'inclusionary argumentation' and this can happen as long as goals are defined in very general terms, when decisions are effected materially, particularist demands are likely to re-emerge.[32] However, even at the discursive level, Habermas (1998) specified certain procedural rules necessary for consensus to be reached, in particular that power differences be neutralized, and these conditions simply do not hold within the development or planning arena. Furthermore, if agreement is established within particular communities, community boundaries are also permeable, so without some widely accepted normative criteria, decisions reached by different communities could lead to a whole range of inconsistent outcomes with no agreed ways of choosing between them. For example in general terms one community may seek global integration and a neighbouring community a more autonomous development trajectory; within each community there is unlikely to be unanimity for either strategy. Thus it is still necessary to have ways of resolving conflicts within and between communities.

One interesting development in this respect is in the participatory budgets (*orçamento participativo* (OP)) which have been introduced mainly by leftist political parties in a number of South American countries. Here participation is associated with control over resources in a way that combines participation at the community level with formal democratic structures stretching from the local level to the nation state. In Brazil OP was introduced by the Workers Party (Partido dos Trabalhadores (PT)) in the cities and states it controls, including Porto Alegre with a population of about 1.5 million, Belo Horizonte, a city with 3 million and Betim, a smaller city of around 300,000 people. In Porto Alegre there is direct participatory control over the 30 per cent of the expenditure, and people have felt empowered because they do have some direct effect on the local government. The idea is that:

> citizens are encouraged to attend neighbourhood meetings to propose, discuss and vote on budget priorities in the area of public works and social services and to elect delegates to subsequent municipal forums where the sum of neighbourhood priorities is put to the final vote. The results are incorporated into the administration's budget proposal and submitted to the city council. An elected council of OP delegates follows subsequent deliberations, as well as the implementation of approved OP projects.
>
> (Nylen 2002: 127)

These participatory budgets have attracted a lot of interest. Porto Alegre has become an icon for participatory democracy and has had real material benefits including increasing the proportion of the population with effective sanitation from 46 per cent to 85 per cent.[33] These budgets however have to operate within the wider national policies, which are constrained by global integration, and similar to the participatory schemes in the UK and those financed by the World Bank the extent of participation is limited. The research on Belo Horizonte and Betim found that 'OP to a great extent preaches to the choir, to the already empowered, and fosters comparatively little new empowerment' in that the majority of OP delegates were already activists in civil society. Thus OP 'may be more efficacious in sustaining and developing existing non elite political activism than in empowering disengaged or alienated citizens' (Nylen 2002: 134 and 127).

While most people would probably wish to have their interests taken into account, the days of the Greek *polis* are long gone and many people have neither the time nor the desire to engage in the detailed discussions required for participatory decision making. Even in Ancient Greece, the fact that men had time to engage in daily decision making rested on the fact that slaves and women took care of all the other work. This is not meant to imply that people are disinterested. The millions who turned out on 15 February 2003 in over 600 cities across the globe to protest against the impending war against Iraq demonstrates the strength of public interest and commitment to world affairs. But participating in these episodic moments is very different from being willing to turn out week by week to meetings concerned with detailed planning or development issues or indeed from organizing and coordinating global protests. While the web enables far greater numbers of people to be informed about protests, committed activists working on a day to day basis are crucial to providing the information and putting it up on the web.

The World Bank seeks to ensure greater participation in decision making in order to get more efficient as well as more inclusive outcomes within the existing unequal economic and social framework, rather than empowerment in the wider sense 'by which the powerless gain greater control over the circumstances of their lives' (Sen and Batliwala 2000 cited by Bisnath and Elson 1999) which would include increased access to material resources and greater control over life at home. More generally, a better system would involve some form of multi level representative democracy that allows for the interdependence of decisions of different groups within and between the spatial tiers. The representatives should also have control over resources and base their decisions on the broader normative criteria on which they were elected. Accountable professionals and technicians could then use their professional expertise to outline alternative scenarios, which approximate these criteria in different ways.

Formal democracy does not seem to be sufficient to guarantee civic or social rights and new systems of participatory governance are generally constrained to work within the existing general model of global market capitalism. Capitalism does however take quite different forms in different countries, leading to significant differences in the well being of people in countries at similar overall levels of development. Thus some of the problems and worst forms of poverty may be amenable to internal resolution. To protect the social rights of poorer people everywhere however, global adherence to some basic economic and social rights or codes of conduct might also be a way forward, and would certainly provide a yardstick against which to assess the implications of nationally and globally funded development projects. Different writers have proposed different schemes. The Zapatistas' demands are for 'work, land, shelter, food, health, education, independence, freedom, democracy' and more specifically for 'an end to centralization and the establishment of municipal self-governance with political, economic and cultural autonomy, justice and peace'.[34] With the exception of the demand for land and arguably the specific demands for local democracy these demands are not so different from the United Nations (1948) Universal Declaration of Human Rights (UDHR).

The UDHR has 30 articles relating to life, liberty, freedom of expression, freedom to assemble and to participate in political and cultural life, the right to work, equal pay, the right to a standard of living adequate to maintain health, education, justice and so on. The first two rights (see Box 9.3) set a framework for the rest. These are also similar to the capabilities approach, that is the freedom to do certain things, proposed by Amartya Sen (1990) and the central human capabilities that are articulated by Martha Nussbaum (2003). This goes further than the UDHR by including capabilities central to sex equality such as the right to bodily integrity, the right to be free from violence in the home and from sexual harassment in the workplace, to overcome feminist concerns that the UN human rights are male centred. Indeed there are references in the UN Declaration to 'brotherhood' and the rights of the 'family' without recognizing that the family itself can be a site of conflict. Referring specifically to gender equality Nancy Fraser (1997: 45) puts forward seven distinct normative principles: anti-poverty; anti-exploitation; income equality; leisure time

---

*Box 9.3* Universal Declaration of Human Rights

*Article 1*
All human beings are born free and equal in dignity and rights. They are endowed with reason and conscience and should act towards one another in a spirit of brotherhood.

*Article 2*
Everyone is entitled to all the rights and freedoms set forth in this Declaration, without distinction of any kind, such as race, colour, sex, language, religion, political or other opinion, national or social origin, property, birth or other status.

Furthermore, no distinction shall be made on the basis of the political, jurisdictional or international status of the country or territory to which a person belongs, whether it be independent, trust, non-self-governing or under any other limitation of sovereignty.

*Source*: UN (2002): http://www.hrweb.org/legal/udhr.html.

---

equality; equality of respect; anti-marginalization; and anti-androcentrism. Most of these principles are self-explanatory but anti-marginalism seeks to prevent women becoming isolated in a domestic sphere and anti-androcentrism requires that current male centred institutions be restructured, so that women achieve comparable levels of participation and social well being, without having to abrogate their reproductive role or concern with care.

The capabilities approach put forward by Sen (1990) has been extremely influential in widening the conception of development from GDP per capita to consider the extent to which people are able to benefit from a country's wealth, as discussed in Chapter 2. More generally these lists of rights, criteria and demands are useful as they provide yardsticks for assessing change and development proposals. However, the UDHR is not a legally binding document but a 'common standard of achievement for all peoples and all nations' towards which people shall strive. Nor are the various systems of rights and capabilities that others have put forward and as soon as they are operationalized complexities arise. To illustrate some of the complexity the development of one set of more specific rights in relation to ethical trade is discussed below.

## ETHICAL TRADING AND CODES OF CONDUCT

Consumers and workers also challenge the global order by seeking greater control over the products they buy or make. Initially these developments were restricted to a tiny minority and organized through NGOs such as Oxfam. More recently, following campaigns from anti-globalization protesters and popular literature, increasing numbers of people in richer countries are demanding natural, healthy and authentic products that are quick to prepare and produced in socially and environmentally friendly conditions. Despite their potentially conflicting nature, these demands have led to fair trade and ethical codes of conduct which have promoted better working conditions, the development of higher value added packaging activities in less developed countries, and created a niche for traditionally made products in the global market. These developments can be contradictory for although some of the positive aspects of globalization may be realised they may also preclude alternative, locally autonomous development strategies.[35]

For over 30 years Oxfam has been working on fair trade and now works with over 160 producer organizations in some 30 countries around the world with the objective of:

> giving poor people power: by paying producers a fair price for their work, helping them gain the skills and knowledge they need to develop their businesses, and challenging ways of trading which keep people poor. Fair Trade means that many of the people who rely on selling crafts and textiles for a living, or who produce food

items such as tea, coffee, honey and chocolate, now have the chance
to work their way out of poverty.

(Oxfam 2002a)

This strategy aims to enable poorer producers who work in volatile markets to
become viable through trade – but *fair* trade – as well as give consumers the
chance to shop fairly. To do so minimum prices are guaranteed, loans are
advanced to help producers avoid getting into debt, and the trade is as direct
as possible, bypassing intermediaries and thereby ensuring that the producer
gets a higher share of the final price. Furthermore assistance is conditional on
fair conditions for people employed by these producers.

One particularly important commodity is coffee, produced primarily in
poorer countries and consumed primarily in richer ones (see Chapter 3). Café
Direct,[36] a fair trade organization, was founded in 1991 by Equal Exchange,
Oxfam, Traidcraft and Twin Trading in response to the sudden collapse of
the price for coffee. The objectives are similar to those of Oxfam above, with a
specific commitment of paying producers 10 per cent above the market price
or the agreed minimum price whichever is the higher and it has undoubtedly
improved the livelihoods of producers as illustrated here: 'We have seen achievements. Now I have money to buy clothes for my children and to build my house.
Day to day things are improving because of the better price' (Mario Hernandez,
Nicaragua (Café Direct 2002)).

There are many similar examples on Café Direct's website and they illustrate
how people can simultaneously earn a living from their integration in the
global market and yet still maintain largely traditional lifestyles, which can be
viewed positively or negatively, depending on the desired development trajectory.[37] Fair trade is expanding as major retailers and supermarkets in the richer
countries have started to stock ethically produced commodities as part of their
place marketing strategies. Price differences have become less important, especially to more affluent consumers, who look for variety and choice, including
the opportunity to buy fair traded goods, when deciding where to shop.

Becoming a fair trade producer however is not without difficulty. In the
Chiapas region, referred to earlier in the chapter, there are a large number of
cooperatives producing organic coffee for organizations linked with fair trade.
However the registration procedure is complex with different importing
countries habitually requiring their own inspection procedures. So it takes
time and resources to gain the organic label even though farmers have been
using organic methods for years, as they have been unable to afford the
chemicals. Overall, however, the proportion of fair traded products is very small
in comparison to the scale of the overall market.

Voluntary, ethical codes of conduct (VCCs) for people working in supply
chains of major corporations also have the promise of improved conditions for
producers and workers. Potentially, they are likely to have a greater impact than
the current fair trade schemes, simply because of their wider scale. At the same
time they do not fundamentally challenge the unequal nature of integration

with the global economy, and are more likely to be associated with changes in traditional ways of life.

NGOs and anti-globalization protesters have made extensive use of contemporary ICTs to raise global awareness about the dire working conditions endured by people working for producers in the supply chains of large corporations.[38] Their protests have had a considerable and sometimes dramatic impact, as adverse publicity can quickly affect sales and share prices.[39] Large corporations have set up voluntary codes of conduct, especially in horticulture, wine and textiles, to improve the conditions of people working in their supply chains, to enhance their corporate image and reduce their vulnerability to consumer boycotts. Indeed so many codes have now been introduced that reference has been made to 'death by a thousand codes' (Blowfield 2001). Altogether, 233 in house codes were counted at Gap in 1999 (Barrientos and Blowfield 2001).

In the UK, the Ethical Trade Initiative (ETI) was introduced in 1998 to harmonize existing company policies and to include some NGO and trade union demands.[40] The ETI was designed to provide a common benchmark or minimum for labour standards based on ILO codes of acceptable working conditions (see Box 9.4). Given the high level of retail concentration in the UK, with supermarkets accounting for 80 per cent of food sales, this initiative leaves producers with little choice but to comply, otherwise they risk losing their UK market.[41]

The codes therefore provide a constraint to the downward pressure on labour standards but are voluntary, so although they do not infringe EU or WTO free trade rules, it is also difficult to ensure compliance, especially when products are sold on general markets or when there are multi tiered supply chains. Sainsbury's, for example, has 450 supermarkets in the UK and buys from over 1500 suppliers throughout the world, who in turn may source from many sub-suppliers (Fullelove 2001). So while the company 'recognises that consumers have a better understanding of what they are buying and is committed to product integrity' it is difficult to ensure compliance throughout the chain. The irony is however that it is these leading companies that create the competitive pressures and downward pressures on wages that lead to the subcontracting and make the monitoring of standards difficult.[42] Unless the codes are monitored effectively, the ETI and VCCs may just be seen as public relations exercises or part of the marketing strategy.

In a study of 20 voluntary codes of conduct Ruth Pearson and Gill Seyfang (2001) point out that the different organizations involved in their formulation – trade associations, corporations, trade unions and NGOs – have different and possibly conflicting motivations for introducing them. VCCs may be accepted by trade associations to pre-empt more stringent ILO standards, so the workers' interests may be neglected. Even when trade unions have been involved in their design, the codes may not protect all workers. In particular, they may overlook the interests of home workers, temporary workers and workers in the informal sector, all of whom have been increasing in the last decade, as well as concerns more specific to women workers, such as sexual harassment and reproductive

*Box 9.4*  UK Ethical Trading Initiative: the base code

**1.**  *Employment is freely chosen*
1.1  There is no forced, bonded or involuntary prison labour.
1.2  Workers are not required to lodge 'deposits' or their identity papers with their employer and are free to leave their employer after reasonable notice.

**2.**  *Freedom of Association and the right to collective bargaining are respected*
2.1  Workers, without distinction, have the right to join or form trade unions of their own choosing and to bargain collectively.
2.2  The employer adopts an open attitude towards the activities of trade unions and their organisational activities.
2.3  Workers' representatives are not discriminated against and have access to carry out their representative functions in the workplace.
2.4  Where the right to freedom of association and collective bargaining is restricted under law, the employer facilitates, and does not hinder, the development of parallel means for independent and free association and bargaining.

**3.**  *Working conditions are safe and hygienic*
3.1  A safe and hygienic working environment shall be provided, bearing in mind the prevailing knowledge of the industry and of any specific hazards. Adequate steps shall be taken to prevent accidents and injury to health arising out of, associated with, or occurring in the course of work, by minimising, so far as is reasonably practicable, the causes of hazards inherent in the working environment.
3.2  Workers shall receive regular and recorded health and safety training, and such training shall be repeated for new or reassigned workers.
3.3  Access to clean toilet facilities and to potable water, and, if appropriate, sanitary facilities for food storage shall be provided.
3.4  Accommodation, where provided, shall be clean, safe, and meet the basic needs of the workers.
3.5  The company observing the code shall assign responsibility for health and safety to a senior management representative.

**4.**  *Child labour shall not be used*
4.1  There shall be no new recruitment of child labour.
4.2  Companies shall develop or participate in and contribute to policies and programmes which provide for the transition of any child found to be performing child labour to enable her or him to attend and remain in quality education until no longer a child; 'child' and 'child labour' being defined in the appendices.
4.3  Children and young persons under 18 shall not be employed at night or in hazardous conditions.

4.4 These policies and procedures shall conform to the provisions of the relevant ILO standards.

**5.** *Living wages are paid*

5.1 Wages and benefits paid for a standard working week meet, at a minimum, national legal standards or industry benchmark standards, whichever is higher. In any event wages should always be enough to meet basic needs and to provide some discretionary income.

5.2 All workers shall be provided with written and understandable information about their employment conditions in respect to wages before they enter employment and about the particulars of their wages for the pay period concerned each time that they are paid.

5.3 Deductions from wages as a disciplinary measure shall not be permitted nor shall any deductions from wages not provided for by national law be permitted without the expressed permission of the worker concerned. All disciplinary measures should be recorded.

**6.** *Working hours are not excessive*

6.1 Working hours comply with national laws and benchmark industry standards, whichever affords greater protection.

6.2 In any event, workers shall not on a regular basis be required to work in excess of 48 hours per week and shall be provided with at least one day off for every 7 day period on average. Overtime shall be voluntary, shall not exceed 12 hours per week, shall not be demanded on a regular basis and shall always be compensated at a premium rate.

**7.** *No discrimination is practised*

7.1 There is no discrimination in hiring, compensation, access to training, promotion, termination or retirement based on race, caste, national origin, religion, age, disability, gender, marital status, sexual orientation, union membership or political affiliation.

**8.** *Regular employment is provided*

8.1 To every extent possible work performed must be on the basis of recognised employment relationship established through national law and practice.

8.2 Obligations to employees under labour or social security laws and regulations arising from the regular employment relationship shall not be avoided through the use of labour-only contracting, sub-contracting, or home-working arrangements, or through apprenticeship schemes where there is no real intent to impart skills or provide regular employment, nor shall any such obligations be avoided through the excessive use of fixed-term contracts of employment.

continued . . .

**9.** *No harsh or inhumane treatment is allowed*

9.1 Physical abuse or discipline, the threat of physical abuse, sexual or other harassment and verbal abuse or other forms of intimidation shall be prohibited.

The provisions of this code constitute minimum and not maximum standards, and this code should not be used to prevent companies from exceeding these standards. Companies applying this code are expected to comply with national and other applicable law and, where the provisions of law and this Base Code address the same subject, to apply that provision which affords the greater protection.

*Source*: http://www.ethicaltrade.org/pub/publications/basecode/eu/index.shtml (last-accessed November 2003).

health, especially as these more marginalized groups rarely participate in the formulation of the codes.

The issues of gender and marginalized workers have been explored empirically in a detailed case study of grape picking in South Africa where progressive national legislation exists and where the UK buyers, the supermarkets, have been following the ETI.[43] Grape picking takes place on remote farms and in a deeply embedded racialized and gendered hierarchy, the interests of the most marginalized – women, temporary workers ('coloured'), and black African migrant workers – are frequently disregarded. Historically the farms were owned and run by Afrikaans male white farmers and the permanent workers were mainly coloured males, who would be given family sized housing on the farms. This arrangement typically assumed that the labourer's wife and children would work on the farm as and when required. Following the end of apartheid, in 1993, the government has introduced progressive labour legislation, which covers all employees, including temporary workers. In practice however it has been hard to enforce because of the low levels of literacy and unionization and because migrant workers, who experience language problems, are uncertain about their rights and even if aware are wary of exercising them for fear of not being rehired in a subsequent season. Problems are intensified for women, because their employment often remains indirect, taking place through their partner, which gives the farmer more discretionary powers. In principle the technical experts from the companies who visit the farms to monitor the quality of the grapes could also monitor the employment codes but they are neither trained nor expected to do so. The way forward therefore probably rests on workers developing their own systems of representation, because although both the legislation and the codes of conduct mean that employees are no longer dependent on potentially despotic farmers, their existence is no guarantee that the standards are met (Barrientos *et al.* 2001).

ETI codes are also criticized when they are effective, because they are seen as yet another way in which rich countries effectively protect their markets by imposing onerous regulations and therefore higher costs on suppliers from elsewhere. More specifically some of the ETI employment regulations (see Box 9.4) are better than standards in the UK. For example, in fruit picking in East Anglia and Kent, labourers from a diverse range of countries, often including illegal immigrants, are bused in to pick the fruit, with conditions reminiscent of gang labour in the eighteenth century.[44] As a consequence the ETI has been extended to the UK, but is also difficult to monitor, owing to this widespread use of subcontracted labour. The ETI is, however, based on ILO codes in which governments from supplier countries have been involved and in the case of South Africa, the post-apartheid government has introduced progressive labour legislation; one aspect of the ETI codes is that they comply with national legislation, thus these measures have built from local legislation and have not simply been imposed from outside (Barrientos 2001).

Designing appropriate universal standards in very different social and economic situations is also difficult as the case of the use of child labour in the production of Nike footballs in Pakistan illustrates. Some well intended measures can be counter-productive. Save the Children Fund (SCF) investigated the Nike case and found that eliminating the children immediately could intensify rather than alleviate their poverty. In a detailed analysis of the lives of the children in the football trade they found that football stitching was not notably hazardous or exploitative for children, it was not bonded work and most children worked to help their families to meet their basic needs. Children were deterred from attending school by poor quality educational provision rather than their work, which was flexible and could be fitted around other activities including school. Further, if the industry was moved into factories then many of the women currently doing the work might no longer be able to work there. Keeping the work in the community centres enabled women to work. Paradoxically the most effective way of preventing child labour would be to raise the adult wage. In this case, SCF persuaded the company not to abolish child labour immediately as its shareholders might have wished, but to phase it out gradually while increasing other opportunities in the area, including education, access to credit and women's stitching centres (Marcus 1997).

ETI and VCC clearly improve conditions for producers and working people when applied. But they are voluntary and thus dependent on the goodwill of the corporations and the potentially transient consumer preferences for fair traded goods. Moreover they are aspirational and something to which firms can be working towards rather than firm standards that are monitored and implemented. These initiatives cannot be seen as a substitute for wider national/ supra national standards or for a comprehensive development strategy which reflects the interests and rights of people affected by them, or perhaps even more radically for a change in the relative prices of primary products.

## CONCLUSION

This chapter has considered different ways in which individuals and social groups have tried to obtain greater control over the areas in which they live and the conditions in which they work. Devolution and increased participation in decision making have become very popular on a world scale. Indeed, the 1990s was perhaps the decade of participation and empowerment. Yet, despite the increasing opportunities for people to exercise their voice, there is nevertheless a sense that people have become increasingly powerless.

Some social movements do obtain their objectives, or at least some of them, especially where, as in the case of Catalonia, there is a wide consensus for their aims and no real threat posed to the neo-liberal agenda, that is where movements are challenging their position within the nation state but not its prevailing economic and social trajectory. By contrast, the Zapatistas are less likely to realize their objectives as they conflict not only with the nation state but also with the neo-liberal agenda. In both cases, however, there is unlikely to be complete uniformity within their regions around their project.

People are also increasingly likely to participate in development and regeneration projects. The motivations for widening participation are not entirely clear, but a wide range of studies, including a very extensive study carried out for the World Bank, indicate that the scope of participation is very limited, confined largely to level 1 on the participation matrix (Box 9.2), that is to comment on proposals only after the main agenda has been determined. This participation is therefore more of a legitimation exercise than one that genuinely seeks to learn from people and draw upon their local knowledge to design programmes to meet their needs.

In order to prevent conflicts within and across spatial tiers or allow them to be resolved, various writers have proposed conventions or rights such as the UN Declaration on Human Rights. These rights or conventions provide a reference point against which to resolve conflicts, which are inevitable in a rapidly changing society. In principle these would allow people to participate in decisions about the broad trajectory of change but leave professionals to implement the policies, with the assurance that some minimum levels of rights would be secure. The difficulty is defining a set of rights around which people can agree, that do not infringe minority interests, that do not generate new conflicts as they become operational and that are adhered to. In this respect there has been some progress in relation to fair and ethical trade and codes of conduct campaigned for by NGOs and trade unionists. But so far codes of conduct are largely voluntary and monitoring is very difficult owing to the widespread use of subcontracting. The advantage of these codes is that they provide a benchmark around which groups can bargain for better conditions.

One of the common themes emerging from participatory rural appraisal or the attempts to achieve greater empowerment within local authorities as well as from the introduction of fair trade and the ETI is that however well meaning, and however effective in securing better futures than otherwise might have been the case, the processes generating the circumstances being challenged remain largely intact. This is why it is necessary to link local issues to national and global ones and back again, as Warren Nyamugasira (1998: 297) concludes:

> NGOs have come to the sad realization that although they have achieved many micro-level successes, the systems and structures that determine power and resource allocations – locally, nationally, and globally – remain largely intact.

The anti-globalization movement or movement for global justice targets these wider processes and there have been a series of major protests coinciding with meetings of various permutations of supra national institutions involving world leaders: Seattle in 1999; Washington, Millau, Melbourne, Prague, Seoul and Nice in 2000; and Quebec City, Gothenburg and Genoa in 2001 where the first demonstrator, Carlo Giuliani, was killed. Following the attack on the World Trade Center in New York on September 11 2001 the protests were more constrained until they joined forces with the anti-war movement in 2002 and 2003. In February 2003 there were the largest demonstrations the world has ever seen as people from all over the globe, including an estimated 2 million in London, that is roughly 1 in every 30 people in the UK, demonstrated against the looming war in Iraq, which nevertheless went ahead (see Figure 9.4). Demonstrations against the war and for the withdrawal of the United States and UK troops from Iraq have continued, as have demonstrations against meetings of world leaders such as the G8 meeting at Evian in France in June 2003, which was accompanied by 100–150,000 protestors.

*Figure 9.4* The anti-war demonstration in London, February 2003.

*Source*: Photograph by Eleanor Phant. Permission given.

As Naomi Klein (2002: xxv) has pointed out, the form of these contestations in which delegates are protected by extremely strong security and demonstrators kept away by high fences is a metaphor for the 'economic system that exiles billions to poverty and exclusion'. In Quebec City in April 2001 for a meeting of the Summit of the Americas the government effectively barricaded the leaders inside the historic city and excluded all others by a high wall. The resonance with the sieges of the middle ages was not lost on the demonstrators who built a 'medieval style wooden catapult and lofted teddy bears over the top' (Klein 2002: xxv). Other actions have included dressing up in funny costumes such as Michelin Men and are designed to emphasize the imbalance of power between the global institutions protected by highly armed guards and that of the people. As time has moved on and world leaders speak at the World Social Forum and NGOs are invited to contribute to the meetings of the supra national institutions there has been a loss of confidence, both in the powers of the supra national institutions and correspondingly in the value of disruptive protests. In the aftermath of the war against Iraq the supremacy of the United States as a world power seems to be largely unchallenged.

## Further reading

A. Blunt and J. Wills (2000) *Dissident Geographies: An Introduction to Radical Ideas and Practice*, London: Prentice Hall.
N. Klein (1999) *No Logo: Taking Aim at the Brand Bullies*, New York: Picador.
T. Ponniah (2003) *Citizen Alternatives to Globalization at the World Social Forum*, London: Zed Books.

## Websites

Oxfam (also has links to fair trade): http://www.oxfam.org.
UK Ethical Trade Initiative: http://www.eti.org.uk/.
World Social Forum (as this is a movement of movements it is best to use http://www.google.com and search for specific sites from there).

## Notes

1 See Sen (1990) and Agarwal (1997).
2 See Bourdieu (1990).
3 See Roy (1999).
4 To see Tulum in Riviera Maya and the ancient Mayan city see http://cancun. rezrez.com/whattoseedo/tours/cultural/tulum_xelha/index.htm (last accessed February 2003).
5 See OECD (2002b).
6 Data on GDP in Mexico comes from Fuente: Instituto nacional de estadística, geografía e informática via the website http://www.inegi.gob.mx/ and for the maquiladoras from the Banco de Información Sectoral website http://www.spice.gob.mx/siem2000/bis/?gpo=1&lenguaje=0. (last accessed February 2003).
7 See for example Global Exchange (2002).

8  See Krugman (2002a).

9  Stiglitz was removed from the IMF while on his way to the WTO trade talks in November 1999 in Seattle, where one of the first major anti-globalization protests took place.

10  See Massey (1995) and Castells (1997).

11  Besides the references this section also draws on personal observations and guided visits from local academics and planners made on four successive annual field courses with students from the LSE. I would particularly like to thank Maria Dolores Garcia-Ramon, Alba Caballe and Anna Ortiz.

12  In 2000 and 2001 bombs were planted in Barcelona, which may reflect how Basque separatists now consider Catalonia's allegiances to lie with rather than in opposition to the Spanish state. One of the bombs in November 2000 was close to where the Prime Minister was due to speak but also close to the Nou Camp stadium which 85,000 fans had left only two hours earlier.

13  In 1987 England defeated Spain 4–2 in a friendly football match. Gary Lineker, an English footballer who played for FC Barcelona, scored all four goals, leading to the headlines in the Barcelona local paper 'Catalonia 4 Spain 2'.

14  See Castells (1997).

15  See Keating (2001).

16  See for example Harvey (2000) on Baltimore and Chapter 5.

17  22@ refers to land zoned for developments based on new technologies.

18  Migrants were encouraged to learn Catalan and adopt Catalan culture and identity although there are tensions between the migrants and the indigenous population.

19  See Ajuntament de Barcelona (2002).

20  Widening participation is not entirely new. As Guijt and Shah (1998) point out, in less developed countries its origins can be traced to the New Deal in India in the 1930s and community development programmes in Latin America in the 1950s. Likewise in the UK in urban planning, schemes for public participation were formalized in the 1960s (Skeffington 1969).

21  See Jackson (1997) on India, Perrons and Skyers (2003) on London, and Nylen (2002) on Brazil.

22  Or in Rousseau's (1968) terms the general will – see Chapter 8.

23  See Audit Commission (1999).

24  Interestingly the World Bank withdrew from the Namada Dam project, following the protests, but the project has been continued by the Indian government (see Roy 2001).

25  See Douglas and Friedman (1998); Perrons and Skyers (2003).

26  See Healey (1998: 1544).

27  See also Forrest and Kearns (1999).

28  See also Sen and Batliwala (2000) and Bisnath and Elson (1999).

29  Jawaharlal Nehru, Prime Minister of India 1947–1964: 'Dams are the temples of modern India' (Roy 2001).

30  See also Kumar and Corbridge (2002).

31  See Perrons and Skyers (2003).

32  See Jessop (1990).

33  See UNDP (2002). Interestingly just after the UNDP refers to Porto Alegre it points out that the UK Women's Budget Group has 'been invited to review the government's budget proposals' (UNDP 2002: 5) and meetings do indeed take place. But this open style of government, which invites feedback, is very different from the participatory budgets in which people are actually given some control over the resources being spent.

34  See http://www.indiana.edu/~jah/mexico/zapmanifest.html (last accessed August 2003).

35 See Escobar, Rocheleau and Kothari (2002).
36 For information on Café Direct see http://www.cafedirect.co.uk/index2.html (last accessed February 2003).
37 See Escobar (1995); Power (2000).
38 See Klein (1999).
39 For example in February 2003 Nestlé had to reduce its claim against the Ethiopian government from $6 million to $1.5 million immediately and agreed to donate the money back to be spent on famine relief following protests by the Ethiopian people and a global campaign by Oxfam amongst other organizations. This resulted in 40,000 letters being written to the company; see http://www.maketradefair.com/go/nestle/settled.
40 See Barrientos and Blowfield (2001) and ETI (2003).
41 For example, 60–70 per cent of South African table grapes are exported to Europe, with a quarter of the total going to the UK (Barrientos, McClenaghan and Orton 2001).
42 See Hale (undated) for an account of the difficulty of applying codes in textiles.
43 See Barrientos, McClenaghan and Orton (2001).
44 See Pinchbeck (1969).

# 10

# CONCLUSION
## Challenging the divided world

Globalization is a fact of contemporary life. People across the globe are connected by flows of information, finance, goods and services and through friends, relations and travel. Contemporary information and communication technologies have transformed many people's lives in positive ways and have the potential for transforming many more but the adherence to neo-liberal policies has undermined economic and social reproduction throughout the world and threatens social sustainability. People in different places are increasingly connected, but their life chances vary profoundly and divisions between the richest and poorest countries are growing. Overall there has been an increase in social wealth but the poorest countries have become poorer, so despite the potential of new technologies, the current model of globalization has reinforced geographically uneven patterns of development.

Relations between countries are now more complex and it is not simply a division between rich and poor countries. Within many countries at medium levels of economic development on the United Nations HDI, there are a small number of cities and regions that are as advanced as anywhere in the world. Shanghai for example has been growing at 18 per cent per annum for the past five years and does not meet the stereotypical descriptions of a third world city in a country that has been under communist rule for the past half century. Such developments reflect the optimistic side of capitalist development, globalization and the new economy and the dynamic drive to extend markets and bring more people into the capitalist sphere of influence – with jobs, consumer goods and in some cases higher material standards of living. The possibilities offered by this system and the way it has been able to make them known all over the world through TV and advertising together with the power of the already rich countries has enabled it to dominate and displace alternative models of development.

At the same time there is a more hidden downside. Not only does the affluence in some areas build upon unfair trade practices and value systems that deprive many people, including primary producers and care workers, from a just return on their endeavours, but there is also intense poverty and deprivation in the superstar regions themselves, in the streets behind the gleaming façades. Thus in addition to spatial divisions in wealth and opportunity there are growing

319

social divisions within countries, cities and regions. Social class, caste, gender, ethnicity and sexual orientation structure social life and under the current model of globalization some of these divisions are also widening, especially social class. Within the superstar regions these social divisions are most extreme, with highly paid knowledge workers at one pole and low paid care and informal sector workers and people without any paid employment at the other, and these class divisions are also marked by ethnicity and by gender. Women and ethnic minorities are overrepresented in the lowest paid jobs but they are not entirely absent from high paid work, so gender and ethnicity cut across class divisions in complex ways. Indeed in the superstar regions there has been an increasing mixing of people from all over the globe, as some are attracted to high level jobs in major corporations and others move there in the hope of finding work and escaping the poverty in the areas from which they come. This leads to vibrant cities, complex patterns of diversity, as well as extreme inequality, issues that were elaborated in Chapter 7. The geographically uneven patterns of development together with the widening social divisions combine to create an increasingly divided world. In this conclusion I briefly recap on some of the contemporary problems of globalization, in particular the growing inequalities, discussed in Part I. I then consider ways in which the framework proposed in this book provides a way of thinking about the interconnectedness between the spatial and social inequalities within and between countries and the implications for future research. Finally I identify some possibilities for social and political change.

Why is it that all the statistics and much of the qualitative research indicates continuing and in some cases widening inequalities between rich and poor countries, between rich and poor people within countries and between men and women, even though there are so many conferences, policies, strategies, resolutions and accords for tackling development, challenging inequality, empowering the poor, mainstreaming gender issues and empowering women? These aspirations are most clearly set out in the millennium development goals (MDGs) (see Box 8.5) to which governments all over the world have subscribed.[1] These goals relate mainly to developing countries and to poverty, hunger, primary education, gender equality, child mortality and access to water and sanitation. The final goal (Goal 8) refers to a 'Partnership for development' and so recognizes that to achieve the other targets changes are required in the policies of developed countries, in particular in relation to aid, debt, unfair trading practices and practices that restrict access to modern technologies and medicines.

Each of the goals has measurable targets, which are monitored, but so far progress towards them has varied between the major regions of the world and in some cases has been slow or non-existent. On the basis of current trends, sub-Saharan Africa will not meet any of the targets while South Asia is set to meet all of them except enrolment in primary education and reducing the proportion of the population that is malnourished. Latin America will meet the targets and East Asia and the Pacific look set to exceed them. Within these broad regions however different countries are proceeding or in some cases receding at different rates. Between 1990 and 2000, 57 countries became poorer and 21 moved backwards on the UNDP HDI measures. Some progress is also being made towards Goal 8 but most countries are not approaching even half the level to be achieved. In particular the subsidies on agriculture continue to dwarf the amount of aid given to people, as cows and cotton in developed countries receive far more than is given in foreign aid. Moreover, the subsidies make it increasingly difficult for producers in less developed countries to maintain a livelihood because the subsidized products are dumped on their markets.[2] As the UNDP (2002: 33) points out:

> The average poor person in a developing country selling into the global markets confronts barriers twice as high as the typical worker in industrial countries, where agricultural subsidies are about $1 billion a day – more than six times total aid. These barriers and subsidies cost developing countries more in lost export opportunities than the $56 billion in aid they receive each year.

Reaching the targets will not necessarily resolve all of the problems however. Aid does not always result in net income transfers to poorer countries as it can be linked implicitly or explicitly to finance imports from richer countries.[3] The UK government, for example, allocated Tanzania £65 million for its poverty reduction programme but also granted an export licence to a British company to supply Tanzania with an air traffic control system designed to military specifications and costing about £40 million when cheaper civilian options were available.[4] Abolishing the debt of poorer countries is also desirable and measures to reduce the debts of the most heavily indebted poor countries (HIPC), mainly in Africa, were introduced in 1996. However, participating countries have to have an established track record of following IMF/World Bank economic adjustment and reform programmes, plus a poverty reduction strategy to ensure some finance for basic health and education.[5] As a consequence debt repayments have fallen and more is being spent on education and health,[6] but progress has been slower than anticipated partly because the incomes of these countries have fallen with the decline in world commodity prices, especially coffee. A third of the countries under the programme will be spending the same amount or more on repayments, and some will still be spending more on debt repayments than on health care; for example in 2002 Zambia was spending 30 per cent more on debt repayments than health, even though one million people were affected by HIV/Aids (Oxfam 2002b).

There are further issues such as whether supra national institutions should be able to dictate the terms on which aid or debt relief is given and whether such transfers are income progressive, given the unequal internal distributions of income within debtor countries. Moreover, these strategies are unlikely to be sufficient to lower inequality in the long term as neither borrowing nor aid form a very large share of the foreign income received by poorer countries. Aid for example amounts on average across developing countries to 0.5 per cent of GDP compared to exports which constitute an equivalent of 26 per cent (UNDP 2002: 31). Given the significance of export earnings poverty cannot be explained by a country's lack of integration with the global economy; it is more a question of the disadvantageous terms on which developing countries engage in trade, so making trade fairer may be a better strategy.

Removing some of these barriers and establishing a more even playing field is clearly important. Indeed the IMF (2002c) has estimated that if all countries removed their agricultural protection, there would be an overall gain of 128 billion dollars, three-quarters of which would accrue to industrialized countries and one-quarter to developing countries. The economic model used to make these estimates is clearly contestable but the point is that countries which generally advocate free trade do so inconsistently. Even when economic models and institutions which embody their beliefs indicate that eliminating agricultural subsidies will be beneficial, they fail to do so suggesting that internal power relations are a clearer guide to policy making than pure economics.

Alternative economic theories, discussed in Chapter 3, predict however that while equal access to world markets may be a necessary condition for increasing economic growth, it is not a sufficient condition. By itself equal access is unlikely to reverse uneven development. Because poorer countries are more confined to routine production in the global division of labour and found at the lower end of value chains. Upgrading is possible and some countries have increased their material well being as a consequence of participating in the global economy so the policies and practices of nation states are important. Nevertheless, market society is a competitive society with winners and losers at all levels and unless states at different levels take action to temper market outcomes, increased integration and increasingly open trade will not resolve inequality and uneven development, only change its form.

Inequality has also been widening within countries. In the United States, which is effectively the richest country in the world,[7] 400 citizens had a joint income equivalent to the combined incomes of Nigeria, Senegal, Uganda and Botswana, and yet life expectancy in the United States is among the lowest of developed countries and for the black male population at 68 years (six years lower than for white men) is equivalent to the average life expectancy for Vietnam or Egypt, countries with much lower GDPs per capita.[8] Furthermore the inequalities have been widening as the incomes of a very rich elite have been expanding rapidly, while those in the middle have been increasing but at a much slower rate and those in the lowest decile have been declining. This shift in income towards the top 1 per cent and even the top 0.1 per cent and 0.01 per cent has taken place over the last 30 years and created an extremely divided society. There has been a return of the 'gilded age'[9] at one pole, while one in four children under 6 years of age are malnourished at the other, conditions reminiscent of the divisions of the late nineteenth and early twentieth centuries, which diminished in the middle of the twentieth century, during the Fordist era. These divisions create the context that allows the United States to rely on a volunteer army to maintain its position worldwide, as it provides one of the few legal routes out of poverty.[10]

Paul Krugman (2002b) reviews a number of possible explanations for widening inequality in the US. First, that the wages of the manual working class have been undermined by competition from low waged regions elsewhere in the world as a consequence of globalization and more specifically by the expansion of world trade and manufacturing in low wage countries, issues illustrated here in Part II by the feminization of employment (Chapter 4) and the decline of male manufacturing employment in (Chapter 5). In contrast to the era of Fordism these workers can no longer earn as much as white collar middle level workers and managers. Second, Krugman examines the idea that new innovations have led to a 'skill based technological change' and increased the demand and hence the incomes of highly skilled and educated workers – or in Reich's (1991) term the symbolic analysts or knowledge based workers – issues discussed in Part III. Third, he looks at the superstar hypothesis, drawing more on Sherwin Rosen (1981) than on Quah (1996) on which I have based my analysis (see Chapters 1 and Part III), but similarly highlighting growing inequalities, as the new technologies allow a minority of superstars to capture a much larger share of the market through TV or other forms of electronic dissemination. Krugman (2002b) argues that while there is something in all of these ideas none fully explains the widening gap in the US, in particular the 2500 per cent increase in chief executive officer incomes and the relative decline in blue collar wages. Instead he puts forward the idea of changing social norms because he argues this also explains what he terms the 'Great Compression' of wages and reduction in inequality between the 1930s and the 1970s. During this era social norms set boundaries to the levels of inequality that would be tolerated. In the current era however he argues that there has been a cultural shift towards greater financial permissiveness, a belief in charismatic leaders and in the effectiveness of incentives so executive earnings are linked to share values, which in combination have created the new social norms which endorse inequality.

Social norms may indeed be part of the explanation but one that raises further questions about how they are determined and how the extraordinarily high incomes of the elite are possible. The ideas in this book suggest ways in which these different explanations for widening inequality in the US can be linked and connected with social and political theories to account for the new social norms. Furthermore it provides scope for examining the relation between inequality within the US and the unequal relations between the US and other countries. The extremely high executive incomes for example cannot be explained simply by new social norms that tolerate inequality, though these are important. They also rest on the economic supremacy of the United States, which in turn can be explained by the power of the large corporations which are global but nonetheless rooted in the US, and the global financial system based on the US dollar that allows the US to sustain a huge trade deficit with the rest of the world and more generally by the unequal trading practices. The US government's desire to retain this dominance explains the military actions in Afghanistan and Iraq, which were designed to control territories with access to key resources, in particular oil, and thereby allow the US government to influence oil prices and to ensure that transactions continue to take place in dollars.[11]

Globalization, understood as the increasing interconnectedness between people and places, means that it is necessary to situate specific events within their wider context. Thus the links between micro and macro levels of analysis always need to be kept in play. The framework presented in Chapter 1 and illustrated throughout the book suggests ways in which different economic, social and political explanations can be interlinked or synthesized in order to help explain social and spatial divisions in contemporary economies. It is not a substitute for explaining specific events but helps to provide a context to aid understanding. Economic and social life is complex just as inequalities and differences are diverse, but capitalist social relations of production lie at the heart of contemporary society and set constraints that impede equalities policies, making targets such as the MDGs largely aspirational. These constraints are not inevitable or unchangeable and exist only so long as people choose to accept them or have them imposed upon them, but until such time as they are changed they set the context of economic and social life and therefore need to be analysed.

In returning to a broadly historical materialist perspective but drawing upon ideas from feminism and the more contemporary mid-range theories of the French Regulation School, welfare regimes and the new economy, I have tried to identify what some of these structural constraints might be and how they operate in terms of influencing the lives of people in different places. In particular two very basic and very enduring structural constraints have been identified that impede moves towards narrowing uneven development, the gap between the rich and poor and between women and men. These inequalities arise from the form taken by capitalism in the contemporary global era and the continuation of patriarchy or processes that continually allow gaps to emerge between women and men even when other circumstances – such as the organization of paid work and in particular the gender balance of paid work – change.

The significance of structural constraints is being increasingly recognized. The protesters for global justice for example have placed the social critique of the capitalist system back on the political and academic agenda.[12] Mercedes González de la Rocha (2003: 3) also highlights the importance of economic and social transformations. Informalization and individualization of work have lowered incomes and undermined the ties of kinship and community as people move away for work. Thus as she argues the foundations of the resources of poverty model that had become a dogma in development studies have been eroded, and 'the lives of the poor are better described today by the opposite: the poverty of resources'. Similarly the UNDP (2003) points out that structural constraints have undermined the three most widely advocated solutions for higher economic growth and development in the 1990s: economic reforms to create macroeconomic stability; institutions of governance to 'enforce the rule of law and control corruption'; and finally widening participation or 'social justice and involving people in decisions that affect them and their communities and countries' (UNDP 2003:1) though it still regards these measures as central to sustainable development. Eliminating corruption is clearly an important issue, though the comparative impact of corrupt individuals is quite hard to estimate. Moreover it might equally be argued that a system in which the rules of free trade are unevenly applied and resources are systematically transferred from poor to rich countries, is itself corrupt. The other two points – stabilization policies and participation – are discussed in more detail below. A return to emphasizing the constraints within which people operate is not to deny the ingenuity of human beings or their agency but to emphasize the significance and power of the limitations within which these choices are made. Returning to the ideas of Marx developed in Chapter 1, human beings do make history, do author their lives, but the context is shaped by processes over which some people and some nations have much more control than others and furthermore that the control that states do have can be used in different ways.

Stabilization in the form of structural adjustment programmes and similar policies of the IMF and World Bank, followed by countries in return for aid, have the longest provenance but have received growing criticism owing to spectacular failures in Russia and Argentina which have both experienced declines in their real incomes. Indeed the 1980s and the 1990s have both been referred to as 'lost decades' owing to the low levels of economic growth or development. Countries where these policies have been followed less avidly, and which have maintained sovereignty by setting conditions for their integration with the global economy, such as China and India, have experienced higher levels of growth. In Brazil, the newly elected Lula[13] government coming into office in January 2003 faced extremely high inflation and high levels of debt, but adhered to the stabilization policies that had been agreed by the previous government. However the new government combined these measures with internal redistribution policies to tackle hunger and poverty. Six months after coming into office inflation levels have fallen and the country is perceived as being more stable economically. There have also been reductions in child illnesses. For example in Guaribas, a city in one of the poorest states of Brazil, Piauí in the north east, coupons for food were distributed. As a consequence the number of children hospitalized with diarrhoea fell from 250 to 15 a month (Lula da Silva 2003). This is just one aspect of the zero hunger programme, which in turn is part of a more inclusive growth strategy.

Despite globalization and the strictures of supra national institutions, nation states still retain some capacity to organize their internal affairs, that is to engage with the global economy in different ways. It is clearly early days for the new Brazilian government but it seems to be making better progress than Argentina, whose crisis in 2001/2002 was undoubtedly linked to years of structural adjustment and stabilization policies but exacerbated by internal corruption on a wide scale. This point is reinforced by the fact that countries with similar levels of GDP per capita can be ranked quite differently on the UNDP's development measures and countries with similar levels of development ranked by GDP per capita or HDI can have different degrees of gender inequality and empowerment, as illustrated in Chapter 2. Together these illustrations suggest that the nation state still retains some power in the global economy and so remains a worthy target for those seeking alternative forms of development.

Despite some deterritorialization of politics, geographical units remain the key organizing element and even though virtually all countries in the world are now capitalist and generally adhere to the norms of neo-liberalism in external relations, there are still significant differences between them in terms of the internal distribution of resources and social well being that are highlighted in the welfare regimes and regulatory frameworks perspectives outlined in Chapter 1. Inequality and absolute poverty are lower in countries where the state continues to support welfare and employment relations less precarious where the social partners play a greater role in their determination. These differences can be starting points for constructing capitalism differently.

The powers of the nation state continue and formal representative democracy has expanded throughout the globe. Yet nation states currently display a very frail form of democracy because although everyone in principle has an equal voice it is a weak voice. One reason for this is that voting figures are often low especially in countries where democracy has a long history. Electorates have become disillusioned or cynical about the role of government; either because the party in power is perceived to make little difference as they all adhere to the current neo-liberal model of development, or because people believe that national governments have little control over the circumstances affecting their lives, which are effectively determined elsewhere. Another reason why democracy is weak is because there is generally a wide gulf between the elected and the electorate, with the leaders only consulting the people in a cursory way at election time, and the electorate generally takes little interest in the day to day workings of the state. Thus there is the curious paradox that just as almost everywhere is formally democratic, democracy itself is in decline.

When opposition/social democratic groups do get into power they face enormous challenges from the supra national institutions and the power of other states within these. To retain a critical agenda, states should try to establish participatory democracy at local levels and in trade unions that can continue to feed into discussions and so retain the momentum for change. Likewise links with oppositional movements on a global scale would probably help these states resist the powers of the US led supra national institutions for conformity around the neo-liberal austerity agenda. In this context Hilary Wainwright (2003) draws an interesting contrast between the Workers Party in Brazil and its democratic participatory structures and the relationship between the ANC and the township civics in South Africa. The civics were based on street by street representation and organized social welfare, public health and environmental protection, resolved local conflicts and provided a foundation on which a new democracy could have been constructed. They could have formed the basis for popular influence over state institutions and provided constant support for the elected officials. Far too often, once elected, leaders deviate from their policy platforms because of externally imposed financial constraints but rather than building upon their grass roots to demonstrate the strength of support for the alternative way of managing the economy that led them to power they apologize to their supporters and argue that the change of direction is necessary to secure their goals in the long term. Sadly this long term rarely materializes. In South Africa once the ANC came to power it succumbed to IMF pressure to introduce political structures similar to the British Labour Party and the role of the civic movements was downgraded. As a consequence problems at this level remain and have led to local criticism of the ANC rather than to local and national harmonization round a collective agenda to raise the position of the poor. Thus while apartheid on racial lines may be waning a new apartheid on the basis of social class has emerged. How events will unfold in Brazil is not clear at the time of writing and caution should be exercised as it is easy to be optimistic about new leftist governments.

Regardless of their limitations elected governments probably offer a more pragmatic way forward. By contrast, Hardt and Negri (2001) envisage the masses rising up against Empire, and there are indeed movements for global justice that have tried to make their voices heard in less conventional ways such as street protests at various gatherings of world leaders and the World Social Forum. These events are characterized by humour, vibrancy and diversity and provide a tantalizing image of how the world could be. The demonstration against the war in Iraq in London in February 2003, of two million people, the largest the country has ever seen, was diverse by gender and ethnicity and consisted of all ages – babies and children with their parents, independent young teenagers, young, middle aged and older people. Judging from appearances and slogans, it consisted of people across the social and political spectrum: new age travellers, football supporters ('Pompey supporters against the war'), the professions ('dentists against plaque') , tea drinkers ('make tea not war'), hairdressers and dancers as well as the 'usual suspects' – the political left, religious groups, anarchists and greens. Overall it was very different from the predominantly white male trade union led demonstrations of earlier decades and very uplifting for those participating. Despite the scale and diversity of this protest the war however went ahead and subsequently protests have been on a smaller scale. However, to harmonize the energies of these diverse groups around common goals except at a very abstract level is much more difficult than combining to oppose a specific event. The idea of these protesters autonomously rising up to secure an alternative future seems rather unlikely.[14] Their diversity is a strength but also a weakness in this respect and while protest movements are invaluable for demonstrating the strength of support for alternative models of globalization and maintaining pressure on governments to change, unless they engage simultaneously with more conventional political institutions, it is not clear how their varied aspirations could be translated into social practice.

Moreover, people are very resilient and seek to some degree to find their own solutions, as for example the nannies discussed in Chapter 4. Even within global capitalism there are partial and occasional winners as well as losers within both poor and rich countries and many losers win sometimes or perceive themselves to do so and thus would be reluctant to abandon their gains in the quest for a more perfect society. Many people earning $1 or $2 a day would prefer to earn more, but would probably still wish to keep their dollar than risk a return to their previous situations. Furthermore women workers often welcome monetary incomes, even when low, as their capitalist oppression at least provides some respite from domestic drudgery and the potential for resisting patriarchal oppression. In this respect it is important to recognize the positive and dynamic side of capitalist development, which does bring improvements in material welfare, through extending social cooperation and the division of labour. In fact one very progressive aspect of contemporary global capitalism is the new information and communications technologies. These have the potential for increasing productivity and raising welfare levels throughout the world. They have also increased and deepened the flow of information across the globe which has widened people's horizons and increased their ability to challenge prevailing ideologies and organize cross-nationally. The problem is that in their current form the technologies have mirrored the contemporary pattern of uneven development, as discussed in Chapter 6, and minorities are allowed to appropriate a disproportionate share of the productivity gains from the increase in social cooperation.

A more realistic way forward would probably be to 'engage with the dominant political system but not be dominated by it' (Wainwright 2003:199), that is to engage with the state but maintain real democratic and participatory links with an expanded active electorate, as discussed above in the case of Brazil. This would be a very different model from the official participatory discourses discussed in Chapter 9, which were largely discursive and effectively embodied the slogan of 1968: 'Je participe, Tu participe . . . Ils décident'[15] largely because the key issues are 'off the agenda'.[16] Conventional democratic procedures are not perfect. Elected representatives fail to match the electorate in terms of gender, ethnicity and age balance. For example in 2000 in north east England men formed 87 per cent of the region's MPs and 77 per cent of its elected councillors and only 3.3 per cent of the total were under 35.[17] Even so perhaps formal democracy has some advantages over ad hoc organizations. Potentially, all legally resident citizens have a voice and mechanisms for dialogue between different levels of government are already in place. The diversity of representatives could be increased via quotas, the success of which has already been demonstrated in relation to gender in the Swedish parliament and, although withdrawn, was probably responsible for the increase in women Labour MPs in the UK. Representative democracy could be used to determine the broad directions of policy and systems of participatory democracy drawn upon to establish the detail of policies affecting particular groups.[18] This would strengthen democracy by making it more engaged with people's preferences without allowing minority groups to use the state as a vehicle for the implementation of their partial interests.

Writing from the standpoint of the middle class in one of the richest countries in the world it is also necessary to recognize that my lifestyle and living standard derive to a significant degree from both local and distant impoverished strangers. That if I desire change in this pattern of inequality, if only from a self-interested perspective of being able to live my life in relative tranquillity, it will be necessary not only to campaign for fairer trade, the cancelling of debt, increased aid to poorer countries and for a change in the overall model of global capitalism, but also to change my own lifestyle. As an individual it is very difficult to live morally in the global economy where all actions are interrelated. As Barbara Ehrenrich comments:

> I tend to buy my jeans at Gap, which is reputed to subcontract to sweatshops. I tend to favour decorative objects, no doubt ripped off, by their purveyors, from scantily paid Third World crafts persons. Like everyone else, I eat salad greens picked by migrant farm workers, some of them possibly children. And so on. We can try to minimize the pain that goes into feeding, clothing, and otherwise provisioning ourselves – by observing boycotts, checking for a union label, and so on – but there is no way we can avoid it altogether without living in the wilderness on berries.
>
> (Ehrenrich 2003: 101)

She goes on to argue that for her, of all the injustices a servant economy is the worst because while it provides some opportunities for poor and immigrant women it also 'breeds callousness and solipsism in the served'. Addressing this inequality would require lateral changes in the division of labour between women and men which may indirectly be a start in tackling some of the deeply rooted gender inequalities that currently seem to be immune to all policy initiatives. If men became engaged in domestic work to a greater degree more people would become aware of its systematic undervaluation in market societies. This would then be a step on the road to challenging the dominance of market valuations in other spheres such as primary commodities and in turn to the unequal exchange of value between rich and poor countries. This may be sheer fantasy but what is clear is that the current exclusive model of development is not socially sustainable.

The world today is very different from the one I grew up in. There has been material progress in a wide range of areas, not least in information technology which has the potential to massively increase human welfare. So 'Another World is Possible' (the slogan of the World Social Forum) but how to get there remains more of a puzzle. The world may currently be concerned with terrorism, but tackling the processes of inequality and injustice is probably more effective than building complex security systems. New ICTs and other technologies contain tremendous potential for productivity increases and transforming lives. ICTs allow knowledge and information to spread around the world much more quickly than ever before and understanding is a necessary condition for change. It is not a sufficient condition however. From an academic perspective contemporary globalization raises questions about appropriate disciplinary boundaries especially the separation of studies between the developed and less developed world when economic and social processes and people and places are increasingly interconnected. The framework proposed in this book together with some of the illustrations, especially in Chapters 4, 5 and 7, provide some indications of how this might be done. From a political perspective, despite the actions of the movement for global justice, it seems unlikely that there will be a mass mobilization against the capitalist system of economic and social regulation. In the context of globalization it would also be extremely difficult for small territorial units to isolate themselves from the current global order. The history of alternative utopias is not promising;[19] to remain pure they tend to have to introduce the very policies of authoritarianism they tried to escape from. The best that can perhaps be hoped is that lessons can be learned from those regions of the world that have managed to achieve a more egalitarian internal distribution of resources and managed to grow with the world economy at the same time. For change to take place societies will need to become more inclusive, which means that the gains from increasing productivity will have to be more fairly shared. New technologies could be harnessed to allow capital widening and deepening to take place simultaneously. These would need to be combined with a new mode of social regulation with global dimensions to allow the gains from human ingenuity to be shared more widely. Similar to the way that a new mode of social regulation was introduced to allow the productivity gains in the Fordist era to be shared more widely with the working classes another mode of social regulation is needed to bring about a more inclusive model of development but this time on a global scale. This in turn suggests some form of global governance but one that would rest upon participatory forms of democracy at different spatial scales in order to overcome the weaknesses of the current supra national institutions and the unequal representation of states within them. Together new technologies and a new system of social regulation would allow the realm of necessity,[20] that is the amount of time and resources that have to be devoted to ensuring that people have their basic needs and wants satisfied, to be rolled back for all and so enable all to share in the realm of freedom that currently only a minority in this divided society enjoy.

## Notes

1 Likewise the member states of the European Union have committed themselves to a strategy for promoting gender equality in economic life, equal participation and representation, equal access to and full enjoyment of social rights for women and men, gender equality in civil rights and a change of gender roles and stereotypes, without really considering how their economic policies and in particular the encouragement of more flexible working practices tends to conflict with some of these goals (European Commission 2000).

2 See UNDP (2003).

3 For an early recognition see Hayter (1971).
4 This project reveals some of the tensions within a nation state and between nation states and supra national institutions as there were divisions within the UK government between the Department for International Trade and Development, which opposed the deal, and the Treasury, which supported it. The World Bank condemned the project and the Tanzanian government tried to get out of the deal fearing that the World Bank would withdraw funding owing to the conflict with Tanzania's anti-poverty strategy (see Hencke, Denny and Elliot 2002). At the time of writing the outcome is not clear.
5 See IMF (2002a).
6 See IMF (2002b) and UNDP (2003).
7 The United States had the highest GDP per capita in 2001 (measured in PPS) apart from Luxembourg, whose figure is distorted by the extremely small population and the presence of EU officials (see UNDP 2003).
8 Figures on average life expectancy are from UNDP (2003). The statistic for the US black male population is from National Center for Health Statistics (2003). In detail the 2000 figure for white males is 74.8, black males 68.3 and for women 80 and 75 years. Figures on income come from the US Inland Revenue Service cited by Sachs (2003).
9 See Krugman (2002b) who likens the present era to the early twentieth century as illustrated in F. Scott Fitgerald's *The Great Gatsby* (1920).
10 Some 20 per cent of the US army are black and are disproportionately in lower ranks and more likely to be involved in fighting. Martin Luther King referred 40 years ago to the irony of poor black and white Americans fighting alongside each other for 'justice' in other parts of the globe, when they could not even eat together at the same table. Now they have the civil rights to eat together but both would be excluded from many restaurants by income poverty.
11 See Chomsky (2003).
12 See Callinicos (2003).
13 His Excellency Luiz Inácio Lula da Silva.
14 See also Corbridge (2003).
15 See Wainwright (2003).
16 See Stiglitz (2002).
17 See Wainwright (2000) and Peck and Tickell (1996).
18 See Wainwright (2003).
19 See Harvey (2000).
20 See Marx (1959).

# REFERENCES

Abrams, F. (2002) *Below the Breadline: Living on the Minimum Wage*, London: Profile Books.

Abu-Lughod, J. (1999) *New York, Chicago, Los Angeles: America's Global Cities*, Minneapolis: University of Minnesota Press.

Aceros, L. (1997) 'Women's work in Brazilian and Argentinian textiles', in S. Mitter and S. Rowbotham *Women Encounter Technology: Changing Patterns of Women's Employment in the Third World*, London: Routledge.

Afshar, H. (ed.) (1998) *Empowering Women: Illustration from the Third World*, Basingstoke: Macmillan.

Agarwal, B. (1997) '"Bargaining" and gender relations: within and beyond the household', *Feminist Economics* 3 (1): 1–50.

Aglietta, M. (1979) *A Theory of Capitalist Regulation*, London: New Left Books.

Ajuntament de Barcelona (2002) '22@ lifestyle work and play the Barcelona way', *Barcelona Bulletin* No. 199, Barcelona: Ajuntament de Barcelona.

Alarcón, R. (1999) 'Recruitment processes among foreign-born engineers and scientists in Silicon Valley', *American Behavioural Scientist* 42 (9): 1381–97.

Alibhai-Brown, J. (1996) 'In defence of diversity', *New Times*, 20 July, p. 2.

Amin, A. and Thrift, N. (1994) *Globalisation, Institutions and Regional Development in Europe*, Oxford: Oxford University Press.

Amin, A. and Tomaney, J. (1995) 'The regional developmental potential of inward investment in the less favoured regions of the European Community', in A. Amin and J. Tomaney (eds) *Behind the Myth of the European Union*, London: Routledge.

Amin, S. (1974) *Accumulation on a World Scale: A Critique of the Theory of Underdevelopment*, London: Monthly Review Press.

Amsden, A. (1993) *Structural Macroeconomic Underpinnings of Effective Industrial Policy: Fast Growth in the 1980s in Five Asian Countries*, Geneva: United Nations Conference on Trade and Development.

Anderson, B. and Phizacklea, A. (1997) *Migrant Domestic Workers: A European Perspective*, Report for the Equal Opportunities Unit, DGV, Brussels: European Commission.

Anderson, K. (1998) 'Sites of difference: beyond a cultural politics of race polarity', in R. Fincher and J. Jacobs (eds) *Cities of Difference*, New York: Guilford Press.

Appadurai, A. (1996) *Modernity at Large: Cultural Dimensions of Globalization*, Minneapolis: University of Minnesota Press.

Appadurai, A. (2001) *Globalization*, Durham NC: Duke University Press.

Asheim, B. (1996) 'Industrial districts as "learning regions": a condition for prosperity', *European Planning Studies* 4 (4): 379–400.

Atkinson, T. (2002) 'Is rising income inequality inevitable? A critique of the "Transatlantic Consensus"', in P. Townsend and D. Gordon (eds) *World Poverty: New Policies to Defeat an Old Enemy*, London: Polity Press.

Audit Commission (1999) *A Life's Work: Local Authorities, Economic Development and Economic Regeneration*, London: Audit Commission Publications.

Austrin, T. and Beynon, H. (1980) 'Global outpost: the working class experience of big business in the North-East of England', Durham: University of Durham mimeo.

Aycrigg, M. (1998) 'Participation and the World Bank. Success Constraints and Responses. Draft for Discussion of the International Conference on Upscaling and Mainstreaming Participation: Lessons Learned and Ways Forward', World Bank: Social Development Papers No. 29.

Bagnasco, A. (1977) *Tre Italia: la problematica territoriale dello sviluppo italiano*, Bologna: Il Mulino.

Bain, P. and Taylor, P. (2000) 'Entrapped by the "electronic Panoptican"? Worker resistance in the call centre', *New Technology Work and Employment* 15 (1): 2–18.

Bain, P. and Taylor, P. (2001) 'Trade unions, workers' rights and the frontier of control in UK call centres', *Economic and Industrial Democracy* 22: 39–66.

Balasubramanyam, V. and Balasubramanyam, B. (2000) 'The software cluster in Bangalore', in J. Dunning (ed.) *Regions, Globalization and the Knowledge-Based Economy*, Oxford: Oxford University Press.

Banerjee, U. (2002) 'Globalisation, Crisis in Livelihoods, Migration and Trafficking of Women and Girls: The Crisis in India, Nepal and Bangladesh', paper presented at the III International Congress of Women, Work and Health in Sweden. Stockholm, July.

Bardhan, K. and Klasen, S. (1999) 'UNDP's gender-related indices: a critical review', *World Development* 27 (6): 985–1010.

Barham, C. (2002) 'Economic inactivity and the labour market', *Labour Market Trends* 110 (2) (February): 69–77.

Barr, N. (1998) *The Economics of the Welfare State*, 3rd edn, Oxford: Oxford University Press.

Barrett, H., Ilbery, B., Browne, A. and Binns, T. (1999) 'Globalization and the changing networks of food supply: the importation of fresh horticultural produce from Kenya into the UK', *Transactions of the Institute of British Geographers* 24 (2): 159–74.

Barrientos, S. (1996) 'Flexible work and female labour: the global integration of Chilean fruit production,' in R. Auty and J. Toye (eds) *Challenging Orthodoxies*, Basingstoke: Macmillan.

Barrientos, S. (2001) 'Gender, flexibility and global value chains', *IDS Bulletin* 32 (3): 83–93.

Barrientos, S. and Blowfield, M. (2001) 'Richer or poorer? Achievements and challenges of ethical trade', ID 21 Insight Issue No. 36, March, University of Sussex: Institute of Development Studies; web reference http://www.id21.org/insights/insights36/index.html (accessed July 2002).

Barrientos, S. and Kritzinger, A. (2003) 'The poverty of work and social cohesion in global exports', in D. Chidester (ed.) *What Holds Us Together. Social Cohesion in South Africa*, Capetown: HSRC.

Barrientos, S. and Perrons, D. (1999) 'Gender and the global food chain: a comparative study of Chile and the UK', in H. Afshar and S. Barrientos (eds) *Women, Globalization and Fragmentation in the Developing World*, London: Macmillan, pp. 150–73.

Barrientos, S., McClenaghan, S. and Orton, L. (2001) 'Stakeholder participation, gender, and codes of conduct in South Africa', *Development in Practice* 11 (5): 575–86.

Barrientos, S., Dolan, C. and Tallontyre, A. (2003) 'A gendered value chain approach to codes of conduct in African horticulture', *World Development* 31 (9): 1511–26.

Baumol, W. (1967) 'Macroeconomics of unbalanced growth: the anatomy of the urban crisis', *American Economic Review* 415–26.

BBC Online (1998) 'Battle over blame for chip plant closure', Saturday 1 August, http://news.bbc.co.uk/hi/english/business/newsid_143000/143390.stm (last accessed January 2003).

Beall, J. (2002) 'The people behind the walls: insecurity, identity and gated communities in Johannesburg', Crisis States Programme Working Paper No. 10, London: LSE DESTIN.

Beatty, C. and Fothergill, S. (2002) 'Hidden unemployment among men: a case study', *Regional Studies* 36 (8): 811–23.

Beck, U. (1992) *Risk Society: Towards a New Modernity*, London: Sage.

Beck, U. (2000a) *The Brave New World of Work*, Cambridge: Polity Press.

Beck, U. (2000b) 'The cosmopolitan perspective: sociology of the second age of modernity', *British Journal of Sociology* 51 (1): 79–106.

Beck, U. (2002) 'Zombie categories: interview with Ulrich Beck', in U. Beck and E. Beck-Gernsheim (eds) *Individualization*, London: Sage.

Beckett, A. (2003) 'Can culture save us?' *The Guardian*, 2 June.

Beek, ver K. (2001) 'Maquiladoras: exploitation or emancipation? An overview of the situation of maquiladora workers in Honduras', *World Development* 29 (9): 1553–67.

Belt, V. and Richardson, R. (2001) 'Saved by the bell? Call centres and economic development in less favoured regions', *Economic and Industrial Democracy* 22 (1): 67–98.

Belt, V., Richardson, R. and Webster, J. (2002) 'Women, social skill and interactive service work in telephone call centres', *New Technology, Work and Employment* 17 (1): 20–34.

Benería, L. (2001) 'Shifting the risk: new employment patterns, informalisation, and women's work', *International Journal of Politics and Culture* 15 (1): 27–53.

Benería, L. (2003) *Gender, Development and Globalization: Economics as if all People Mattered*, London: Routledge.

Bennett, K., Beynon, H. and Hudson, R. (2000) *Coalfields Regeneration: Dealing with the Consequences of Industrial Decline*, Bristol: Policy Press.

Berger, J. and Mohr, J. (1975) *A Seventh Man: A Book of Images and Words about the Experience of Migrant Workers in Europe*, Harmondsworth: Penguin.

Beynon, H., Hudson, R. and Sadler, D. (1993) *A Place called Teeside: A Locality in a Global Economy*, Edinburgh: Edinburgh University Press.

BGMEA (2002) Bangladesh Garment Manufacturers and Exporters Association website: http://www.bgmea.com/ (last accessed December 2002).

BIE (2002) Banco de Información Económica INEGI http://dgcnesyp.inegi.gob.mx/bie.html-ssi (accessed October 2002).

Bisnath, S. and Elson, D. (1999) 'Women's Empowerment Revisited', Background Paper for Progress of the World's Women 2000, a UNIFEM Report, New York: UNIFEM.

BJS (2003) Prison and Jail Inmates at Midyear 2002, Justice Department's Bureau of Justice Statistics: http://www.ojp.usdoj.gov/bjs/abstract/pjim02.htm (last accessed August 2003).

Blindel, J. (2003) 'The price of a holiday fling', *Guardian Weekend*, 5 July.

Blowfield, M. (2001) 'Death by a thousand codes: Is harmonisation possible?' ID 21 Insight Issue No. 36, March, University of Sussex: Institute of Development Studies; web reference http://www.id21.org/insights/insights36/index.html (accessed July 2002).

Blunt, A. and Wills, J. (2000) *Dissident Geographies: An Introduction to Radical Ideas and Practice*, London: Prentice Hall.

Bonacich, E. and Appelbaum, R. (2000) *Behind the Label: Inequality in the Los Angeles Apparel Industry*, Los Angeles: University of California Press.

Borenstein, S. and Saloner, G. (2001) 'Economics and electronic commerce', *Journal of Economic Perspectives* 15 (1): 3–12.

Bossy, J. and Coleman, S. (2000) 'Womenspeak Parliamentary Domestic Violence Internet Consultation', Report of the main findings, Bristol: Women's Aid Federation of England.

Bourdieu, P. (1990) *The Logic of Practice*, Cambridge: Polity Press.

Bourdieu, P. (1998) 'Utopia of endless exploitation: The essence of neoliberalism', from *Le Monde Diplomatique*, December 1998, translated by Jeremy J. Shapiro: http://www.forum-global.de/soc/bibliot/b/bessenceneolib.htm (last accessed August 2003).

Bourdieu, P. (1999) *The Weight of the World: Social Suffering in Contemporary Society*, Cambridge: Polity Press.

Bourdieu, P. and Wacquant, L. (1999) 'On the cunning of imperialist reason', *Theory, Culture and Society* 16 (1): 41–60.

Brannen, J. (2002) 'The work–family lives of women: autonomy or illusion?' Paper presented at the first ESRC seminar on work life and time in the new economy, LSE Gender Institute, October.

Braverman, H. (1974) *Labor and Monopoly Capital: The Degradation of Work in the Twentieth Century*, London: Monthly Review Press.

Brennan, D. (2003) 'Selling sex for visas: sex tourism as a stepping-stone to international migration', in B. Ehrenrich and A. Hochschild (eds) *Global Women*, London: Granta Books.

Brenner, N. (1998) 'Global cities, global states: global city formation and state territorial restructuring in contemporary Europe', *Review of International Political Economy* 5 (1): 1–37.

Brough, D. (2003) Italy gets power back after massive blackout, Planet Ark: http://www.planetark.org/dailynewsstory.cfm/newsid/22398/story.htm (last accessed November 2003).

Buck, N., Gordon, I., Hall, P., Harloe, M. and Kleinman, M. (2002) *Working Capital: Life and Labour in Contemporary London*, London: Routledge.

Budlender, D., Elson, D., Hewitt, G., Mukhopadhyay, T. *et al.* (2002) *Gender Budgets make Cents: Understanding Gender Responsive Budgets*, London: Commonwealth Secretariat.

Bulman, J. (2002) 'Patterns of pay: results of the 2002 New Earnings Survey', *Labour Market Trends* 110 (12): 643–54.

Bunnell, T. (2002) 'Cities for nations? Examining the city–nation-state relation in the information age', *International Journal of Urban and Regional Research* 26 (2): 284–98.

Cabinet Office (2003) *Ethnic Minorities and the Labour Market Final Report*, London: Cabinet Office.

Café Direct (2002) website: http://www.cafedirect.co.uk/index2.html (last accessed August 2003).

CAFOD (2002) 'End the debt relief mirage, UK aid agencies urge World Bank and IMF' (30 August): http://www.cafod.org.uk/news/debtrelief20020830.shtml (last accessed August 2003).

Callinicos, A. (2003) *An Anti-Capitalist Manifesto*, Cambridge: Polity Press.

Carnoy, M. (2000) *Sustaining the New Economy: Work, Family and Community in the Information Age*, New York: Harvard University Press.

Carroll, R. (1998) 'The chill east wind at your doorstep', *The Guardian*, 28 October, p. 2.

Castells, M. (1989) *The Informational City: Information Technology, Economic Restructuring, and the Urban–Regional Process*, Oxford: Blackwell.

Castells, M. (1996) *The Rise of the Network Society*, Oxford: Blackwell.

Castells, M. (1997) *The Power of Identity*, Oxford: Blackwell.

Castells, M. (2001) *The Internet Galaxy: Reflections on the Internet Business and Society*, Oxford: Oxford University Press.

Castles, S. (2000) *Ethnicity and Globalization: From Migrant Worker to Transnational Citizen*, London: Sage.

Causer, P. and Williams, T. (2002a) Region in Figures, London: Office for National Statistics.

Causer, P. and Williams, T. (2002b) Region in Figures, South East, Winter, No. 6, London: Office for National Statistics; available at http://www.statistics.gov.uk/downloads/theme_compendia/region_in_figures_winter02/London.pdf (last accessed February 2003).

Cavendish, R. (1982) *Women on the Line*, London: Routledge & Kegan Paul.

CCMA (2002) 'The south-east dominates the UK call centre market: 38% of operations are in or around London': http://www.ccma.org.uk/Articles/southeast.htm.

Chaddock, G. (2003) 'US notches world's highest incarceration rate', Report on findings from a recent Bureau of Justice report *The Christian Science Monitor*, 18 August.

Chambaz, C. (2001) 'Lone-parent families in Europe: a variety of economic and social circumstances', *Social Policy and Administration* 35 (6): 658–71.

Chambers, R. (1983) *Rural Development: Putting the Last First*, London: Longman.

Chambers, R. (1997) *Whose Reality Counts? Putting the Last First*, London: Intermediate Technology Publications.

Chant, S. (1997) *Women-headed Households: Diversity and Dynamics in the Developing World*, Basingstoke: Macmillan.

Chant, S. (2001) 'Families on the verge of breakdown? Views on contemporary trends in family life in Guanucuste, Costa Rica', in C. Jackson (ed.) *Men at Work: Labour, Masculinities, Development*, London: Frank Cass.

Chant, S. (2002) 'Researching gender, families and households in Latin America: from the 20th into the 21st century', *Bulletin of Latin American Research* 21 (4): 545–75.

Chant, S. with Craske, N. (2003) *Gender in Latin America*, London: Latin American Bureau.

Chant, S. and McIlwaine, C. (1995) *Women of a Lesser Cost: Female Labour, Foreign Exchange, and Philippine Development*, London: Pluto Press.

Chaplin, C. (1936) *Modern Times* (film); see http://www.filmsite.org/mode2.html for a review (last accessed January 2003).

Charles, D. and Benneworth, P. (2001) 'Situating the north east in the European space economy', in J. Tomaney and N. Ward (eds) *A Region in Transition: North East England in the Millennium*, Aldershot: Ashgate.

Charles, N. (1993) *Gender Divisions and Social Change*, Hemel Hempstead: Harvester Wheatsheaf.

Chaudhuri, A. (2000) 'Work unlimited', *The Guardian*, 30 August.

Chomsky, N. (2003) *Rogue States: The US Force in World Affairs*, London: Pluto Press.

Christerson, B. and Lever-Tracy, C. (1997) 'The third China? Emerging industrial districts in rural China', *International Journal of Urban and Regional Research* 21 (4): 569–90.

Ciccolella, P. and Mignaqui, I. (2002) 'Buenos Aires: Social spatial impacts of the development of global city functions', in S. Sassen (ed.) *Global Networks, Linked Cities*, London: Routledge.

Clancy, M. (2002) 'The globalization of sex tourism and Cuba: a commodity chains approach', *Studies in Comparative International Development* 36 (4): 63–88.

Clarke, W. (1999) 'Mass migration and local outcomes', *Urban Studies* 35 (3): 371–83.

Clos, J. (2002) 'Barcelona mayor proactive in Porto Alegre and New York forums', *Barcelona Bulletin* No. 199, Barcelona: Ajuntament de Barcelona.

Cockburn, C. (1991) *In the Way of Women: Men's Resistance to Sex Equality in Organizations*, Basingstoke: Macmillan.

Connell, R. (1995) *Masculinities*, Cambridge: Polity Press.

Connell, R. (2002) *Gender*, Cambridge: Polity Press.

Cooke, B. and Kothari, U. (2001) *Participation: The New Tyranny*, London: Zed Books.

Cooke, P. (1995) *Revitalising Older Industrial Regions: North-Rhine Westphalia and Wales Contrasted*, London: Anglo-German Foundation.

Corbridge, S. (2003) 'Countering empire', *Antipode* 35 (1): 184–90.

Corporate Watch (2002) What's wrong with supermarkets?: http://www. corporate watch.org.uk/pages/whats_wrong_suprmkts.htm (accessed September 2002).

Coyle, D. and Quah, D. (2002) *Getting the Measure of the New Economy*, London: Isociety, The Work Foundation.

Cravey, A. (1997) 'The politics of reproduction: households in the Mexican industrial transition', *Economic Geography* 73 (2): 166–86.

Cross, S. and Bagilhole, B. (2002) 'Girls' jobs for the boys? Men, masculinity and non-traditional occupations', *Gender, Work and Organisation* 9 (2): 204–26.

Culture North East (2001) *Regional Cultural Strategy for the North East of England 2001–2010*, Newcastle upon Tyne: Culture North East.

Cutler, D. and Glaeser, E. (1997) 'Are ghettoes good or bad', *Quarterly Journal of Economics* 112 (30): 827–72.

Dabrowski, M. and Gortat, R. (2002) 'Political and economic institutions, growth and poverty: experiences of transition countries', Human Development Report Office, Occasional Paper, Background Paper for HDR 2002/2.

Daniels, S. and Rycroft, S. (1993) 'Mapping the modern city: Sillitoe, Alan, Nottingham novels', *Transactions of the Institute of British Geographers*, 18 (4): 460–80.

Davis, M. (1990) *City of Quartz: Excavating the Future in Los Angeles*, London: Verso.

Davis, M. (2000) *Magical Urbanism: Latinos Reinvent the US City*, London: Verso.

Davis, M. (2001) 'The flames of New York', *New Left Review* 12: 34–50.

Day, H. (2003) 'Permanently Temporary Migrations: Contemporary Spatial Evidence of Filipina Domestic Workers in Hong Kong's Public Domain', unpublished Masters Dissertation, LSE.

Denman, C. (2002) 'Health care practices during pregnancy of maquiladora workers in northern Mexico', Paper presented at the Women, Work and Health International Congress, Stockholm, July.

Denny, C. (2001) 'Lifestyle fixers take the strain for City workers', *The Guardian*, 16 July, p. 5.

DETR (1999) *Towards an Urban Renaissance: Final Report of the Urban Task Force*, London: E & F N Spon.

Diagonal mar (2002) Information: http://212.9.72.172/html_eng/home_pivote.htm (accessed August 2003).

Diamantopoulou, A. (2002) 'Fighting modern slavery: the EU's role in supporting victims of trafficking', speech to the European conference on preventing and combating trafficking in human beings, organized by the International Organisation for Migration, Brussels, 19 September.

Dicken, P. (1998) *Global Shift: Transforming the World Economy*, New York: Guilford Press.

Dicken, P. (2003) *Global Shift: Reshaping the Global Economy in the 21st Century*, London: Sage.

Dodge, M. and Kitchen, R. (2000) *Mapping Cyberspace*, Harlow: Addison-Wesley.

Dolan, C. and Tewari, M. (2001) 'From what we wear to what we eat: upgrading in global value chains', *IDS Bulletin* 32 (3), University of Sussex: Institute of Development Studies.

Doogan, K. (2001) 'Insecurity and long-term employment', *Work, Employment and Society* 15 (3): 419–41.

Donkor, K. (2002) 'Structural adjustment and mass poverty in Ghana', in P. Townsend and D. Gordon (eds) *World Poverty: New Policies to Defeat Old Enemies*, Bristol: Policy Press.

Douglas, M. and Friedman, J. (eds) (1998) *Cities for Citizens*, Chichester: Wiley.

DTI (1998) *Our Competitive Future: Competitiveness in the Knowledge Driven Economy*, Department of Trade and Industry, London: Stationery Office.

Duffield, M. (2002) 'Trends in female employment', *Labour Market Trends* 110 (11): 605–16.

Duncan, R. (2003) *The Dollar Crisis: Causes and Consequences, Cures*, Chichester: Wiley.

Duncan, S. (1996) 'The diverse worlds of European patriarchy', in M. D. Garcia-Ramon and J. Monk (eds) *Women of the European Union: The Politics of Work and Daily Life*, London: Routledge, pp. 74–110.

Duncan, S. and Edwards, S. (1997) *Single Mothers in an International Context: Mothers or Workers?* London and Bristol: UCL Press.

Dunford, M. (1994) 'Winners and losers: the new map of economic inequality in the European Union', *European Urban and Regional Studies* 1 (2): 95–114.

Dunford, M. (2002) 'Inequality and transition', paper presented to the British Association, Leicester, September.

Dunford, M. and Fielding, A. (1997) 'Greater London, the south-east region and wider Britain: metropolitan polarization, uneven development and inter-regional migration', in H. Blotvogel and A. Fielding (eds) *People, Jobs and Mobility in the New Europe*, Chichester: Wiley.

Dunford, M. and Greco, L. (forthcoming) *After the Three Italies*, Oxford: Blackwell.

Dunford, M. and Perrons, D. (1983) *Arena of Capital*, Basingstoke: Macmillan.

*Economist* (2001) 'Chile's badly treated fruit pickers: what help for the Temporeras', *The Economist*, 5 February.

Edin, P., Fredriksson, P. and Åslund, O. (2000) 'Ethnic enclaves and the economic success of immigrants – evidence from a natural experiment', Working Paper No. 9, Uppsala: IFAU Institute for Labour Market Policy Evaluation.

Ehrenrich, B. (2001) *Nickel and Dimed: Undercover in Low Wage USA*, London: Granta Books.

Ehrenrich, B. (2003) 'Maid to order', in B. Ehrenrich and A. Hochschild (eds) *Global Women*, London: Granta Books.

Ehrenrich, B. and Hochschild, A. (2003) *Global Women*, London: Granta Books.

Elson, D. (ed.) (1995) *Male Bias in the Development Process*, Manchester: Manchester University Press.

Elson, D. (1999) 'Labour markets as gendered institutions: equality, efficiency and empowerment issues', *World Development* 27 (3): 611–27.

Elson, D. and Pearson, R. (1981) 'Nimble fingers make cheap workers: an analysis of women's employment in third world manufacturing', *Feminist Review*, spring: 87–107.

Elson, D. and Pearson, R. (1998) 'Nimble fingers revisited', in C. Jackson and R. Pearson (eds) *Feminist Visions of Development*, New York: Routledge.

Emmanuel, A. (1972) *Unequal Exchange: A Study of the Imperialism of Trade*, London: Monthly Review Press.

Escobar, A. (1995) *Encountering Development: The Making and Unmaking of the Third World*, Princeton: Princeton University Press.

Escobar, A., Rocheleau, D. and Kothari, S. (2002) 'Environmental social movements and the politics of place', *Development* 45 (1): 28–36.

Esping-Andersen, G. (1990) *The Three Worlds of Welfare Capitalism*, Cambridge: Polity Press.

Esping-Andersen, G. (1999) *Social Foundations of Post-industrial Economies*, Oxford: Oxford University Press.

ETI (2003) ETI base code: ethical trading initiative: http://www.ethicaltrade.org/pub/publications/basecode/en/index.shtml (last accessed November 2003).

European Commission (2000) *Towards a Community Framework Strategy on Gender Equality 2001–2005* (Com (2000) 335 final), Brussels.

European Commission (2001) *Employment in Europe 2001: Recent Trends and Prospects*, Luxembourg: European Commission.

European Foundation (2002) 'Quality of women's work and employment: tools for Change', Foundation Paper No. 3.

Fagan, C. and Burchell, B. (2002) *European Foundation for the Improvement of Living and Working Conditions: Gender, Jobs and Working Conditions in the European Union*, Luxembourg: Office for Official Publications of the European Communities.

Fagnani, J. (1998) 'Helping mothers to combine paid and unpaid work – or fighting unemployment? The ambiguities of French family policy', *Community, Work and Family* 1 (3): 297–312.

Fagnani, J. (2002) 'The French 35 hour working law and the work–life balance of parents: friends or foes?' Paper presented at the first ESRC seminar on work life and time in the new economy, LSE Gender Institute, October.

Fainstein, S., Gordon, I. and Harloe, M. (eds) (1992) *Divided Cities: New York and London in the Contemporary World*, Oxford: Blackwell.

Federal Reserve System (2003) *Flow of Funds Accounts of the United States*, Washington DC: Federal Reserve System.

Fernie, S. and Metcalf, D. (1997) *Hanging on the Telephone*, London: LSE Centre for Economic Performance.

Fincher, R. and Jacobs, J. (eds) (1998) *Cities of Difference*, The Guildford Press, London.

Finquelievich, S. (2001) 'Electronic democracy, Buenos Aires and Montevideo', *Cooperation South* 1: 61–81.

Fitter, R. and Kaplinsky, R. (2001) 'Who gains from product rents as the coffee market becomes more differentiated? A value chain analysis', in G. Gereffi and R. Kaplinsky (eds) *Value of Value Chains*, University of Sussex IDS Bulletin 32 (3): 46–59.

Fleck, S. (2001) 'A gender perspective on maquila employment and wages in Mexico', in E. Katz and M. Correia (eds) *The Economics of Gender in Mexico: Work, Family, State and Market*, Washington DC: World Bank.

Fletcher, V. (2002) 'Revealed: East End sweatshops where they pay workers only £3.75 an hour to make clothes for Topshop', *Evening Standard*, 19 November, pp. 16–17.

Flyvbjerg, B. (1998) 'Empowering civil society: Habermas, Foucault and the question of conflict', in M. Douglas and J. Friedman (eds) *Cities for Citizens*, Chichester: Wiley.

Folbre, N. and Nelson, J. (2000) 'For love or money – or both?' *Journal of Economic Perspectives* 14 (4): 123–40.

Fölster, K. (1999) 'Paid domestic work in Germany', paper presented at the Arbetslivinstitutet workshop on labour market and social policy: gender relations in transition, Brussels, May.

Forester, J. (1998) 'Rationality, dialogue and learning: what community and environmental mediators can teach us about the practice of civil society', in M. Douglas and J. Friedman (eds) *Cities for Citizens*, Chichester: Wiley.

Forrest, R. and Kearns, A. (1999) *Joined-up Places? Social Cohesion and Neighbourhood Regeneration*, York: Joseph Rowntree Foundation.

Francis, P. (2001) 'Participatory development at the World Bank: the primacy of process', in B. Cooke and U. Kothari (eds) *Participation: The New Tyranny*, London: Zed Books.

Frank, A. (1971) *Capitalism and Underdevelopment in Latin America: Historical Studies of Chile and Brazil*, Harmondsworth: Penguin.

Fraser, N. (1997) *Justice Interruptus: Critical Reflections on the Postsocialist Condition*, London: Routledge.

Fraser, N. (2000) 'Rethinking recognition', *New Left Review* 3 (May/June): 107–20.

Freeman, C. (1987) *Technology Policy and Economic Performance: Lessons from Japan*, London: Frances Pinter.

Freeman, C. and Louçã, F. (2001) *As Time Goes By: From the Industrial Revolution to the Information Revolution*, Oxford: Oxford University Press.

Friedan, B. (1963) *The Feminine Mystique*, London: Gollancz.

Friedman, J. (1986) 'The world city hypothesis', *Development and Change* 17 (1): 69–83.

Friedman, J. (2001) 'Intercity networks in a globalizing era', in A. Scott (ed.) *Global City-Regions*, Oxford: Oxford University Press.

Fröbel, F., Heinrichs, J. and Kreye, O. (1980) *The New International Division of Labour: Structural Unemployment in Industrialised Countries and Industrialisation in Developing Countries*, Cambridge: Cambridge University Press.

Fujita, M., Krugman, P. and Venables, A. (1999) *The Spatial Economy: Cities, Regions and International Trade*, Cambridge, Mass: MIT Press.

Fullelove, L. (2001) 'Sainsbury's and ethical trade', ID 21 Insight Issue No.36, March, University of Sussex: Institute of Development Studies.

Fussell, E. (2000) 'Making labour flexible: the recomposition of Tijuana's maquiladora female labour force', *Feminist Economics* 6 (3): 59–80.

Gamburd, M. (2003) 'Breadwinner no more', in B. Ehrenrich and A. Hochschild (eds) *Global Women*, London: Granta Books.

Garcia-Ramon, M.D. and Albet, A. (2000) 'Pre-Olympic and post-Olympic: a "model" for urban regeneration today'? *Environment and Planning A* 32: 1331–4.

Garreau, J. (1991) *Edge City: Life on the New Frontier*, New York: Doubleday.

Gateshead Council (2003) Millennium Bridge Factsheet: http://www.gateshead.gov.uk/bridge/facts.htm (last accessed June 2003).

Gdaniec, C. (2000) 'Cultural industries, information technology and the regeneration of post-industrial urban landscapes: Poblenou in Barcelona – a virtual city'? *GeoJournal* 50: 379–87.

Gereffi, G. (1999) 'International trade and industrial upgrading in the apparel commodity chain', *Journal of International Economics* 48: 37–70.

Gereffi, G. and Kaplinsky, R. (eds) (2001) *Value of Value Chains*, University of Sussex *IDS Bulletin* 32 (3).

Gereffi, G. and Korzeniewicz, M. (eds) (1994) *Commodity Chains and Global Capitalism*, Westport: Praeger.

Gereffi, G., Humphrey, J., Kaplinsky, R. and Sturgeon, T. (2001) 'Introduction: globalisation, value chains and development', *IDS Bulletin* 32 (3): 1–8.

Geske, A. and Plantenga, J. (1997) *Gender and Economics: A European Perspective*, New York: Routledge.

Giddens, A. (1999) *Runaway World: How Globalization is Reshaping our Lives*, London: Profile Books.

GLA (2002) *London Divided: Income Inequality and Poverty in the Capital*, London: GLA.

Glassman, J. (1999) 'State power beyond the "territorial trap": the internationalisation of the state', *Political Geography* 18: 669–96.

Global Exchange (2002) Plan Puebla Panama (PPP): http://www.globalexchange.org/campaigns/mexico/ppp/ppp.html (last accessed August 2003).

Glucksmann, M. (1990) *Women Assemble: Women Workers in the New Industries in Interwar Britain*, London: Routledge.

Glucksmann, M. (2002) 'Call connections: call centres and varieties of socio-economic intermediation', Paper for Workshop on Polanyian Perspectives on Instituted Economic Processes, Development and Transformation, ESRC Centre for Research on Innovation and Competition, University of Manchester, 23–25 October.

Goldthorpe, J. (1968) *The Affluent Worker: Industrial Attitudes and Behaviour*, Cambridge: Cambridge University Press.

GONE (2002) Key Facts and Figures, June, Newcastle: Government Office of the North East.

González de la Rocha, M. (2003) 'The construction of the myth of survival', Paper prepared for the International Workshop on Feminist Fables and Gender Myths: Repositioning Gender in Development Policy and Practice, Institute of Development Studies, Sussex, 2–4 July.

Gordon, I. (2002) 'Global cities, internationalisation and urban systems', in P. McCann (ed.) *Industrial Location Economics*, Cheltenham: Edward Elgar.

Gordon, I. (2003) 'Capital needs, capital growth and global city rhetoric in Mayor Livingstone's London plan', Paper presented in session on The Production of Capital Cities, Association of American Geographers Annual Meeting, New Orleans, 7 March.

Gordon, I. and McCann, P. (2000) 'Industrial clusters: complexes, agglomeration and/or social networks?' *Urban Studies* 37 (3): 513–32.

Gowan, P. (1995) 'Neo-liberal theory and practice for Eastern Europe', *New Left Review*, 216: 129–40.

Graham, S. and Marvin, S. (2001) *Splintering Urbanism: Networked Infrastructures, Technological Mobilities and the Urban Condition*, London: Routledge.

Gramsci, A. (1971) 'Americanism and Fordism', from A. Gramsci, *Selections from the Prison Notebooks*, New York: International Publishers.

Green, J. (1998) 'Save seconds: teach them how to say goodbye', *CallCentre Europe* 20: 28–9.

Greenspan, A. (1998) 'Is there a new economy?' *California Management Review* 41 (1): 74–85.

Gregson, N. and Lowe, M. (1994) *Servicing the Middle Classes*, London: Routledge.

Grossman, R. (1979) 'Women's place in the integrated circuit', *Southeast Asia Chronicle* – Pacific Research SRC 66 (9) No. 5.

Guijt, I. and Shah, M. (eds) (1998) *The Myth of Community: Gender Issues in Participatory Development*, London: Intermediate Technology Publications.

Gwynne, R. (1998) 'Globalization, commodity chains and fruit exporting regions in Chile', *Tijdschrift voor Economische en Sociale Geografie* 90 (2): 211–25.

Habermas, J. (1998) *On the Pragmatics of Communication*, Cambridge, Mass.: MIT Press.

Haggett, P. (1965) *Locational Analysis in Human Geography*, London: Edward Arnold.

Hale, A. (undated) 'What hope for "ethical" trade in the globalised garment industry?' Women Working Worldwide: http://www.poptel.org.uk/women-ww (accessed August 2003).

Hall, P. (1966) *The World Cities*, London: Weidenfeld & Nicolson.

Hall, P. (1998) 'Globalization and world cities', in F. Lo and Y. Yeung (eds) *Globalization and the World of Large Cities*, Tokyo: United Nations University Press.

Hall, P. (2001) 'Global city-regions in the 21st century', in A. Scott (ed.) *Global City-Regions*, Oxford: Oxford University Press.

Hamnett, C. (2003) *Unequal City: London in the Global Arena*, London: Routledge.

Hardt, M. and Negri, A. (2001) *Empire*, Cambridge, Mass.: Harvard University Press.

Harvey, D. (1985) *The Urbanization of Capital*, Oxford: Blackwell.

Harvey, D. (1989) 'From managerialism to entrepreneurialism: the transformation of urban governance in late capitalism', *Geografiska Annaler* 71B: 3–17.

Harvey, D. (2000) *Spaces of Hope: Towards a Critical Geography*, Edinburgh: Edinburgh University Press.

Harvey, D. (2003) 'The new imperialism' (Clarendon lectures in geographical and environmental studies), Oxford: Oxford University Press.

Hayek, F. (1945) 'The uses of knowledge in society', *American Economic Review* XXXV (4): 519–30.

Hayter, T. (1971) *Aid as Imperialism*, Harmondsworth: Penguin.

Healey, P. (1996) 'The communicative turn in planning theory', *Environment and Planning B: Planning and Design* 23: 217–34.

Healey, P. (1998) 'Building institutional capacity through collaborative approaches to urban planning', *Environment and Planning A* 30: 1531–46.

HELA (2001) Call Centres Local Authority Circular, LAC No. 94/1 rev, Health and Safety Executive Local Authority Enforcement Liaison Committee: http://www. hse.gov.uk/lau/lacs/94-1.htm (last accessed June 2003).

Held, D., McGrew, A., Goldblatt, D. and Perraton, J. (1999) *Global Transformations: Politics, Economics and Culture*, Cambridge: Polity Press.

Hencke, D., Denny, C. and Elliot, L. (2002) 'Tanzania air traffic control deal condemned as "waste of money"', *Guardian*, 14 June.

Hirdman, Y. (1990) 'Women from possibility to problem? Gender conflict in the welfare state – the Swedish model', Research Report Series No. 3, Stockholm: Arbetslivscentrum.

Hirschman, A. (1958) *The Strategy of Economic Development*, New Haven: Yale University Press.

Hirst, P. and Thompson, G. (1996) *Globalization in Question: The International Economy and the Possibilities of Governance*, Cambridge: Polity Press.

Hirst, P. and Thompson, G. (2002) 'The future of globalization', *Cooperation and Conflict: Journal of the International Nordic International Studies Association* 37 (3): 247–65.

Hobbes, T. (1983) *De Cive*, Oxford: Clarendon Press.

Hobbes, T. (1996) *Leviathan*, Oxford: Oxford University Press.

Hochschild, A. (1997) *The Time Bind*, New York: Metropolitan Books.

Hochschild, A. (2000) 'Global care chains and emotional surplus value', in W. Hutton and A. Giddens (eds) *On the Edge: Living with Global Capitalism*, London: Jonathan Cape.

Hollands, R. (2001) 'The restructuring of young Geordies' employment, household and consumption identities', in J. Tomaney and N. Ward (eds) *A Region in Transition: North East England at the Millennium*, Aldershot: Ashgate.

Holman, D. and Fernie, S. (2000) 'Can I help you? Call centres and job satisfaction', *Centrepiece* 5 (1): spring.

Hotelling, H. (1929) 'Stability and competition', *Economic Journal* 39 (1): 41–57.

HRW (2002) Incarceration and Race, Human Rights Watch: http://www.hrw.org/reports/2000/usa/Rcedrg00-01.htm#P167_28183 (last accessed December 2002).

Hudson, R. (2000) *Production, Places and Environment: Changing Perspectives in Economic Geography*, Harlow: Prentice Hall.

Humphrey, J. (2002) 'Business to business e-commerce and access to global markets: exclusive or inclusive outcomes? IDS mimeograph, Sussex University: Institute of Development Studies.

Humphrey, J. and Schmitz, H. (2002) 'How does insertion in global value chains affect upgrading in industrial clusters?' *Regional Studies* 36 (9):1017–27.

Humphrey, J., Mansell, R., Paré, D. and Schmitz, H. (2003) *The Reality of E-Commerce with Developing Countries*, Sussex, London: Media@lse.

Hymer, S. (1975) 'The multinational corporation and the law of uneven development', in H. Radice (ed.) *International Firms and Modern Imperialism: Selected Readings*, Harmondsworth: Penguin.

ICO (2001) International Coffee Organization: http://www.ico.org/ (last accessed December 2002).

IDS (2000) *Pay and Conditions in Call Centres*, London: Income Data Services.

IDS (2003) *Pay and Conditions in Call Centres*, London: Income Data Services.

IDS (2001) 'E-commerce: accelerator of development', IDS Policy Briefing 14, Sussex University: Institute of Development Studies.

ILO (1998) Export Processing Zone Project: http://www.ilo.org/public/english/dialogue/govlab/legrel/tc/epz/ (last accessed February 2003).

ILO (2001a) 'Seeking socially responsible tourism', magazine of the ILO *World of work* 39 (June).

ILO (2001b) International Labour Organization history, available at http://www.ilo.org/public/english/about/history.htm (accessed June 2002).

ILO (2003a) Laborsta Labour Statistics Database, Geneva: http://www.ilo.org/public/english/bureau/stat/ (last accessed November 2003).

ILO (2003b) About the ILO: http://www.ilo/public/english/about/index.htm (last accessed November 2003).

IMF (2001) 'What is the International Monetary Fund?' available at http://www.imf.org/external/pubs/ft/exrp/what.htm#glance (accessed June 2002).

IMF (2002a) 'Debt relief under the heavily indebted poor countries (HIPC) initiative: a factsheet', August 2002: http://www.imf.org/external/np/exr/facts/hipc.htm (last accessed November 2002).

IMF (2002b) 'Debt relief for poor countries (HIPC): progress through September 2002: a factsheet', September 2002: http://www.imf.org/external/np/exr/facts/povdebt.htm (accessed November 2002).

IMF (2002c) *World Economic Outlook, Trade and Finance*, September: http://www.imf.org/external/pubs/ft/weo/2002/02/pdf/chapter2.pdf (accessed November 2002).

IMF (2003) About the IMF: http://www.imf.org/external/about.htm (last accessed August 2003).

Jackson, C. (1997) 'Sustainable development at the sharp end: field-worker agency in a participatory project', *Development in Practice* 7 (3): 237–47.

Jackson, C. (ed.) (2001) *Men at Work: Labour, Masculinities, Development*, London: Frank Cass.

Jackson, P. and Penrose, J. (1993) *Constructions of Race, Place and Nation*, London: UCL Press.

James, D. (2001) 'Starbucks' programs show improvement in commitment to fair trade, but not nearly enough', Global Exchange, Coffee Campaign: http://www.globalexchange.org/economy/coffee/news2001/gx102201.html (last accessed December 2002).

Jarvis, H. (2002) '"Lunch is for wimps": what drives parents to work long hours in "successful" British and US cities?' *Area* 34 (4): 340–52.

Jensen, M. (2001) 'Afriboxes, telecenters, cybercafes: ICT in Africa', *Cooperation South* 1: 97–109.

Jessop, B. (1990) *State Theory: Putting Capitalist States in their Place*, Cambridge: Polity Press.

Jessop, B. (1991) *The Politics of Flexibility: Restructuring State and Industry in Britain, Germany and Scandinavia*, Aldershot: Edward Elgar.

Jessop, B. (2002a) *The Future of the Capitalist State*, Cambridge: Polity Press.

Jessop, B. (2002b) 'Liberalism, neoliberalism, and urban governance: a state-theoretical perspective', *Antipode* 33 (4): 452–72.

Jevons, W. (1911) *The Theory of Political Economy*, 4th edn, London: Macmillan.

Kabeer, N. (1995) 'Necessary, sufficient or irrelevant? Women, wages and intra-household power relations in urban Bangladesh', IDS Discussion Paper No. 25, Sussex: Institute of Development Studies.

Kabeer, N. (2000) *The Power to Choose: Bangladeshi Women and Labour Market Decisions in London and Dhaka*, London: Verso.

Kabeer, N. (2003) *Gender Mainstreaming in Poverty Eradication and the Millennium Development Goals: A Handbook for Policy-makers and Stakeholders*, Ottawa: Commonwealth Secretariat, Canadian International Development Agency.

Kaldor, N. (1970) 'The case for regional policies', *Scottish Journal of Political Economy*, 17: 337–48.

Keating, M. (2001) 'Rethinking the region: culture, institutions and economic development in Catalonia and Galicia', *European Urban and Regional Studies* 8 (3): 217–34.

Keating, M. and Loughlin, J. (1997) *The Political Economy of Regionalism*, London: Frank Cass.

Kelkar, G. and Nathan, D. (2002) 'Gender relations and technical change in Asia', *Current Sociology* 50 (3): 427–42.

Kelso, P. (2002) 'Angry women find a voice over pay that doesn't add up', *The Guardian*, 18 July, p. 5.

Kempadoo, K. (2001) 'Freelancers, temporary wives, and beach-boys: researching sex work in the Caribbean', *Feminist Review* 67: 39–62.

Keynes, J. (1947) *The General Theory of Employment Interest and Money*, London: Macmillan.

Keynote (2000) KN52150 Keynote supermarkets and superstores, January: http://www.the-list.co.uk/acatalog/kn52150.html (last accessed February 2003).

King, R. (1998) 'From guest workers to immigrants: labour migration from the Mediterranean periphery', in D. Pinder (ed.) *The New Europe: Economy, Society, and Environment*, New York: Wiley.

Klein, N. (1999) *No Logo: Taking Aim at the Brand Bullies*, New York: Picador.

Klein, N. (2002) *Fences and Windows: Dispatches from the Frontlines of the Globalization Debate*, London: Flamingo.

Klein, N. (2003) 'Cut the strings', *Guardian*, 1 February.

Kommolvadhin, N. (2002) 'Women and the new economy in Thailand', unpublished PhD thesis in progress, London School of Economics.

Krugman, P. (1998) 'What's new about economic geography?' *Oxford Review of Economic Policy* 14 (2): 7–17.

Krugman, P. (2002a) 'For richer', *New York Times*, 20 October.

Krugman, P. (2002b) 'The lost continent', *New York Times*, 9 August.

Kumar, S. and Corbridge, S. (2002) 'Programmed to fail? Development projects and the politics of participation', *Journal of Development Studies* 39 (2): 73–103.

Leborgne, D. and Lipietz, A. (1991) 'Two social strategies in the production of new industrial spaces', in G. Benko and M. Dunford (eds) *Industrial Change and Regional Development*, London: Belhaven.

Lewis, J. (ed.) (1997) *Lone Parents in European Welfare Regimes*, London: Jessica Kingsley.

Lewis, J. (2001) 'The decline of the male breadwinner model: implications for work and care', *Social Politics* 8 (2):152–69.

Lewis, T. (2002) 'Argentina's revolt', *International Socialist Review* 21: January–February.

Leyshon, A. (2003) 'Scary monsters? Software formats, peer-to-peer networks and the spectre of the gift', *Environment and Planning D: Society and Space* 21.

Lim, L. (ed.) (1998) *The Sex Sector: The Economic and Social Bases of Prostitution in Southeast Asia*, Geneva: International Labour Office.

Lim, L. (2002) *Female Labour-Force Participation: Gender Promotion Programme*, Switzerland: ILO; available from http://www.un.org/esa/population/publications/completingfertility/RevisedLIMpaper.PDF (last accessed June 2003).

Lipietz, A. (1987) *Mirages and Miracles: The Crises of Global Fordism*, London: Verso.

Lipietz, A. (1992) *Towards a New Economic Order: Postfordism, Ecology and Democracy*, Oxford: Polity Press.

List, F. (1909) *The National System of Political Economy*, London: Longman.

Litske, R. (2002) 'Migration and mobility in the EU', Presentation, 9 September, Dublin: European Foundation for Living and Working.

Lojkine, J. (1976) 'Contribution to a Marxist theory of capitalist urbanization', in C. Pickvance (ed.) *Urban Sociology: Critical Essays*, London: Tavistock.

Lovering, J. (1999) 'Theory led by policy: the inadequacies of the "new regionalism" (illustrated from the case of Wales)', *International Journal of Urban and Regional Research* 23: 379–95.

Lula da Silva, L. (2003) Speech delivered by His Excellency Luiz Inácio Lula da Silva, President of the Federative Republic of Brazil at the London School of Economics, July.

Lundberg, U. and Berntsson, L. (2002) 'Has the total workload of Swedish men and women become more equal during the last 10 years?' Paper presented at the Women Work and Health Third International Congress, Stockholm, June.

Lutz, H. (2002) 'At your service madam! The globalization of domestic service', *Feminist Review* 70: 89–104.

Luxembourg Income Study (2002): http://www.lisproject.org/keyfigures/ineqtable.htm (last accessed December 2002).

L–EWRG (1980) *In and Against the State*, London–Edinburgh Weekend Return Group, London: CSE.

McCrone, G. (1969) *Regional Policy in Britain*, London: Allen & Unwin.

McDowell, L. (1991) 'Life without father and Ford: the new gender order of post-Fordism', *Transactions of the Institute of British Geographers* 16 (4): 400–19.

McDowell, L. (1997) *Capital Culture: Gender at Work in the City*, Oxford: Blackwell.

McDowell, L. (2002) 'Masculine discourses and dissonances: strutting "lad", protest masculinity, and domestic respectability', *Environment and Planning D: Society and Space* 20: 97–119.

MacLeod, M. and Goodwin, M. (1999) 'Space, scale and state strategy: rethinking urban and regional governance', *Progress in Human Geography* 23 (4): 503–27.

Madanipour, A., Cars, G. and Allen, J. (eds) (1998) *Social Exclusion in European Cities: Processes, Experiences and Responses*, London: Jessica Kingsley.

Mahdi, S. and Barrientos, A. (2003) 'Saudization and employment in Saudi Arabia', *Journal of Career Development International* 8 (3): 70–7.

Maguire, K. (2003) 'BT strike threat over Indian call centres', *The Guardian*, 3 June, p. 9.

Malecki, E. (2002) 'The economic geography of the Internet's infrastructure', *Economic Geography* 78 (4): 399–424.

Mann, M. (2000) 'Has globalization ended the rise of the nation state', in S. Corbridge (ed.) *Development: Critical Concepts in the Social Sciences*, London: Routledge.

Mansell, R. (2001) 'Digital opportunities and the missing link for developing countries', *Oxford Review of Economic Policy* 17 (2): 282–95.

Mansell, R. (2002) 'From digital divides to digital entitlements in knowledge societies', *Current Sociology* 50 (3): 407–26.

Marcus, R. (1997) 'Stitching footballs: the voices of children in Pakistan', in R. Marcus and C. Harper (eds) *Small Hands: Children in the Working World*, London: SCF Publication.

Marcuse, P. (1997) 'The enclave, the citadel, and the ghetto: what has changed in the post-fordist U.S. city', *Urban Affairs Review* 33 (2): 228–64.

Marcuse, P. (2002) 'Urban form and globalization after September 11th: The view from New York', *International Journal of Urban and Regional Research* 26 (3): 596–606.

Markusen, A. (1996) 'Sticky places in slippery space: a typology of industrial districts', *Economic Geography* 72: 293–313.

Markusen, A. (1999) 'Fuzzy concepts, scanty evidence, policy distance: The case for rigour and policy relevance in critical regional studies', *Regional Studies* 33 (9): 869–83.

Marshall, A. (1961) *Principles of Economics*, 9th (variorum) edn, London: Macmillan.

Martin, R. and Sunley, P. (2001) 'Deconstructing clusters: chaotic concept or policy panacea?' Paper presented at the Regional Studies Association Conference on Regionalizing the Knowledge Economy, London, 21 November.

Marx, K. (1959) *Capital*, Vol. 3, London: Lawrence & Wishart.

Marx, K. (1973a) *Grundrisse: Introduction to Political Economy*, Harmondsworth: Penguin.

Marx, K. (1973b) *A Contribution to the Critique of Political Economy*, Preface, London: Lawrence & Wishart.

Marx, K. (1976) *Capital*, Vol. 1, Harmondsworth: Penguin.

Marx, K. and Engels, F. (1983) *Communist Manifesto*, London: Lawrence & Wishart.

Massey, D. (1978) 'In what sense a regional problem?' *Regional Studies* 13: 233–44.

Massey, D. (1984) *Spatial Divisions of Labour: Social Relations and the Geography of Production*, Basingstoke: Macmillan.

Massey, D. (1995) 'Politicising space and place', *Scottish Geographical Magazine* 112: 117–23.

Massey, D. (1996) 'Masculinity, dualisms, and high technology', in N. Duncan (ed.) *Bodyspace*, London: Routledge, pp. 109–26.

Meier, V. (1999) 'Cut-flower production in Columbia – a major development story for women'? *Environment and Planning A* 31: 273–89.

Metcalf, D. (2003) 'British Unions: What Future?' Future of Unions in Modern Britain, Mid-Term Report on Leverhulme Trust-Funded Research Programmes 2000–2002, London: CEP.

Milanovic, B. (2002) 'True world income distribution, 1988 and 1993: first calculation based on household surveys alone', *The Economic Journal* 112 (January): 51–92.

Milanovic, B. and Yitzhaki, S. (2001) 'Decomposing World Income Distribution: Does the World Have a Middle Class?' Papers 2562, World Bank – Country Economics Department.

Mill, J. S (1989) *On Liberty*, Cambridge: Cambridge University Press.

Miller, R. (1991) 'Selling Mrs Consumer's advertising and the creation of suburban social spatial relations 1910–1930', *Antipode* 23 (3): 263–81.

Mir, A., Mathew, B. and Mir, R. (2000) 'The codes of migration: contours of the global software labor market', *Cultural Dynamics* 12 (1): 5–33.

Mirza, H. (1997) 'Mapping a genealogy of black British feminism', in H. Mirza (ed.) *Black British Feminism: A Reader*, London: Routledge.

Mishel, L., Bernstein, J. and Schmitt, J. (2003) *The State of Working America, 2002/2003*, Ithaca: ILS Press.

Mitter, S. and Rowbotham, S. (1997) *Women Encounter Technology: Changing Patterns of Employment in the Third World*, London: Routledge.

Momsen, J. (ed.) (1999) *Gender, Migration and Domestic Service*, London: Routledge.

Monbiot, G. (2002a) 'At the seat of empire: Africa is forced to take the blame for the devastation inflicted on it by the rich world', *The Guardian*, 25 June, p.15.

Monbiot, G. (2002b) 'While Lula's Brazil kowtows to the free market, Blair's Britain only pretends to do so', *The Guardian*, 29 October, p.15.

Moragas Spà, M., Rivenburg, N. and Garcìa, N. (1995) 'Television and the construction of identity: Barcelona, Olympic host', in M. Moragas Spà and M. Botella (eds) *The*

*Keys of Success: The Social, Spatial, Economic and Communications Impact of Barcelona '92*, Bellaterra Servei de Publicacions de la Universitat Autònoma de Barcelona.

Morgan, K. (1997) 'The learning region: institutions, innovation and regional renewal', *Regional Studies* 31 (5): 491–503.

Morris, L. (1995) *Social Division, Economic Decline and Social Structural Change*, London: UCL Press.

Morris, S. (2003) 'Sisters deceived hundreds into prostitution', *The Guardian*, 4 June, p.8.

Murray, C. (1999) *The Underclass Revisited*, Washington DC: AEI Press.

Murray, R. (1991) *Local Space – Europe and the New Regionalism*, Stevenage: South East Economic Development Strategy.

Myrdal, G. (1963) *Economic Theory and Under-developed Regions*, London: Methuen.

National Center for Health Statistics (2003) Health, United States, 2002: http://www.cdc.gov/nchs/hus.htm (accessed August 2003).

National Statistics (2003a) *E-commerce: The Business Facts*, NEMW (07), London: National Statistics.

National Statistics (2003b) 'Internet Access: Household and Individuals', April, London: National Statistics; http://www.statistics.gov.uk/pdfdir/int0403.pdf (last accessed May 2003).

Nelson, M. and Smith, J. (1999) *Working Hard and Making Do: Surviving in Small Town America*, London: University of California Press.

NES (2002) New Earnings Survey First Release, October: http://www.statistics.gov.uk/pdfdir/nes1002.pdf (last accessed December 2002).

Ng, C. and Mohamad, M. (1997) 'The management of technology and women in two electronic firms in Malaysia', *Gender, Technology and Development* 1 (2): 176–204.

Nolan, P. and Slater, G. (2002) 'The labour market: history, structure and prospects', in P. Edwards (ed.) *Industrial Relations: Theory and Practice*, Oxford: Blackwell.

Norris, P. (2001) *Digital Divide: Civic Engagement, Information Poverty and the Internet Worldwide*, Cambridge: Cambridge University Press.

Northover, H., Ladd, D., Drapkin, J., Kline, S. and Lemoine, F. (2002) A Joint Submission to the World Bank and IMF Review of HIPC and Debt Sustainability: http://www.christian-aid.org.uk/indepth/environm.htm (accessed November 2002).

NUA (2002) NUA Internet Surveys: http://www.nua.ie/surveys/how_many_online/ (last accessed August 2003).

Nussbaum, M. (2003) 'Capabilities as fundamental entitlements: Sen and social justice', *Feminist Economics* 9 (2–3): 33–59.

Nyamugasira, W. (1998) 'NGOs and advocacy: how well are the poor represented?' *Development in Practice* 8 (3): 297–308.

Nylen, W. (2002) 'Testing the empowerment thesis: the participatory budget in Belo Horizonte and Betim, Brazil', *Comparative Politics* 34 (2):127–45.

O'Brian, R. (1992) *Global Financial Integration: The End of Geography?* London: Pinter.

O'Connell Davidson, J. and Sanchez Taylor, J. (1999) 'Fantasy islands: exploring the demand for sex tourism', in K. Kempadoo (ed.) *Sun, Sex and Gold: Tourism and Sex Work in the Caribbean*, Oxford: Rowman & Littlefield.

O'Connor, J. (1973) *The Fiscal Crisis of the State*, New York: St Martin's Press.

OECD (2000) *Economic Outlook* Chapter VI 'E-commerce: impacts and policy challenges' 67 (1) Paris: Organization for Economic Co-operation and Development.

OECD (2002a) *Measuring the Information Economy*, Paris: OECD; http://www.oecd.org/pdf/M00036000/M00036089.pdf (last accessed July 2003).

OECD (2002b) Economic Surveys: Mexico Issue 7.

OED (2000) Review of World Bank participation schemes, World Bank: Operations Evaluation Department; for report and updates see http://www.worldbank.org/participation/.

Ohmae, K. (1990) *The Borderless World: Power and Strategy in the Interlinked Economy*, London: Collins.

Ohmae, K. (1995) *The End of the Nation State: The Rise of Regional Economies*, London: HarperCollins.

ONE North East (2002) Regional profile: http://www.onenortheast.com.

Ong, A. and Peletz, M. (1995) *Bewitching Women, Pious Men: Gender and Body Politics in Southeast Asia*, Berkeley: University of California Press.

ONS (2003) Regional profiles, regional trends on line: http://www.statistics.gov.uk/statbase/explorer.asp? (last accessed November 2003).

Oxfam (2002a) *Rigged Rules and Double Standards: Trade, Globalisation and the Fight against Poverty*, Make Trade Fair: Oxfam. See also http://www.oxfam.org.uk/fair_trade.html (accessed July 2002) and http://www.maketradefair.com (accessed August 2002).

Oxfam (2002b) Debt relief and the HIV/Aids crisis in Africa, Oxfam briefing paper 25: http://www.oxfam.org.uk/what we do/issues/debt aid/downloads/bp25 debt hivaids.pdf (last accessed November 2003).

Pahl, J. (1989) *Money and Marriage*, London: Macmillan.

Palloix, C. (1976) 'The labour process: from Fordism to neo-Fordism', in *The Labour Process and Class Strategies*, London: CSE Pamphlets and Stage 1.

Palmer, G., Mohibur, R. and Kenway, P. (2002) *Monitoring Poverty and Social Exclusion*, York: Joseph Rowntree Foundation.

Parkin, J. (2002) 'Throwing away the key: the U.S. as the world's leading jailer', *International Socialist Review*, January/February.

Parkinson, M. (1998) *Combating Social Exclusion: Lessons from Area-based Programmes in Europe*, Bristol: Policy Press.

Parreñas, R. (2001) *Servants of Globalization: Women, Migration and Domestic Work*, California: Stanford Press.

Patel, D. (2003) 'ICT for Women's Empowerment', Self-Employed Women's Association (SEWA) Paper presented at the Workshop on Engendering the Digital Opportunities, Bridging the Gender Divide in the ICT Sector K.L. Management Development Centre, Indian Institute of Management, Ahmedabad, January; http://www.iimahd.ernet.in/ctps/GenderandICT/Report%20of%20the%20Workshop.PDF (last accessed July 2003). (See also http://www.iimahd.ernet.in/ctps/GenderICT/Presentations/sewa.pdf (last accessed January 2004).

Pateman, C. (1988) *The Sexual Contract*, Cambridge: Polity Press.

PDN (2001) 'InfoSoc 2001 highlights importance of K-Society', *Penang Development News*, September, p.4; http://www.pdc.gov.my/pdn/.

Pearson, R. (1997a) 'Gender perspectives on health and safety in information processing', in S. Mitter and S. Rowbotham (1997) *Women Encounter Technology: Changing Patterns of Women's Employment in the Third World*, London: Routledge.

Pearson, R. (1997b) 'Renegotiating the reproductive bargain: gender analysis of economic transition in Cuba in the 1990s', *Development and Change* 28: 671–705.

Pearson, R. (2000) 'Moving the goalposts: gender and globalisation in the twenty-first century', *Gender and Development* 8 (1): 10–19.

Pearson, R. (2001) 'All change? Men, women and reproductive work in the global economy', in C. Jackson (ed.) *Men at Work: Labour, Masculinities, Development*, London: Frank Cass.

Pearson, R. and Seyfang, G. (2001) 'New hope or false dawn? Voluntary codes of conduct, labour regulation and social policy in a globalizing world', *Global Social Policy* 1(1): 49–78.

Peck, J. (2001) 'Neoliberalizing states: thin policies/hard outcomes', *Progress in Human Geography* 25 (3): 445–55.

Peck, J. (2003) 'Geography and public policy: mapping the penal state', *Progress in Human Geography* 27 (2): 222–32.

Peck, J. and Tickell, A. (1996) 'The return of Manchester men: men's words and men's deeds in the remaking of the local state', *Transactions of the Institute of British Geographers* 21: 595–616.

Peck, J. and Tickell, A. (2002) 'Neoliberalizing space', *Antipode* 34 (3): 380–404.

Peck, J. and Yeung, H. (2003) *Remaking the Global Economy*, London: Sage.

Peet, R. and Hardwick, E. (1999) *Theories of Development*, London: Guildford Press.

Perez, C. (2000) 'Change of paradigm in science and technology policy', *Cooperation South* 1: 43–9.

Perrons, D. (1981) 'The role of Ireland in the new international division of labour', *Regional Studies* 15 (2): 81–100.

Perrons, D. (1999a) 'Reintegrating production and consumption or why political economy still matters', in R. Munck and D. O'Hearn (eds) *Critical Development Theory: Contributions to a New Paradigm*, London: Zed Books.

Perrons, D. (1999b) 'Flexible working patterns and equal opportunities in the European Union: conflict or compatibility?' *European Journal of Women's Studies* 6 (4): 391–418.

Perrons, D. (2000a) 'Care, paid work and leisure: rounding the triangle', *Feminist Economics* 6 (1): 105–14.

Perrons, D. (2000b) 'Deconstructing the Maastricht myth? Economic and social cohesion in Europe: regional and gender dimensions of inequality', in R. Hudson and A. Williams (eds) *Divided Europe*, London: Sage, pp.186–209.

Perrons, D. (2001) 'Towards a more holistic approach to economic geography', *Antipode* 33 (2): 208–15.

Perrons, D. (2003) 'The new economy and the work life balance: a case study of the new media sector in Brighton and Hove', *Gender Work and Organisation* 10 (1): 65–93.

Perrons, D. (2004) 'Understanding social and spatial divisions in the new economy: new media clusters and the digital divide', *Economic Geography* 80 (1): 45–61.

Perrons, D. and Skyers, S. (2003) 'Empowerment through participation? Conceptual explorations and a case study', *International Journal of Urban and Regional Research* 27 (2): 265–85.

Perroux, F. (1950) 'Economic space: theory and applications', *Quarterly Journal of Economics* 64 (1): 89–104.

Phelps, N., Makinnon, D., Stone, I. and Briadford, P. (2003) 'Embedding the multinationals? Institutions and the development of overseas manufacturing affiliates in Wales and north east England', *Regional Studies* 37 (1): 27–40.

Phillips, A. and Taylor, B. (1980) 'Sex and skill', *Feminist Review* 6: 79–88.

Piattoni, S. (1994) 'Regions and economic growth: the experience of Abruzzo and Puglia in the 1970s and 1980s', in U. Buttman (ed.) *Die Politik der dritten Ebene*, Baden-Baden: Nomas Verlagsgesellschaft, pp. 173–96.

Pike, A. (1999) 'The politics of factory closures and task forces in the north east of England', *Regional Studies* 33 (6): 567–75.

Pike, A. and Tomaney, J. (1999) 'Far Eastern FDI and the political economy of local development in north east England', *Asia-Pacific Business Review* 6 (2): 132–61.

Pinchbeck, I. (1969) *Women Workers and the Industrial Revolution, 1750–1850*, London: Frank Cass (originally published by G. Routledge, 1930).

Pineda, J. (2001) 'Partners in women-headed households: emerging masculinities?' in C. Jackson (ed.) *Men at Work: Labour, Masculinities, Development*, London: Frank Cass.

Piore, M. and Sabel, F. (1984) *The Second Industrial Divide: Possibilities for Prosperity*, New York: Basic Books.

Pirez, P. (2002) 'Buenos Aires: fragmentation and privatization of the metropolitan city', *Environment and Urbanization* 14 (1): 145–58.

Polanyi, K. (1957) *The Great Transformation: The Political and Economic Origins of our Time*, Boston: Beacon Press.

Pollard, J. (2000) 'The global financial system: worlds of monies', in P. Daniels, M. Bradshaw, D. Shaw and J. Sidaway (eds) *Human Geography: Challenges of a New Millennium*, London: Prentice Hall.

Pollard, S. (1993) *The Development of the British Economy 1914–1990*, London: Edward Arnold.

Ponniah, T. (2003) *Citizen Alternatives to Globalization at the World Social Forum*, London: Zed Books.

Porter, M. (1998a) *The Competitive Advantage of Nations*, London: Collier Macmillan.

Porter, M. (1998b) 'Clusters and the new economics of competition', *Harvard Business Review*, November–December: 77–90.

Porter, M. (2000) 'Location, clusters, and company strategy', in G. Clark, M. Gertler and M. Feldman (eds) *Oxford Handbook of Economic Geography*, Oxford: Oxford University Press.

Power, M. (2000) 'Alternative geographies of uneven development', in P. Daniels, M. Bradshaw, D. Shaw and J. Sidaway (eds) *Human Geography: Challenges of a New Millennium*, London: Prentice Hall.

Power, M. (2003) *Rethinking Development Geographies*, London: Routledge.

Pratt, G. (1999) 'From registered nurse to registered nanny: discursive geographies of Filipina domestic workers in Vancouver, B.C.,' *Economic Geography* 75 (3): 215–36.

Prism Research (2000) 'Homecare Workers Recruitment and Retention Study', a report by Prism Research for Brighton and Hove Social Services, Draft Final Report.

Quah, D. (1996) 'The invisible hand and the weightless economy', Centre for Economic Performance Occasional Paper No. 12, London: LSE.

Quah, D. (1999) 'The weightless economy in economic development', Centre for Economic Performance Discussion Paper No. 417, London: LSE.

Quah, D. (2001) 'Technology dissemination and economic growth: some lessons for the new economy', Public Lecture, University of Hong Kong (available from the author's website).

Quah, D. (2003) 'Digital goods and the new economy' (available from the author's website).

Rabellotti, R. (1997) *External Economies and Cooperation in Industrial Districts: A Comparison of Italy and Mexico*, London: Macmillan.

Rajasekarapandy, R. (2003) 'Case studies on gender aspects of community management of knowledge centres', Paper presented at the Workshop on Engendering the Digital

Opportunities, Bridging the Gender Divide in the ICT Sector K.L. Management Development Centre, Indian Institute of Management, Ahmedabad, January; http://www.iimahd.ernet.in/ctps/GenderandICT/Report%20of%20the%20Workshop.PDF (last accessed July 2003). See also http://www,iimahd.ernet.in/ctps/GenderICT/Presentations/sewa.pdf (last accessed January 2004).

Rake, K., Davies, H., Joshi, H. and Alami, R. (2000) *Women's Income over the Lifetime*, London: Stationery Office.

Rammohan, K. and Sundaresan, R. (2003) 'Socially embedding the commodity chain: an exercise in relation to coir yarn spinning in southern India', *World Development* 31 (5): 903–23.

Reeves, R. (2001) *Happy Mondays: Putting the Pleasure Back into Work*, London: Momentum.

Regional Trends (2002) Electronic Tables, London: ONS; http://www.statistics.gov.uk/downloads/theme_compendia/Regional_trends_37/Regional_Trends_37_contents.pdf (last accessed February 2003).

Reich, R. (1991) *The Work of Nations: Preparing Ourselves for the 21st Century Capitalism*, London: Simon & Schuster.

Reich R. (2001a) *The Future of Success: Work and Life in the New Economy*, London: Heinemann.

Reich, R. (2001b) 'American Capitalism in the 21st Century: Does Britain want to go there?' CEP Public Lecture, London School of Economics, 14 May.

Reimer, S. (1998) 'Working in a risk society', *Transactions of the Institute of British Geographers* 23 (1): 116–27.

Rex, J. and Tomlinson, S. (1979) *Colonial Immigrants in a British City: A Class Analysis*, London: Routledge & Kegan Paul.

Richardson, R., Belt, V. and Marshall, N. (2000) 'Taking calls to Newcastle: the regional implications of the growth in call centres', *Regional Studies* 34 (4): 357–70.

Richman, D. (2000) Gates rejects idea of e-utopia: http://seattlepi.nwsource.com/business/gate19.shtml (accessed August 2003).

Rimmer, P.J. and Morris-Suzuki, T. (1999) 'The Japanese Internet: visionaries and virtual democracy', *Environment and Planning A* 31: 1189–206.

Robertson, R. (1992) *Globalization: Social Theory and Global Culture*, London: Sage.

Robinson, F. (1994) 'Something old, something new? The Great North in the 1990s', in P. Garrahan and P. Stewart (eds) *Urban Change and Renewal: The Paradox of Place*, Aldershot: Avebury.

Robinson, J. (2002) 'Global and world cities: a view from off the map', *International Journal of Urban and Regional Research* 26 (3): 531–54.

Rocha, J. (2002) 'Democracy dawns for Bolivia's first people', *The Guardian*, 3 August, p. 11.

Rodríguez-Pose, A. and Gill, N. (2003) 'The global trend towards devolution and its implications', *Environment and Planning C: Government and Policy* 21 (3): 333–51.

Roland, R. (1991) *Globalization: Social Theory and Global Culture*, London: Granta Books.

Rosen, S. (1981) 'The economics of superstars', *American Economic Review* 71 (5): 845–58.

Rostow, W. (1960) *The Stages of Economic Growth: A Non-Communist Manifesto*, Cambridge: Cambridge University Press.

Rousseau, J. (1968) *The Social Contract*, Harmondsworth: Penguin Books.

Roy, A. (1999) *The Cost of Living: The Greater Common Good and the End of Imagination*, London: Flamingo.

Roy, A. (2001) Interview with David Bar, *The Progressive* 5 (April): http://www. progressive.org/intv0401.html (accessed August 2003).

Rubery, J., Ward, K., Grimshaw, D. and Beynon, H. (2003) 'Time and the New Employment Relationship', Paper presented at the second ESRC seminar on work life and time in the new economy, University of Manchester, February.

Sachs, J. (2003) 'A rich nation, a poor continent', *New York Times*, 9 July.

Sanchez-Jankowski, M. (1999) 'The concentration of African-American poverty and the dispersal of the working class: an ethnographic study of three inner-city areas', *International Journal of Urban Regional Research* 23 (4): 619–37.

Sandercock, L. (1998) *Towards Cosmopolis: Planning for Multicultural Cities*, Chichester: Wiley.

Sartre, J. P. (1976) *A Question of Method: Introduction to Critique of Dialectical Reason*, London: New Left Books.

Sassen, S. (1991 and 2001a) *The Global City: New York, London, Tokyo* (1st and 2nd edns), Princeton: Princeton University Press.

Sassen, S. (1998) *Globalization and its Discontents*, New York: New York Press.

Sassen, S. (1999) 'Embedding the global in the national: implications for the role of the state', in D. Smith, D. Solinger and S. Topik (eds) *States and Sovereignty in the Global Economy*, London: Routledge.

Sassen, S. (2000) *Cities in a World Economy*, 2nd edn, London: Pine Forge Press.

Sassen, S. (2001b) 'Global cities and global city-regions: a comparison', in A. Scott (ed.) *Global City-Regions*, Oxford: Oxford University Press.

Sassen, S. (2002a) 'Towards a sociology of information technology', *Current Sociology* 50 (3): 365–88.

Sassen, S. (ed.) (2002b) *Global Networks, Linked Cities*, London: Routledge.

Sassen, S. (2003) 'Global cities and survival circuits', in B. Ehrenrich and A. Hochschild (eds) *Global Women*, London: Granta.

Saxenian, A. and Hsu, J. (2001) 'The Silicon Valley–Hsinchu connection: technical communities and industrial upgrading', *Industrial and Corporate Change* 10 (4): 893–919.

Schmitz, H. (1999) 'Global competition and local co-operation: success and failure in the Sinos Valley, Brazil', *World Development* 27 (9): 1627–50.

Scott, A. (1980) *The Urban Land Nexus and the State*, London: Pion.

Scott, A. (1998) *Regions and the World Economy: The Coming Shape of Global Production, Competition and Political Order*, Oxford: Oxford University Press; see also http://www. oxfam.org.uk/fair_trade.html (accessed July 2002) and http://www.maketrade fair.com (accessed August 2002).

Scott, A., Agnew, J., Soja, E. and Storper, M. (2001) 'Global city-regions', in A. Scott (ed.) *Global City-Regions*, Oxford: Oxford University Press.

Sen, A. (1990) 'Gender and cooperative conflicts', in I. Tinker (ed.) *Persistent Inequalities, Women and World Development*, Oxford: Oxford University Press.

Sen, A. (1992) 'The missing women', *British Medical Journal* 304 (March): 587–8.

Sen, A. (2000) *Development as Freedom*, New York: Anchor Books.

Sen, G. and Batliwala, S. (2000) 'Empowering women's reproductive rights', in H. Presser and G. Sen (eds) *Women's Empowerment and Democratic Processes: Moving Beyond Cairo*, Paris: IUSSP.

Sennett, R. (1998) *The Corrosion of Character*, London: WW Norton.

SEU (2000) *Closing the Digital Divide: ICT in Deprived Areas*, London: Stationery Office.

SEWA (2002) (Self-Employed Women's Association, India): http://www.sewa.org.

Shatkin, G. (1998) '"Fourth world" cities in the global economy: the case of Phnom Penh, Cambodia', *International Journal of Urban and Regional Research* 22 (3): 378–93.

Sikka, P., Wearing, B. and Nayak, A. (1999) *No Accounting for Exploitation*, Basildon: Association for Accountancy and Business Affairs.

Silverstone, R. (ed.) (1996) *Visions of Suburbia*, New York: Routledge.

Singh, A. (1999a) 'Financial liberalization and globalization: implications for industrial and industrializing economies', in J. Michie and J. Grieve Smith (eds) *Global Instability: The Political Economy of World Governance*, London: Routledge.

Singh, A. (1999b) 'What really happened in Asia', *Economic Bulletin, National Institute for Economic Policy* 1 (1): 13–15.

Skeffington, A. (1969) *People and Planning: Report of the Committee on Public Participation in Planning*, London: HMSO.

Sklair, L. (2000) 'The transnational capitalist class and the discourse of globalisation', *Cambridge Review of International Affairs* XIV: 67–85.

Sklair, L. (2002) 'The transnational capitalist class and global politics: deconstructing the corporate-state connection', *International Political Science Review* 23 (2): 159–74.

Skyers, S. (2003) 'Reconnecting People and Communities. Participation in Partnerships and Local Labour Markets: The Impact of Local Area Regeneration Strategies', unpublished PhD Thesis, London School of Economics.

Smith, A. (1976) *An Inquiry into the Nature and Causes of the Wealth of Nations*, Chigaco: University of Chicago Press.

Smith, N. (1984) *Uneven Development: Nature, Capital and the Production of Space*, Oxford: Blackwell.

Smith, N. (1996) *The New Urban Frontier: Gentrification and the Revanchist City*, London: Routledge.

Smith, N. (2003) 'New globalism, new urbanism: gentrification as global urban strategy', *Antipode* 34 (3): 427–50.

Solow, R. (1987) 'We'd better watch out', *New York Times*, Book Review, 12 July.

Spivak, G. (1999) *A Critique of Postcolonial Reason: Toward a History of the Vanishing Present*, London: Harvard University Press.

Standing, G. (1989) 'Global feminisation through flexible labour', *World Development* 17 (7): 1077–95.

Standing, G. (1999) 'Global feminisation through flexible labour: a theme revisited', *World Development* 27 (3): 583–602.

Standing, G. (2002) *Beyond the New Paternalism: Basic Security as Equality*, London: Verso.

Stanley, K. (2001) 'Let some people get rich first,' said Deng; 'sure enough, they have', *Star Tribune*, 17 June.

Stiglitz, J. (1998) 'Redefining the role of the state: What should it do? How should it do it? And how should these decisions be made?' Public lecture presented on the tenth anniversary of MITI Research Institute (Tokyo), 17 March; available at http://www.worldbank.org/html/extdr/extme/jssp031798.htm (accessed June 2002).

Stiglitz, J. (2002) *Globalization and its Discontents*, London: Allen Lane/Penguin Books.

Storper, M. (1995) 'The resurgence of regional economies, ten years later: the region as a nexus of untraded interdependencies', *European Urban and Regional Studies* 2 (3): 191–222.

Sturgeon, T. (2001) 'How do we define value chains and production networks?' in G. Gereffi and R. Kaplinsky (eds) *Value of Value Chains*, University of Sussex IDS 32 (3).

Swift, R. (1998) 'The NI interview with Angelica Alvarez Cerda: Richard Swift talks with a Chilean activist fighting for fruit pickers', *New Internationalist* Issue 297.

Tacoli, C. (1999) 'International migration and the restructuring of gender asymmetries: continuity and change among Filipino migrants in Rome', *International Migration Review* 33 (Fall): 658–9.

Talbot, J. (1997) 'Where does your coffee dollar go? The division of income and surplus along the coffee commodity chain', *Studies in Comparative International Development* 32 (1): 56–91.

Tangkitvanich, S. (2001) 'Global e-commerce policies seen from the south', *Cooperation South* 1: 16–28.

Taylor, F. W. (1967) *The Principles of Scientific Management*, New York: W. W. Norton.

Taylor, P., Catalano, G. and Walker, D. (2002) 'Exploratory analysis of the world city network', *Urban Studies* 39 (13): 2377–94.

Tomaney, J. and Ward, N. (eds) (2001) *A Region in Transition: North East England at the Millennium*, Aldershot: Ashgate.

Toynbee, P. (2003) *Hardwork: Life in Low-pay Britain*, London: Bloomsbury.

TUC (2001) 'Calls for change', second TUC report of call centre workers campaign, London: TUC.

TUC (2002) *About Time: A New Agenda for Shaping Working Life*, London: Trades Union Congress.

Turok, I. (1993) 'Inward investment and local linkages – how deeply embedded is Silicon Glen?' *Regional Studies* 27 (5): 401–17.

Turok, I. and Webster, D. (1998) 'The New Deal: jeopardised by the geography of unemployment'? *Local Economy* 12 (4): 309–28.

Tzannatos, Z. (1999) 'Women and labour market changes in the global economy: growth helps, inequalities hurt and public policy matters', *World Development* 27 (3): 551–69.

UN (2002) United Nations World Migration Data, UN Population Division see http://www.un.org/esa/population/publications/ittmig2002/ittmig2002.htm (accessed February 2003).

UNCTAD (2001) E-commerce and Development Report 2001: http://r0.unctad.org/en/pub/ps1ecdr01.en.htm (last accessed November 2003).

UNCTAD (2002) *World Investment Report 2001: Promoting Linkages*, Geneva: United Nations.

UNDP (1996) *Human Development Report 1996*, Oxford: Oxford University Press.

UNDP (1997) *Human Development Report 1997*, Oxford: Oxford University Press.

UNDP (1999) *Human Development Report 1999*, Oxford: Oxford University Press.

UNDP (2001) *Human Development Report 2001: Making New Technologies Work for Human Development*, Oxford: Oxford University Press.

UNDP (2002) *Human Development Report 2002: Deepening Democracy in a Fragmented World*, Oxford: Oxford University Press.

UNDP (2003) *Human Development Report 2003 Millennium Development Goals: A Compact among Nations to End Human Poverty*, Oxford: Oxford University Press.

UNHCR (2000) Refugees by numbers: http://www.unhcr.ch/un&ref/numbers/numb2000.pdf (last accessed January 2004).

UNICEF (1997) *Children at Risk in Central and Eastern Europe: Perils and Promises*, London: UNICEF.

UNICEF (2000) 'A League Table of Child Poverty in Rich Nations', Innocenti Report Card Issue No.1, Florence: Innocenti Research Centre.

UNIFEM (2000) *Progress of the World's Women: 2000 Biennial Report*, New York: United Nations.

UNIFEM (2002) Progress of the World's Women: Gender Equity and the Millennium Development Goals: http://www.unifem.undp.org/resources/progressv2/ (accessed August 2003).

UNISON (2002) UNISON submission to Low Pay Commission, September: http://www.unison.org.uk/acrobat/B509.pdf (last accessed August 2003).

U.N. Wire (2001) United Nations Foundation Independent News briefing, 3 October: http://www.unfoundation.org/unwire/index.asp (last accessed February 2003).

United Nations (1948) Universal Declaration of Human Rights: http://www.un.org/Overview/rights.html (accessed August 2002).

United Nations (2003) Basic facts about the United Nations: http://www.un.org/aboutun/basicfacts/unorg.htm (last accessed February 2003).

United States Bureau of Justice (1996) Correctional Populations in the United States, Bureau of Justice Statistics: http://www.ojp.usdoj.gov/bjs/abstract/cpius96.htm (last accessed November 2003).

US (2002) *A Nation Online: How Americans are Expanding Their Use of the Internet*, United States Department of Commerce, Economics and Statistics Administration: http://www.esa.doc.gov/508/esa/nationonline.htm (accessed August 2002).

Vera-Sanso, P. (2001) 'Masculinity, male domestic authority and female labour participation in Southern India', in C. Jackson (ed.) *Men at Work: Labour, Masculinities, Development*, London: Frank Cass.

Wacquant, L.J. (1997) 'Three pernicious premises in the study of the American ghetto', *International Journal of Urban and Regional Research*, 21 (2): 341–56.

Wacquant, L.J. (1999) 'How penal common sense comes to Europeans: notes on the transatlantic diffusion of the neoliberal *doxa*', *European Societies* 1: 319–52.

Wade, R. (1990) *Governing the Market: Economic Theory and the Role of Government in East Asian Industrialization*, Princeton, NJ: Princeton University Press.

Wade, R. (1998) 'The Asian debt-and-development crisis of 1997–?: causes and consequences', *World Development* 26 (8): 1535–53.

Wade, R. (2001) 'Is globalization making world income distribution more equal?' London School of Economics DESTIN Working Paper No. 01–01.

Wainwright, H. (2003) *Reclaim the State: Experiments in Popular Democracy*, London: Verso.

Wainwright, M. (2000) 'Men again', *Guardian*, 27 June.

Walby, S. (1990) *Theorising Patriarchy*, Oxford: Blackwell.

Walby, S. (1997) *Gender Transformations*, London: Routledge.

Walcott, S. (2002) 'Chinese industrial and science parks: bridging the gap', *The Professional Geographer* 54 (3): 349–64.

Wallerstein, I. (1991) *Geopolitics and Geoculture: Essays on the Changing World-System*, Cambridge: Cambridge University Press.

War on Want (2003) Win–win for the world's poor: http://www.waronwant.org/?lid=1443 (accessed February 2003).

WDI (2001) World Development Indicators: http://www.worldbank.org/data/wdi2001/ (accessed April 2002).

Wehling, J. (1995) Zapatismo: What the EZLN is fighting for: http://flag.blackened.net/revolt/mexico/comment/why.html (last accessed February 2002).

Weiss, L. (1999) 'Managed openness: beyond neoliberal globalism', *New Left Review* 238: 126–40.

Westwood, S. (1984) *All Day, Every Day: Factory and Family in the Making of Women's Lives*, London: Pluto Press.

Wetherall, M. and Edley, N. (1999) 'Negotiating hegemonic masculinity: imaginary positions and psycho-discursive practices', *Feminism and Psychology* 9 (3): 335–56.

Whitehorn, K. (1997) Interview with Tessa Jowell, *Observer*.

Wilkinson, B., Gamble, J., Humphrey, J., Morris, J. and Anthony, D. (2001) 'The new international division of labour in Asian electronics: work organisation and human resources in Japan and Malaysia', *Journal of Management Studies* 38 (5): 675–95.

Wilkinson, H. (1997) *Time Out: The Costs and Benefits of Paid Parental Leave*, London: Demos.

Willis, K. and Yeoh, B. (eds) (2000) *Gender and Immigration*, London: Edward Elgar.

Willott, S. and Griffin, C. (1996) 'Men, masculinity and the challenge of long term unemployment', in M. Mac An Ghaill (ed.) *Understanding Masculinities*, Buckingham: Open University Press.

Wilson, W. (1998) *When Work Disappears: New Implications for Race and Urban Poverty in the Global Economy*, London: Centre for Analysis of Social Exclusion.

Women's Unit (2000) *More Choice for Women in the New Economy: The Facts*, London: Cabinet Office.

Work Foundation (2003) 'End Child Poverty Once and For All', Work and Child Poverty Briefing Paper, London: Work Foundation; http://www.theworkfoundation. com/pdf/Child%20Poverty.pdf (last accessed August 2003).

World Bank (1996) *Participation Source Book*, Washington DC: World Bank.

World Bank (2000) Poverty in an Age of Globalization: http://www.worldbank.org/ html/extdr/pb/globalization/povertyglobalization.pdf (accessed December 2002).

World Bank (2003a) *Global Development Finance 2003 – Striving for Stability in Development Finance*: http://www.worldbank.org/prospects/gdf2003/gdf_ch07_ web.pdf (accessed August 2003).

World Bank (2003b) About Us: http://www.worldbank.org/ (accessed February 2003).

World Bank Group (2001) Participation Newsletter: http://lnweb18.worldbank.org/ essd/essd.nsf/Participation/ParticipationNewsletter10–01–2?OpenDocument& ExpandSection=3.1 (last accessed February 2003).

World Bank Group (2002) Urban Development website: http://www.worldbank.org/ urban/ (last accessed February 2002). Population figures: http://www.worldbank. org/urban/env/population-regions.htm (accessed February 2003).

WTO (2002) World Tourism Organization Facts and Figures: http://www.world-tourism.org/market_research/facts&figures/menu.htm (last accessed October 2002).

WTO (2003) The WTO in Brief: http://www.wto.org/english/thewto_e/whatis_e/ inbrief_e/inbr00_e.htm (accessed August 2003).

Wu, F. (2003) 'Globalization, place promotion and urban development in Shanghai', *Journal of Urban Affairs* 25 (1): 55–78.

Zhang, X. and Zhang, K. (2003) 'How does globalisation affect regional inequality within a developing country? Evidence from China', *Journal of Development Studies* 39 (4): 47–67.

Zook, M. (2001) 'Old hierarchies or new networks of centrality? The global geography of the Internet content market', *American Behavioural Scientist* 44 (10).

Zook, M. (2004) *The Geography of the Internet Industry*, Oxford: Blackwell.

# INDEX

Note: Page numbers in **bold** type refer to **figures**. Page numbers in *italic* type refer to *tables*. Page numbers followed by 'n' refer to notes.

domestic violence 181
domestic work 288
Dominican Republic 114
Doxford International Business Park
    153
dual adult households: USA 16
dual earning households 22, 222–4

e-commerce 170–80
e-democracy 180–3
earnings differentials 219
East Asia 320
Eastern Europe 244
economic integration **260**
economic migrants 221
economic restructuring: and
    employment change **138**;
    regulationist perspective 130–6
economic and social restructuring:
    employment 136–42
economies of scale 60
economy *see* new economy
education 44
efficiency gains 219
Ehrenreich, B. 67, 111, 123, 329
elderly: caring 17
Elson, D. 65, 87–8n, 95, 96, 120,
    121–2, 124n, 305, 317n
embourgeoisement 132
employment: and economic and social
    restructuring 136–42; female 23–4;
    and gender 141; gender stereotypes
    85; manufacturing 70; masculinity
    86, 156–9; rate 162n; self- 86;
    *see also* feminization of employment;
    maquiladoras
empowerment: and participation
    297–307; through participation 98
enclaves 226, 229
Engels, F.: and Marx, K. 254
England: NE 142–55, *143*, 266–7, 328;
    regional development *144*
Enron affair 254, 279n
Esping-Andersen, G. 7–8, 18, 21, 30n,
    137
Ethical Trade Initiative (UK) 309,
    310–12
ethical trading 307–13
European Regional Development Fund

(ERDF) 268
European Union (EU) 23, 58;
    condemnation of trafficking 115;
    integration in 261
export processing zone (EPZ) 99–102,
    104

fair trade 307, 308
female: employment 23–4;
    non-agricultural employment **83**;
    supervision 85
feminism 6
feminization of employment 16, 69, 81,
    82, 84, 86, 89, 109, 121, 128,
    134–42; limits of challenge to
    patriarchy 122
fertility 84; levels 159
Filipinas: 105–8
financial services 139
financial speculation: destabilizing effect
    254
Fitter, R.: and Kaplinsky, R. 72
flexi contracts: supermarkets 94
flexible contracts 91
Folbre, N. 8, 14
Fordism 6, 123, 130, 131, 134–5, 243,
    246
foreign direct investment (FDI) 58, 68
Foucault's panoptican 152
France 106, 190, 274
Fraser, N. 111, 306
free markets 242
free trade areas 100, 259, 261
freelancers 16
French Regulation School 6, 12, 128,
    130, 136, 324
fruit chain: global 90–5
*Full Monty, The* 156

G7/8 261, 280n
gated communities 227, 229
Gates, B. 191
GDP 35, 44, 273; changing and
    widening inequality *43*; and internet
    access 188–90
gender 24; and development *50*;
    distribution of roles 109; inequality
    49; relations and employment
    117–23; stereotypes 85